The Harness Book

Work Horse and Mule
Harness Design & Function

Lynn R. Miller

The Harness Book
Lynn R. Miller
copyright © 2022 Lynn R. Miller

Publisher
Davila Art & Books LLC.
in conjunction with Small Farmer's Journal
PO Box 1627 Sisters, Oregon 97759
541-549-2064

authored by Lynn R. Miller

First Edition November 2022, Soft Cover

Library of Congress Catalog Number
ISBN 978-1-885210-36-4

Also by the author

Work Horse Handbook
Training Workhorses / Training Teamsters
Horsedrawn Plows and plowing
Haying with Horses
Horsedrawn Mower Book
Horsedrawn Tillage Tools
Art of Working Horses

Horses in Harness (out of print)
Complete Barn Book (out of print)
Starting Your Farm (out of print)
Small Farm Bookkeeping (out of print)

Essays
Why Farm
Farmer Pirates and Dancing Cows
Old Man Farming

Poetry/Prose
Thought Small
The Elastic Signature

Fiction
The Glass Horse
The Brown Dwarf
Talking Man

dedication

This book is dedicated to future teamsters, especially those committed to the comfort and efficiency of their equine workmates.

And it is also dedicated to one of my best partners, the extraordinary Belgian mare Cali. She taught me, in dramatic fashion, to pay attention to her comfort so that she in turn might pay attention to what I needed of her.

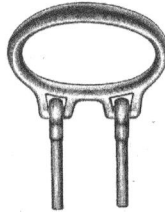

harness (n.)

c. 1300, "personal fighting equipment, body armor," also "armor or trappings of a war-horse," from Old French *harnois*, a noun of broad meaning: "arms, equipment; harness; male genitalia; tackle; household equipment" (12c.), of uncertain origin, perhaps from Old Norse *hernest* "provisions for an army," from herr "army" (see harry (v.)) + nest "provisions" (see nostalgia). Non-military sense of "fittings for a beast of burden" is from early 14c. German *Harnisch* "harness, armor" is the French word, borrowed into Middle High German. The Celtic words are believed to be also from French, as are Spanish *arnes*, Portuguese *arnez*, Italian *arnese*.

The Harness Book

Work Horse and Mule

Harness Design & Function

TABLE OF CONTENTS

Sidebars

This volume is packed with a heap of information and wagon loads of imagery. Some of the material has appeared previously in other titles within the LRM *Work Horse Library* series, also various SFJ articles and as a consolidation preview article in four successive issues of the *Small Farmer's Journal*. Some of this material is brand new to this book. The lion's share has come from our extensive archives.

Though design of harness systems and hardware so clearly follows function married to materials (and targeted knowledge tied to its day) this author has long seen a lasting and serviceable beauty to all the paraphernalia and setups. As said elsewhere, we do hope that gathering this material together, making it available in print form with what we hope will be a long shelf life, serves to keep this viable and remarkably telling craft accessible as a positive option for the future.

The realm or world of working horses and mules (in our case a North American adventure) once incorporated thousands of individual craftsmen, manufacturers and distributors servicing millions of teamsters, farmers, miners, oilfield workers, loggers, freighters, firefighters and commercial carriage operators. Today the true teamsters community is scattered and diminished but with pockets of strong numbers. Sadly, within the last 25 years we have seen a marked decline in the number of horsepowered Amish farmsteads. Who can tell if this trend will continue? My sense is that the economic, social and climatic challenges ahead could mean a dramatic increase in the numbers of people who choose to depend on animal power. The remaining active Amish horse and mule numbers could be most useful to such a trend.

Thank you for your interest. Lynn R. Miller, Singing Horse Ranch 2022

It's not all in here

When I first had the idea for this project my notion was that it would write itself with captions wrapped around all of the photos, drawings, and illustrations I had amassed. But the deeper I got into it, reminding myself from my fifty years with the horses of all the myriad variations, adjustments, measurements, combinations, questions, applications and such, I realized that it was not only a much bigger project than imagined, it was also an impossible task. I would not be able to remember or find or figure out every single detail I had been privy to, let alone find all those far flung unknowns. This book would, in other words, be cursed as perpetually incomplete. For this I must apologize. I can only hope that there is enough here, and in a form, that excites you to go for what you suspect is missing, find your own answers and inventions. If that be you, thanks for taking on this book as a searcher and a reader. And to those young people brand spankin' new to the magical world of working animals, I am ecstatic for you. LRM

Introduction

In the late sixties, when I first looked intently at harnessed mules and horses longing to understand how the system worked, ignorant of what was most important, it was the harness that confused me more than the anatomy and movements of the animals, even more than the overall system. I saw a tangled basket of straps, chains, ropes, all seeming to have purpose. Yes, there were some diagrams in dusty libraries and old books and these did offer basic explanations of the structural design of some harness varieties. But those didn't help me to understand in a truly useful way. It would be a few years before I would have my own first team and a pile of old harness to figure out. The little bit of book-learning and diagram-scanning I did failed to educate me. I have told the story before of how my innocence and arrogance got me into big trouble the first time I harnessed and tried to drive a team. Some of that tragedy came from the harness being put on all wrong, making it impossible to function properly. That does not need to be the case with newcomers today. There is a lot of good information available, some of it through our own books. But the subject of work harness is vast and various as evidenced by the hundreds of design variables we see in the many old catalogs in our archives.

How'd it go in the beginning? It wasn't obvious at first, took some time for me to figure out the subtleties of fitting animal to harness. Sure, at a glance you can see that the collar goes around the neck and the harness hangs over the back and hips, attaching forward to hames and collar, and backwards to the pulling apparatus or singletree/doubletree, but there's more to it than that.

The subtleties I spoke of have to do with function and fit. As the mule or horse steps forward, into the collar, the

comfort with which it displaces the load, drawn from behind, is directly proportionate to how well the collar fits. If the collar is too tight it pinches, chaffs and shuts off the breathing. If the collar is too loose, either in length or width at point of draft or contact, or 'argues' with the shoulder, top of neck and perhaps even sides, it will create serious discomfort which can lead to sores and a baulky horse. Moving back over the animal, in a basket-brichen (or breeching) harness style, the belly band should fit loosely as should the quarter straps running from brichen forward to pole strap. The brichen or breeching should be neither too high nor too low, preferably running in a line to the connections with the two quarter straps. If too high, you lose efficiency in backing and braking. If too low the same is true and the animal feels something is wrong, as the motion of the hip is impaired. (There are exceptions, explained further along in the volume, as represented by Yankee or hip brichen harness design.)

Most of what we are talking about is either leather based harness, or a synthetic harness built on the same structural principles. (There are variables that employ a combination of leather and nylon webbing, biothane or nylon based materials, ropes as lines and even sometimes chain tugs or traces.)

By definition, we might be forgiven thinking of harness as a restraint system for equines. We may be, but I won't, because fifty years plus of experience and observation has shown me that most of us as teamsters are the limited and limiting element in the teamster's craft. Given a well-fit and comfortable harness, the best working equines frequently exalt. While as teamsters we seldom do. Add a soft voice and certain yet soft touch and horses and mules will gladly work for us. Half a century of experience has taught me that a lifetime is not long enough to fully appreciate and absorb all the mysteries and peculiarities of working horses and mules in harness. Uppermost in that realm is the equine's capacity for appreciating how it's handled. To return to my now familiar excuse as apology, "the longer I do it, the less I know and the easier it gets."

The harness, rather than a restraint, is best and most usefully seen as a translation and enabling device; it translates the motion of the willing equine into a motive power the teamster might direct. When we insist that it is primarily a restraint system we limit what is possible in the relationship with the working animal(s). While it is certainly true that, with beginners, safety must be paramount, if we insist that the animals are not to be trusted without restraint, there is the common risk that we carry that limiting attitude and damaging set of demands forward with us throughout our teamster life.

For at least 5,000 years, back before the civilizations of Mesopotamia, humans have employed animals as motive power. This required some sort of rigging that allowed forward motion to be translated to pulling action. This objective was accomplished, sometimes crudely, in various ways such as with tied yoking around cattle horns.

The modern work harness we are concerned with, used for farming, drayage, mining and woodswork, evolved from a beginning that some believe occurred in China around 500 AD. The rudimentary basics of modern harness design don't show up until the early European Middle Ages.

As for the word 'harness' some believe it first appeared around 1300 AD as a noun denoting the armor or trappings of a warhorse with origins back to the Old French 'harnois.' It may go further back to the Old Norse 'hernest' as in provisions for an army. But most specifically it doesn't appear to show up until 14th century Germany and the word 'Harnisch.' As is the case with the origin of words, experts guess and the science quivers in apprehension of new diggings that will likely stir yet another set of conclusions.

As for the architects of harness design, I leave it to those in the future to do the detective work; work that may unveil the names of inventors and ingenious harness makers through the ages, those who put their minds and hands to harness structure and to shaping each piece to greater efficiency and comfort.

As to the objective of understanding, we may look to the mechanics of the implements, hitchings and work itself to see how these elements pushed harness design to its current structure.

But first, some effort needs to go into a rudimentary grasp of the mechanics of draft. The 19th and early 20th century saw the application of science to the mysteries of efficient pulling systems employing equines. The practical aspects were frequently derailed by romantic notions of a horse moving, as children might

The Fancy & Mechanics of Draft

imagine, as a gravity defying four-legged acrobat floating forward with pumpkin carriages and magical skeleton structures in tow, rolling through space with scarcely an indication of restraint or weight. But real and fascinating science ignored popular fantasy and looked into the

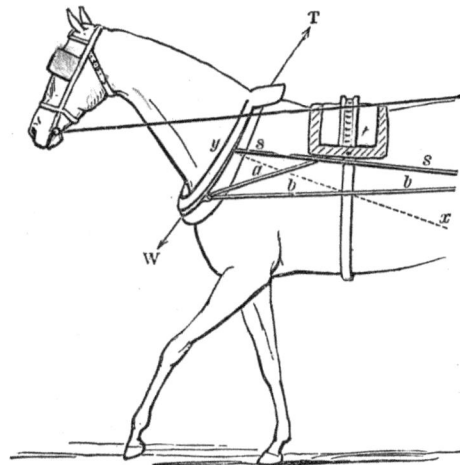

equine skeletal and muscle structure to appreciate how natural forward motion might best be applied to the displacement of a drawn load. These two drawings are from the 1859 treatise entitled "On Seats and Saddles..." by Major Francis Doyne Dwyer.

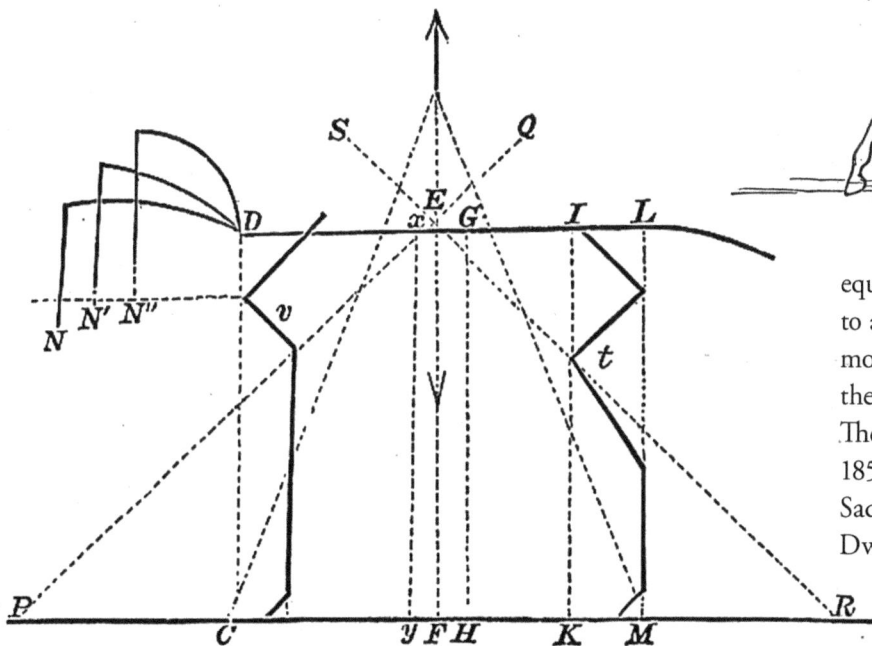

**How the implements and work affected
the evolution of harness design.**

When the work is all about dragging some weight, object or implement across the ground - a 'dead drag' some might say - where there is little or no chance of that walking plow, spike tooth harrow or log running up on the heels of the draft animals - the required harness does not need a brichen assembly. But when the work features a rolling implement or vehicle, and includes possible precise adjustments as to depth of shovels, or planting discs, a brichen assembly might be essential. Highly unique and specialized applications of animal power, such as mining and working in ocean surf, may benefit from suitable materials and surfaces.

The steerage of implements and wagons, and in some cases the exacting work it may be called to do, affected some structural aspects of harness design. For ex-

ample: auto-steer front ends on four-wheeled manure spreaders can make it a challenge to back up precisely. When a teamster/farmer needs to back up the spreader through a narrow door and down a narrow lane (to get closer to the manure piles), leather or chain breast straps attached to the neckyoke ends can make the job impossible. The slightest bump or swing of the breast straps will tweak the front wheels and cause the backing spreader to go wrong.

In the New England states, a high degree of design went into answering the needs of teams in tight situations on farms, city streets and in the woods. Side backer and D-ring harnesses were developed to give a far more precise control over the vehicle or implement tongue, and, because of this, allow very precise backing. This is accomplished with short tugs running forward to all four ends of single trees off the side backer neckyoke.

Conventional brichen harness design, with a single simple neckyoke bar, anchored by tension or otherwise, allows that the neckyoke may sway and this will

influence auto-steer and fifth-wheel stability.

All of this acknowledged, in North America the conventional western-style basket brichen work harness is far and away the most popular design.

And the day after today?

The working horse and mule almost disappeared from North America in the mid-twentieth century. A resurgence of interest and application began in the 1970's peaking around 2005 when, once again, general interest fell off. But not before the development of a whole range of cottage businesses manufacturing HD farm and logging equipment and harness. These companies and individuals developed exciting innovations for animal power, some so fresh and revolutionary that they haven't had time to be fully tested in the field. For example, synthetic materials are still evolving in the construction of harness and hitch gear, as well as ground-drive gearing and linkage. Innovations today from European circles feature some outstanding advances.

When the cycle comes full 'round once again, and people choose to move into animal power, information and supplies should be more abundant.

The heavy human foot print upon this planet is being sorely tested by the laws of nature. Every single day people are being forced to find new and better ways to live in union with biological imperatives. But sometimes *new* isn't the answer, sometimes *better* lies with revisiting the principles of the best of what was so hastily left behind.

Note: Throughout this volume there are hundreds of illustrations appearing from the old harness catalogs in our archive library. In some cases we have left the descriptors including pricing as they originally appeared. Simple adjustments or conversions for inflation do not work here. In 1900, with millions of people dependent upon draft animals for farm and transport, demand for harness dictated there be many harness-makers, large and small, spread all across North American. Efficiencies of scale coupled with an abundance of labor and raw materials resulted in very low prices. Conditions today, with a handful of harness-makers, limited supplies, and a shrinking labor pool, have resulted in disproportionately higher prices. Supply and demand today is on the far end of the pendulum's swing.

The whole "next big thing," those leaps into plastics, toxic chemistry, digital realities, imagined valuations, electric interface and augmented reality have led us to where we are. Enough said? I don't think so.

Whether in 2022, or 2052 or 2100, it's past time to bring the horses in, curry them, put harness on and return to the woods, fields, roads and profit columns.

Chapter One

What Makes It A Work Harness?

To answer the question in the chapter's title, first a look at the job the gear, or harness, is expected to perform.

If horses and/or mules are to displace a load, to drag things across the ground, to pull a wheeled conveyance, or to artfully work an adjustable implement through and across land, they need some sort of rigging for the job. In North America, a popular and effective harness design has evolved over many decades. For our purposes we identify this rigging as a western basket brichen 'work' harness. We say 'work' to differentiate it from light driving, carriage and show harness designs. The distinctions are:

1. a simpler, more basic overall design with utility the governing feature.

2. heavier or suitable construction with few frills.

3. adjustable features synchronized with work requirements and draft efficiencies.

4. the opportunity for maximum comfort for the working animal.

Can work be done with a light carriage harness or a fancy show harness? Yes, but... in either case it entirely defeats the purpose. Examples: If the carriage harness employs a breast collar, rather than a neck collar, then protracted time spent with heavy pulling will result in sore-shouldered horses. And, typically most fancy parade or hitch harness is not designed for optimum work efficiency fit. Some of the worst examples of poor fitting collars and brichens can be seen on expensive show horses with incredibly expensive harness. To my eye, a well fit, well built work harness, fully employed and worn by willing and well behaved animals, is a thing of useful beauty.

The horse or mule steps into the collar, pushing as it walks. Rib-like hames, set into the groove of the collar, allow a way to attach tugs or traces which translate the push to pull. In this illustration (page 17) of a chain-tug plowing harness (without brichen), the image of the horses is stripped away and the barest essentials of the harness are in view. Here it is perhaps

easier to 'see' what the work harness needs to do and how it rests and rides in relation to the animal. But this is a staged posture for the horses to show off the harness. If these *ghost* horses were actually pulling, their heads might be down some and necks arched, tipping the collar angle up steeper to match the leaning angle of the shoulders.

Converting, comfortably, the forward motion of a trained four-legged animal to motive power is the job of the harness. The illustration of the Robson dog harness, while notably different from a work horse or mule harness, nevertheless gives a clear sense that the work animal, stepping forward and pushing into the collar, pulls on, by virtue of the two parallel tug or trace straps, the cross piece, or single tree, which then pulls the load.

Today, new work harnesses are being manufactured by many small harness shops across North America and Europe. It's a good thing but there is certainly no guarantee that this important rigging, these harnesses, will always be readily available.

Times change and regretably we are beginning to witness, once again, the craft of working horses and mules ***falling*** from favor. There will be consequences. Many small harness shops may vanish. Harness will be difficult to come by. For this reason I am trusting

that the work that we have put into this volume will give people, long in the future, some of the information necessary to retool, build harness, and return in large numbers to the symbiotic relationship with draft animals.

Where's the information? When I started working horses in the early 1970's I was cautioned that there were good reasons why people had quit working horses and mules. Conventional wisdom: *It was very hard work* (I learned that was necessarily so), *there weren't the trained animals available to use* (I taught myself to train

THE ROBSON

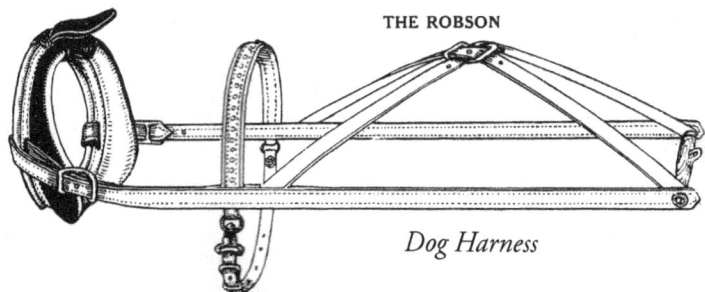

Dog Harness

them), *the harness and equipment was no longer being made* (not true, it was just that small shops were hidden, tucked away under shade trees on remote farms), and *no one knew how to work the animals anymore* (also very much not true.) So, early on, the path was somewhat clear to me. I needed to find where the information was

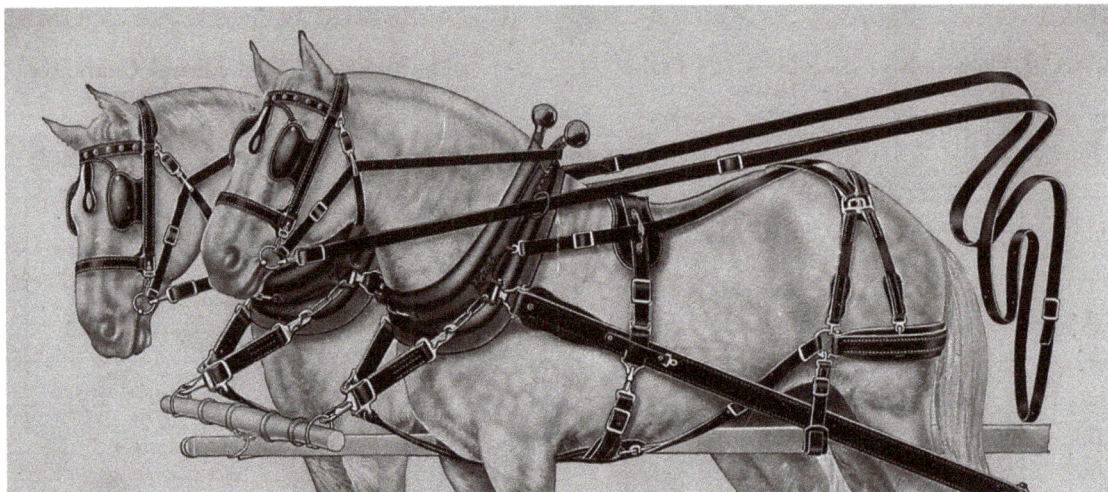

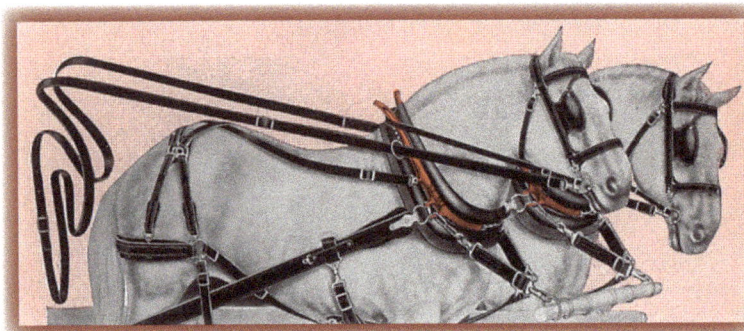

Cart (Show) Harness
Coach Harness
Buggy Harness
Carriage Harness
Brewery or Todd style Show Harness

Parade Harness is an interesting subvariable as the name suggests. Any harness is a candidate to use in a parade. Harness shops do sometimes offer Parade Harness as a type and it is customarily work harness that has been dressed up with patent leather touches, spotting, and fancy elements incorporated into the hardware. Sometimes 'horse brasses' are tacked on, as well as tassels and

hiding. That has been part of my life's work: finding the information and figuring ways to keep it fresh and on a clearly marked, accessible shelf for others to find. The result, to date, is forty-six years of publishing **Small Farmer's Journal,** an international quarterly which features practical horsefarming. And the continuing construct of the **Work Horse Library**, a series of books which include: *Work Horse Handbook, Horsedrawn Plows and plowing, Haying with Horses, Training Workhorses – Training Teamsters, The Horsedrawn Mower Book* and several more. (Please see back of this volume for a complete list.)

Returning to the question, 'what makes it a work harness?' It is appropriate to give a list of the exemptions. These styles of harness are not necessarily suited for work:

An auction display of spotted parade harness.

ringed and/or spotted lead straps.

Strictly speaking, *delivery* or *market harness*, usually for a single animal, is considered for work. (Imagine milk wagons or stud carts, for example.) Along those lines, in their day breweries and freight companies utilizing big hitches, while dressing for commercial purposes, still featured work harness.

The future of the craft of working horses and mules belongs to those who love it. They are the ones who will keep the flame burning. LRM

No. 2765

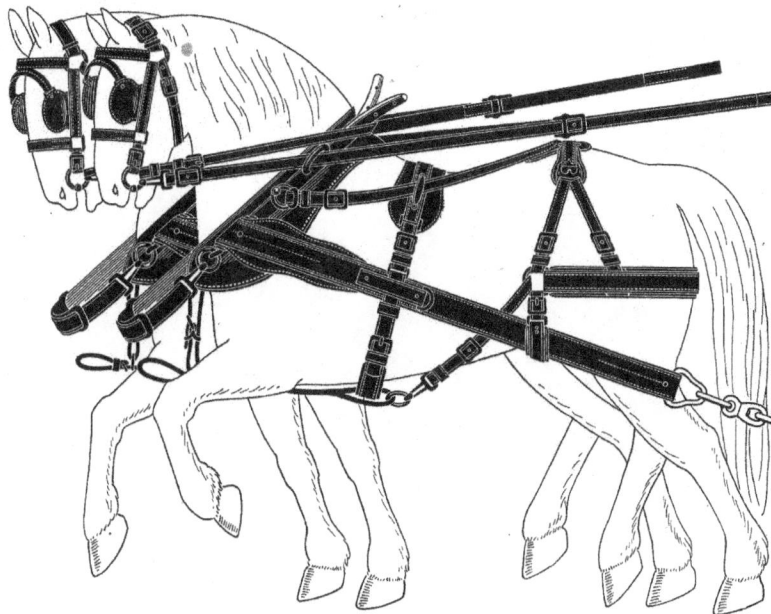

No. 2765

X C TRIMMED. CREASED.

No. 2765A—1½-inch Harness with No. 5 Oiled Bolt Hames. Per Set $107.90
No. 2765C—1½-inch Harness with Nic. Ball Steel Hames. Per Set 110.00
If wanted with 1¾-inch Traces, Breast Straps and Martingales, add 4.70

BRIDLES—Per set $11.70. Ring crown, 1-inch cheeks, Concord blinds.
LINES—1⅛-inch by 20 feet with Conway loop and snap.
TRACES—1½-inch by 6 feet with 6-link heel chain. Sewed bolt ends, 1½-inch billets.
BACK BANDS—4¾-inch Swell harness leather housings, leather lined, 1½-inch layer with dee, 1½-inch market straps to reverse. Colorado bridges.
BREECHING—Per set $28.70. 2¼-inch single strap with 1½-inch full length layer, 1½-inch split turnbacks, 1⅛-inch reverse hipstraps, 1¼-inch reverse sidestraps, 1-inch lazy strap with wear leather and dee.
BELLY BANDS—2-inch single strap with 1½-inch layer—1½-inch buckles.
BREAST STRAPS—1½-inch with snaps and slides.
MARTINGALES—1½-inch with ring.
COLLAR STRAPS—⅞-inch.

No. 2766A—1½-inch with No. 5 Oiled Bolt Hames. Per Set $114.70
No. 2766C—1½-inch with Nickel Ball Tubular Steel Hames. Per Set 116.80
The Number 2766 harness is the same as Number 2765, only has long round side check bridles and third turnbacks.
No. 2766—Bridles. Per Set $13.90

A good heavy serviceable harness where one is wanted at a lower price than our Gopher Brand.

WHAT FORM THIS BOOK TAKES: *Once, while looking for an illustration in our archival library, I found myself marveling at the all the old harness catalogs we had amassed. I mentioned to Eric that I had an idea for a series of basic articles on work harness for the SFJ and mused about putting that together with hundreds of engravings from the old catalogs. It was that simple. The idea for this book was born. Then came decisions about the selection process. Trusting my instincts I knew that I enjoyed comparing the sometimes miniscule differences in harness design, the nomenclature, and all the information about dimensions, materials and pricing. So I decided to include that stuff often. There you have it, and here you have it. Trust it works. LRM*

'*All manner of things*' applies when we talk about the pieces and possibilities with work harness.

If there seems to be something missing in the approach of this book let it be this: I, as the author, am less concerned with the long view to 'why' – these pieces – this way or that. I am, instead, totally enamoured of the short view of '*how's this go together?*' And '*ain't that curious?*' And '*my goodness but that is elegant and beautiful.*' And ultimately '*now this works for me in just about every way.*'

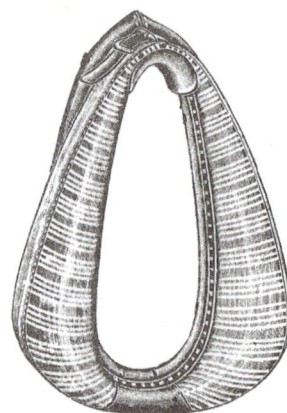

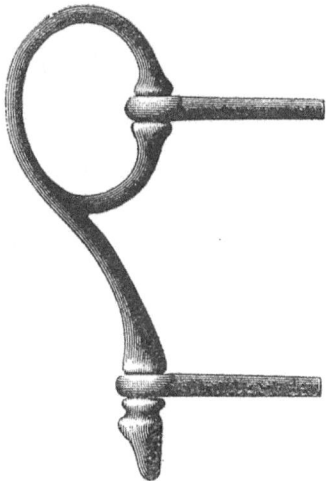

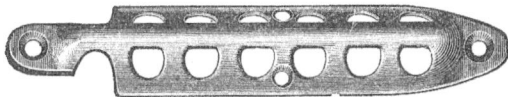

Chapter Two
Collars, Hames, Hame Straps, Chain Binders

The work harness prevalent in North America over the 20th and early 21st centuries evolved slowly to its unique design. Stemming in the middle ages from European engineering, the design basics may have their origins reaching back to Greco-Roman and even Egyptian and Phoenician ages. The primary influence has been the demands of function. Rather than get into arguments about what harness type or design is best, the purpose of this volume is to build an introduction worthy of harness makers and armchair historians.

There are a number of key mechanical differences within North American harness and hitch design, such as the plug neckyoke versus the ring neckyoke, side backer systems versus polestrap style, back pad styles versus market tug styles. Those are best explained after a straightforward introduction to the more common two or three-strap western brichen-style harness.

Half Sweeney.   Full Sweeney.

COLLARS: Horses and mules may be made to work in ill-fitting, improperly adjusted harness – for a while. Please be assured, however, that long hours of hard work in poor fitting collars WILL result in problems for your animals and reduced efficiency.

Western-style team basket brichen work harness.

Features: Back pad or saddle – back strap spider assembly with hip drops which carry the brichen strap. Brichen attaches forward to two quarter straps which fasten to pole strap affixing to neckyoke. Hanging from hame bottoms each harness uses a breast strap or chain which carries the weight of the neckyoke and tongue. Harness may or may not employ an overcheck or check rein.

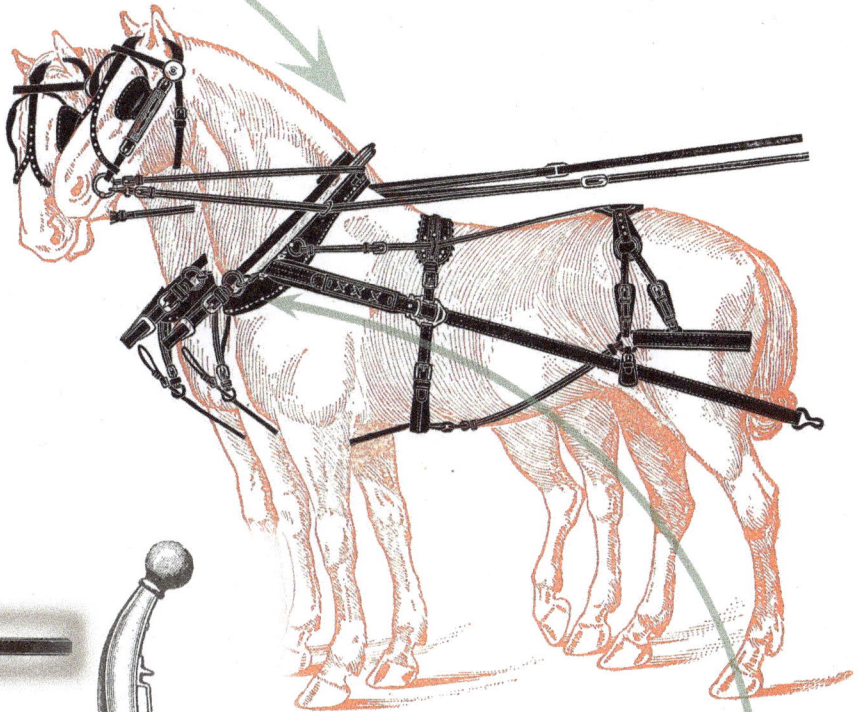

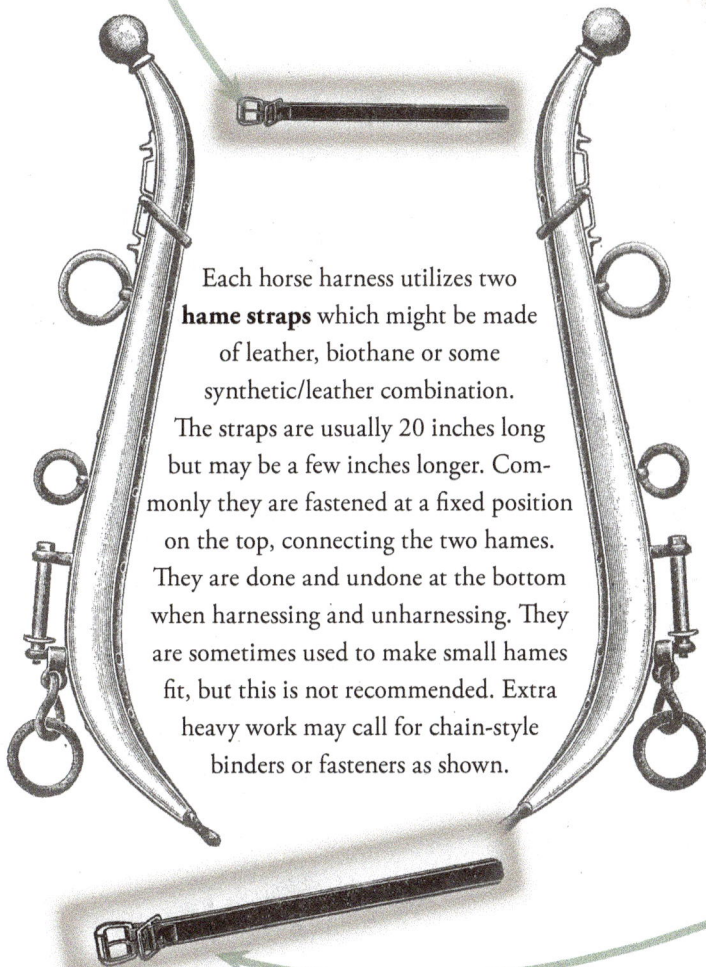

Each horse harness utilizes two **hame straps** which might be made of leather, biothane or some synthetic/leather combination. The straps are usually 20 inches long but may be a few inches longer. Commonly they are fastened at a fixed position on the top, connecting the two hames. They are done and undone at the bottom when harnessing and unharnessing. They are sometimes used to make small hames fit, but this is not recommended. Extra heavy work may call for chain-style binders or fasteners as shown.

HAME STRAPS, or some ways to fasten the hames to the collar, are critical to the harness. It is impossible to make a collar-style harness function without them. And **hame straps need to be adjusted tight to avoid mishaps.** A loose hame strap can jump out of the collar groove and choke a pulling horse or tangle the harness.

Collars are measured, in inches, from the inside top front edge to the inside bottom front edge. The more common sizes are from 19" through 26". Larger and smaller do occur.

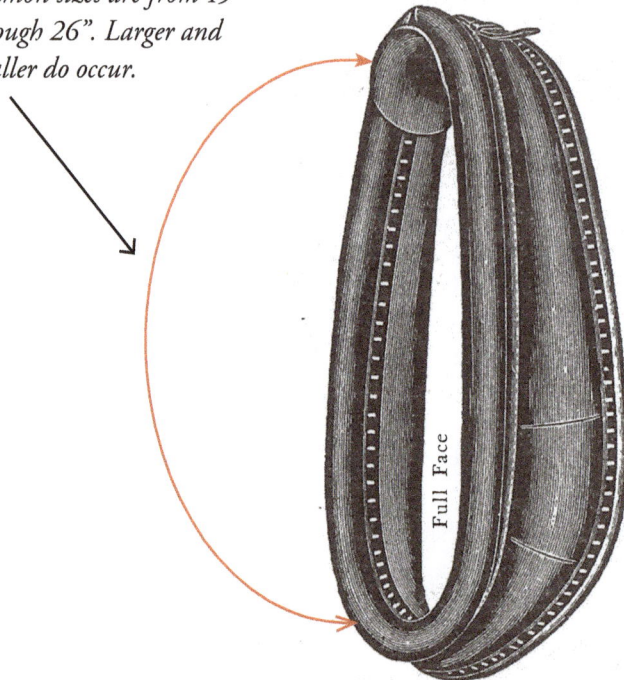

Full Face

Two styles of hame chain binders.

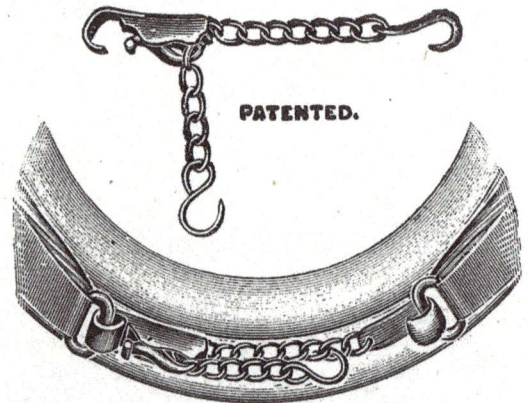

PATENTED.

Sore Shoulders Preventable

Here's the anecdote: You work your three-abreast 8+ hours a day for two weeks, at plowing, discing, harrowing. Next comes planting. Coming into the stalls on this morning are two of your three with problem conditions you can see. One has what looks like a soft raised area under the skin on the point of the shoulder. Touching it reminds you of touching a boil. The second horse, on a similar location, has a small pink flap of loose skin, pink because most of the hair is gone and it is sore. This had been a boil feature just like the former horse but now it has moved into new territory, it is now a full fledged collar sore. You know that this horse is out of the work string until the shoulder heals. And the other horse? Perhaps. There are things you can do and you might be able to avoid a longer layoff. But for someone who depends on these work animals to keep going there is a more important question. Could this have been prevented? Yes, by proper collar and hame fit along with daily attentiveness.

Half Sweeney.

Full Sweeney.

COLLARS: Horses and mules may be made to work in ill-fitting, improperly adjusted harness – for a while. Please be assured, however, that long hours of hard work in poor fitting collars WILL result in problems for your animals and reduced efficiency.

Arguably the most critical harness piece, when it comes to proper fit, is the collar. There are four common types (notwithstanding the Irish collar with

FLEXIBLE THROAT COLLARS.

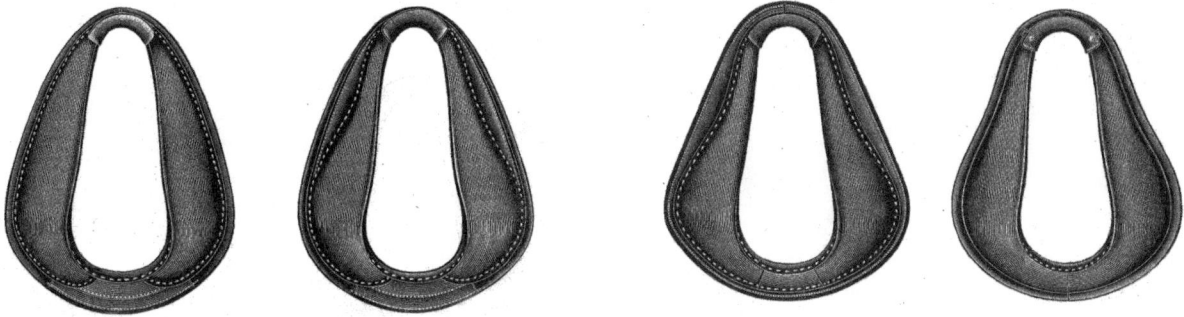

Note: *The two collars on the left have flat, unstuffed throats to allow more room for breathing.*

ticking face). They are: 1. Full face 2. Half sweeney 3. Full sweeney and 4. Mule. These speak to the design of the inside bearing surface of the collar, where it makes contact with the neck and shoulders of the work animal. Full face collars are suited for slab-sided necks whereas the half and full sweeney have concaved shapes which accommodate thicker necks.

Far and away the most common construction features a sewn, shaped, pair of tubes of leather. The collar is stuffed with chopped straw. At one time the most prized collars were handmade by hand-stuffing with long straw. Chopped straw can in certain environs and situations become lumpy. That problem is less common with long straw types.

There are at least three different areas of concern for proper collar fit: 1. Overall length 2. Width and shape 3. Marriage of hames to collar. Separately, it is important that the collar and/or the collar and pad combo have no sharp or abrupt edges or features that might rub the horse or mule's neck the wrong way.

On the first score: The collar, seated against the shoulder should be just loose enough at the throat to avoid cutting off the wind when the animal steps forward to pull.

The customary way to measure this is to see if the flat of your fingers will slip between throat and collar. The most common mistake is to use a collar that is too long. When this happens the draft of the collar (the widest portion) slips below the point of the shoulder. The collar then rocks on the point of the shoulder and moves up and down causing sores at the shoulder and at the withers. When given the choice the best teamsters understand that a snug fit is almost always superior to a loose fit.

There are two ways that higher end collar design might be employed to help with fit. Flexible throat collars, which feature a flattened flap of doubled leather rather than a continuation of the collar's tube structure, allow that a snug fit may occur without cutting off the wind.

And the other older design was the *pipe throat collar* employing a metal (usually bronze) throat piece that was shaped to make it easier for the horse to breath.

On the second score: If the collar is too thin at the sides it will rub the sides of the horse's neck. Choosing a correct design of collar should alleviate this. The full sweeney is best suited for a thick-necked horse. If it is used on a normal neck it could cause problems with extra wear at the top and a rocking action while pulling.

On the third score: Collars are measured from the inside throat up to the inside top. Hames are measured from the bottom up to the mid range of the strap ratchet. If designed with the adjustable top hame strap ratchet, a pair of hames might be made to fit three to four sizes of collars. A 26" hame can fit, for instance, a 25" - 27" collar. Hames need to fit the collar. If the hames are too long or too short they can affect the shape of the collar and the way it fits the animal. All retail hames feature a straight run for most of their inside edge. Custom hames are especially made to bend slightly for a snugger fit in the collar groove. It is not common to see these employed and they are hard to find, but they do point to an uncommonly attentive teamster, as that matching curve will deliver extra comfort through a perfect fit of the collar to the horse's shoulder.

So much of this discussion would leave you to be-

HOW TO MEASURE A HORSE FOR A COLLAR

It does not matter how well a horse collar is made, if it is not properly fitted to the horse, it is sure to hurt him; it is therefore necessary to use the greatest care in selecting the proper size of collar. For this purpose we especially recommend the E V E R E T PATENT COLLAR MEASURE as the best method of measuring a horse for a collar; you will find this measure listed and illustrated among our tools. If you do not have the Everet measure place your hand on the top of the horse's neck, and measure in a straight line with a rule to the bottom of the neck, as shown in above cut.

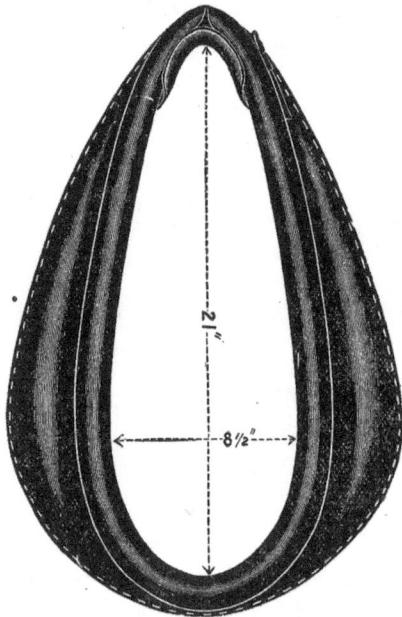

HOW TO MEASURE A HORSE COLLAR

If you have an old collar that fits the horse, and you wish to order one the same size, please measure as shown in the accompanying cut. Inside measurements only are wanted. Never measure a collar on the face, but always on the rim side, as shown by arrowheads in cut. Should the horse require an extra wide collar at the top, please specify the width needed, in a similar manner as shown in cut at the draft of the collar; but state that it is the width wanted at the top.

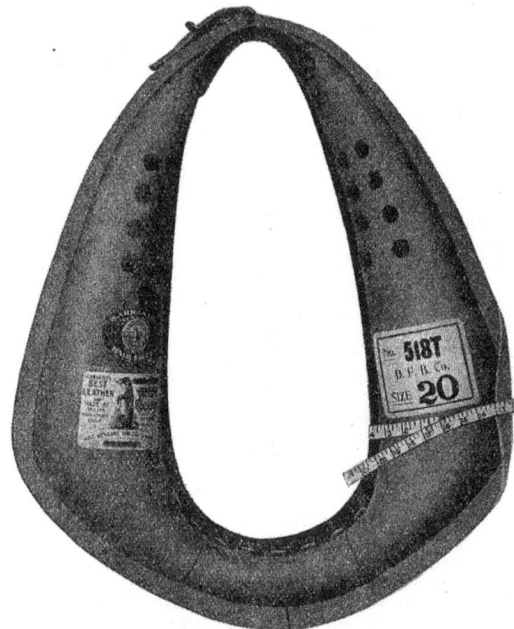

THE ABOVE CUT SHOWS HOW WE MEASURE A COLLAR FOR THE SIZE OF THE DRAFT

If you have no tape measure an ordinary string may be used in finding the circumference of the collar, marking the string where it meets and then measuring the string on a rule.

Final assessment of a collar fit: *Every horse and mule shoulder is different. And this anatomy can shift and surprise when put to work in a collar. That is why, in an ideal world, you should have an opportunity to judge how the collar 'sits' as well as fits when the animal is actually leaning in and pulling. You may find, in spite of your measurements, that it is too short, or too long, or too wide, or too thin, or entirely the wrong shape. Make the change and enjoy the result.*

lieve that a collar fit is a collar fit, end of story. But there are variables. The young horse in harness, for example a two or three-year-old, is still growing up and out. In the spring this horse will have a different size neck than six months later. Collar fit will need to be adjusted accordingly. On some farms with more horses than necessary it is possible to see dramatic weight gain on pasture that will have the collar be too small.

On my farm, from early on, I prized and prize having a wide assortment of collar sizes and shapes. I keep an eye open at farm auctions for any usable collars selling at reasonable or cheap prices. Horses' necks change over time and circumstance. To keep going with a string of six to ten draft horses or big mules, you need to have plenty of collars of all sizes and shapes from 20" on up to 28".

I learned the hard way that keeping my work string properly fitted in collars and harness gave me a leg up in being able to stay in the field.

COLLAR FIT 101, 102, 103 & 104

101: *From withers to the throat, the inside length is critical and can be estimated without a collar, BUT to assure a fit, the collar must be on the horse and the horse leaning into it.*

102: *The width and shape of the collar is important to assure that it does not pinch the sides of the neck and allows the collar to sit against the shoulder in a pull. Full Sweeney, Half Sweeney and Straight Face (or 'Norman') are the three primary shapes. The full sweeney is for the fattest profile neck.*

103: *When used with a stuffed, tufted and sewn sweat pad, the collar needs to be two inches longer than the animal's true measure. And the pad needs to be one to two inches longer than the collar. For example, a 26" collar would need a 27" or 28" pad. The reason for this is that the collar is measured for length straight up and down. While the pad length for the collar is measured along the inside curvature of the collar. The fastened pad needs to have its draft or widest portion align with the same area of the collar, and at the bottom the pad needs to be separated by two to four inches to allow freedom for the animal's windpipe.*

104: *The collar 'style' and construction should match the application. For example, heavy duty use for logging or pulling benefits from a collar that is wider at the draft and of considerably heavier construction giving a more stable 'bed' for the horse to lay into when exerting itself.*

Notice with these horses below that the sweat pads fit with a separation at the bottom to allow freedom for the windpipes.

Ichthammol ointment for preventing sore shoulders: *From long experience I have learned that catching potential shoulder sores early and applying Ichthammol ointment as an overnight drawing salve will shrink and toughen loose skin, reduce or alleviate the formation of boils, and remove itchiness. I keep it handy at all times. I check every shoulder at night, and if the working animal has developed a hot spot and/or loose skin and/ or raised areas with hair loss under the collar or pad, I apply a thick coat of the ointment to that area.*

A newer, short straw 'lumpy' collar.

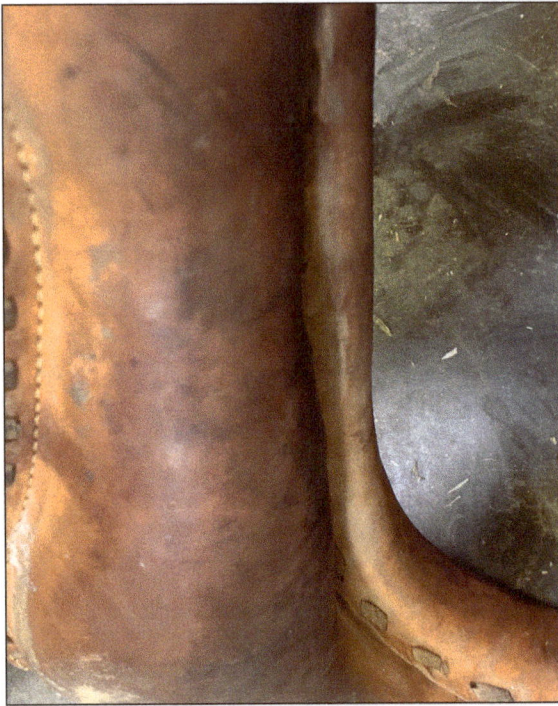

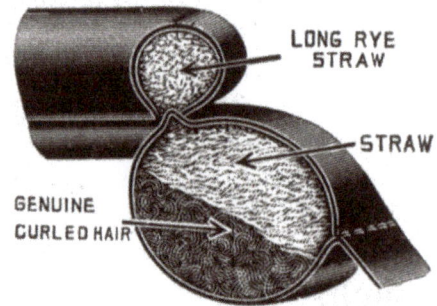

LONG RYE
STRAW

STRAW

GENUINE
CURLED HAIR

chopped straw
versus
long straw collars

The majority of modern horse and mule work collars are made by the mechanical process of stuffing chopped straw (or similar stuffing materials) into shaped leather tubes. The eccentric tubes are expertly made to result, on completion, in the particular shapes and sizes represented on these pages.

The cheaper chopped straw collars are prone to be lumpy, something which can be somewhat cured by soaking the collar in water and then very carefully and softly pounding the lumpy surface with a wooden or rubber mallet. But even if successful with such manipulation, you will likely find that the collar, over time and use, finds new ways to return to lumpiness.

The very best collars, those with smooth hard surfaces on the inside, are handmade and stuffed with long stem rye grass (see illustration below), each stem being laid parallel. This does away with lumpiness and provides a smoother and much more solid shape.

Collars With Hand Stuffed Throats Fit Better and Wear Longer.

The throat is the weakest point in most horse collars.

In making horse collars only extreme care in selecting materials and in putting these materials together properly produces good collars.

By good collars we mean those that fit your horse—that do not make sore necks and shoulders—that stand up year after year without constant repair costs—that do not break or weaken in the throat.

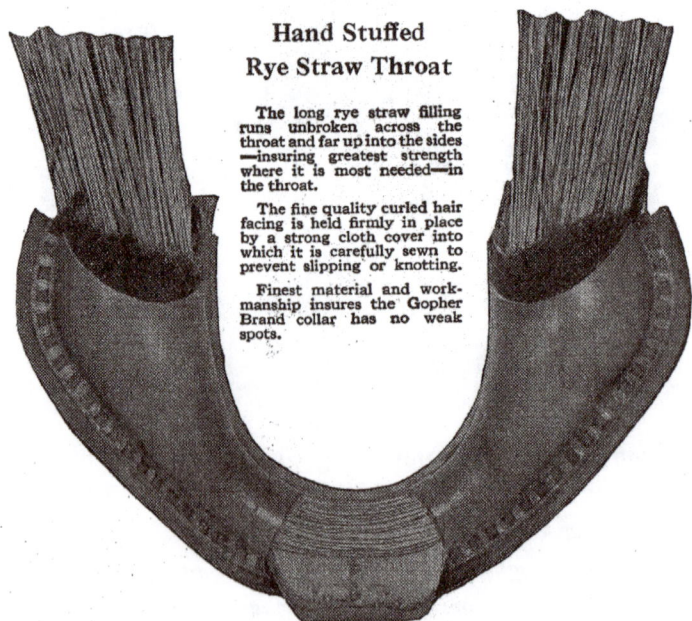

Hand Stuffed
Rye Straw Throat

The long rye straw filling runs unbroken across the throat and far up into the sides —insuring greatest strength where it is most needed—in the throat.

The fine quality curled hair facing is held firmly in place by a strong cloth cover into which it is carefully sewn to prevent slipping or knotting.

Finest material and workmanship insures the Gopher Brand collar has no weak spots.

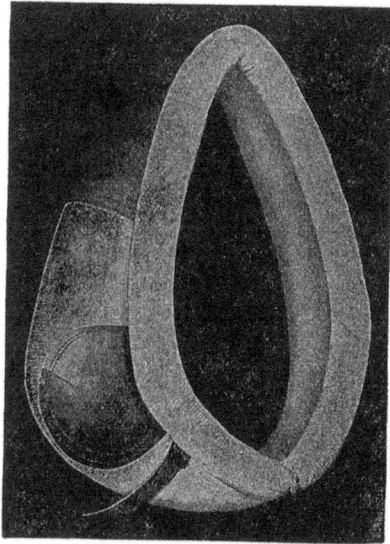

SURE CURE

Specialty 'medical' or 'remedial' collars were made by several manufacturers around a hundred years ago. 'Cure' collars, such as the one described below from a Dodson-Fisher catalog, employed different materials for stuffing and cover.

SURE CURE

SURE CURE—Extra heavy sail duck with reinforced waterproof covering over the top, heavy leather bearings and leather straps near bottom to secure the hame, filled with specially treated cotton that will not pack nor become hard and is well stuffed over the top, which forms a cushion on top of horse's neck.
Per dozen $28.00

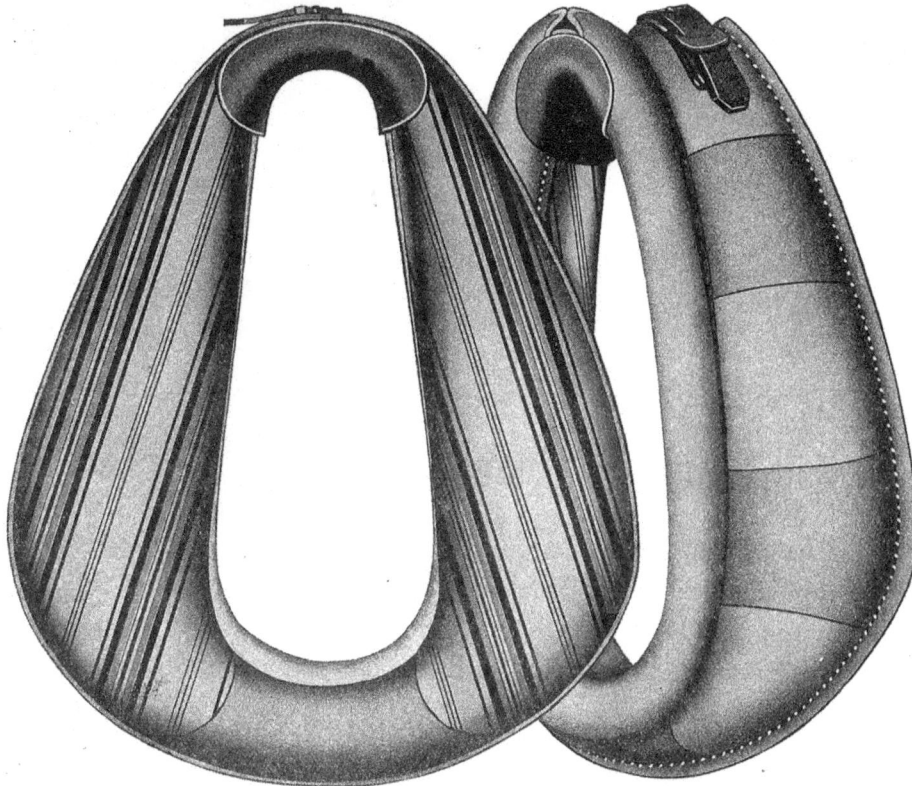

No. 12 DUCK COLLARS

'Irish' collars, or 'duck' collars (see above) feature an inside face that is canvas or 'ticking' cloth covered. These surfaces are considered to be better for horses with sensitive skin issues. It is somewhat like combining a sweat pad and collar all into one. While they may be 'healthy' for the working horse, they do wear out far quicker, and they are nearly impossible to get these days.

DODSON-FISHER CO.

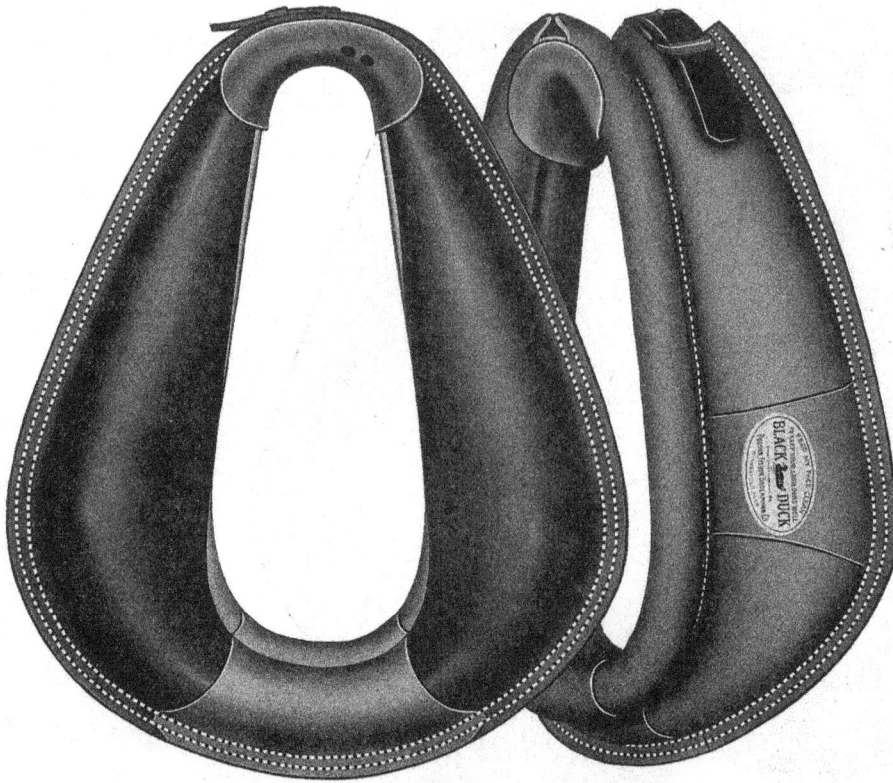

No. 54

No. 54—Per doz. $58.90

No. 54S—Sweeney
    Per doz. 58.90

Double texture black rubber face, split leather back and rim, sole leather collar cap, 18-inch draft. Always cool and dry. Sweat does not penetrate.

A rubber-faced collar. Doubtless we will see, over time, other experimental synthetic surfaces.

No. 58

No. 58—Per doz. $56.30

No. 58S—Sweeney
    Per doz. 56.30

Crack-a-jack striped ticking face, split leather back and rim hair felt face, sole leather collar cap, 18-inch draft.

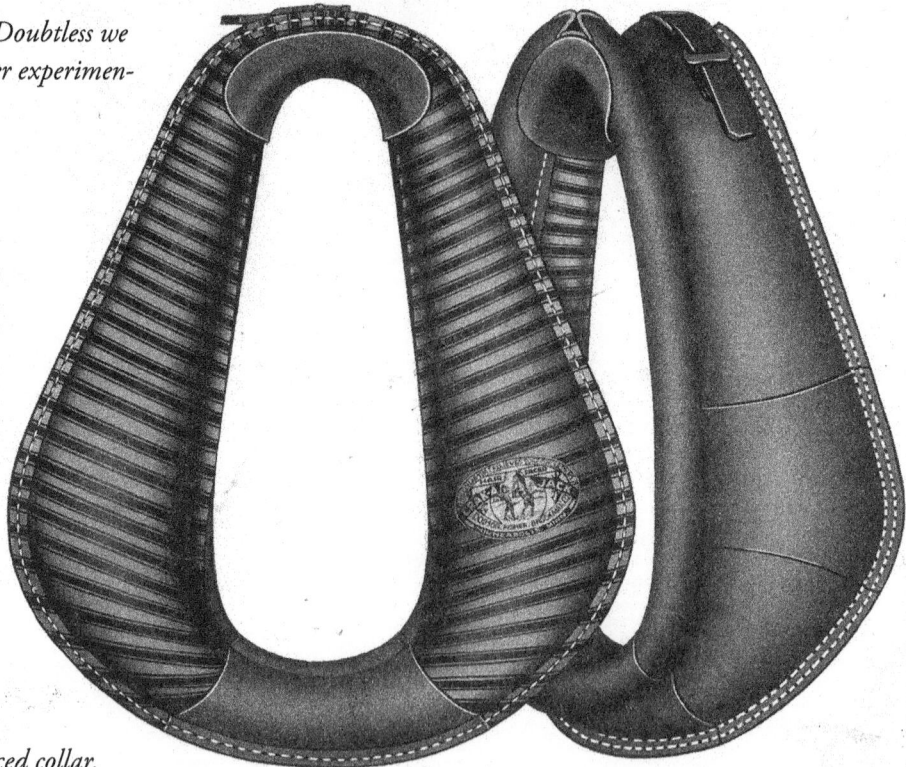

A ticking-faced collar.

Notice the stitching: If it is clean, uniform and substantive this usually connotes a superior collar.

No. 80

No. 80R Norman shape. Per doz. $89.50

No. 80SR—Sweeney shape, Per doz. **90.50**

Russet split leather back and rim, Russet collar leather f a c e, pieced back, thong middle seam, wide ribbon outside s e a m, sole leather collar cap, 16-inch draft.

No. 495R MULE

No. 495R—Mule Collars Per dozen....$133.10

Russet collar leather back and face, pieced back, t h o n g middle seam, wide ribbon outside seam, sole leather collar c a p s, 16-inch draft.

In some instances, a hundred years ago, a 'straight' or 'full' faced collar was referred to as a 'Norman.' Not sure the origin of that term in this case.

No. 335R

No. 335R—Norman Shape, per dz. $151.50.

Russet collar leather back and face, pieced back, t h o n g middle seam, wide ribbon outside seam, sole leather collar c a p s, 17-inch draft.

No. 336SR

No. 336SR—Sweeney Shape, per dz. $157.40

Russet collar leather back and face, pieced back, t h o n g middle seam, wide ribbon outside seam, sole leather collar c a p s, 17-inch draft.

No. 338SR — Same as No. 336SR, only made full Sweeney.

Per dozen . . . $153.20

Imagine being able to purchase a dozen new high quality collars for $153.20?

DODSON-FISHER CO.
GOPHER BRAND HORSE COLLARS
Have Proven Their Value-- They are Trade Builders

(GOPHER BRAND)

Per Dozen

No. 327R $174.30

Norman shape.

Russet collar leather back and face.

Heavy full shoulders.

Thong middle seam.

Wide ribbon outside seam.

Genuine curled hair face.

Pressed sole leather collar caps.

17 inch draft.

Buckle pieces and billets thonged on.

(GOPHER BRAND)

Per Dozen

No. 375SR $170.00

Sweeney shape.

Russet collar leather back and face.

Heavy full shoulders.

Thong middle seam.

Wide ribbon outside seam.

Genuine curled hair face.

Pressed sole leather collar caps.

17 inch draft.

Buckle pieces and billets thonged on.

GOPHER BRAND COLLARS FIT THE HORSE AND HE WORKS MORE WILLINGLY

DODSON-FISHER CO.
GOPHER BRAND HORSE COLLARS
Have Hand Stuffed Throats Insuring Longer Wear

(GOPHER BRAND)

Per Dozen
No. 385SR $181.60
Sweeney shape.
Russet collar leather
back and face.
Heavy full shoulders.
Thong middle seam.
Wide ribbon outside
seam.
Genuine curled hair
face.
Pressed sole leather
collar pads strapped
on.
18 inch draft.
Buckle pieces and bil-
lets thonged on.

(GOPHER BRAND)

Per Dozen
No. 387SR $169.50
Sweeney shape.
Russet collar leather
back and face.
Heavy full shoulders.
Thong middle seam.
Wide ribbon outside
seam.
Hair faced.
Pressed sole leather
collar caps sewed on.
18 inch draft.
Buckle pieces and bil-
lets thonged on.

DODSON-FISHER CO.
GOPHER BRAND HORSE COLLARS
You Make Steady Satisfied Customers When You Sell Gopher Brand Collars

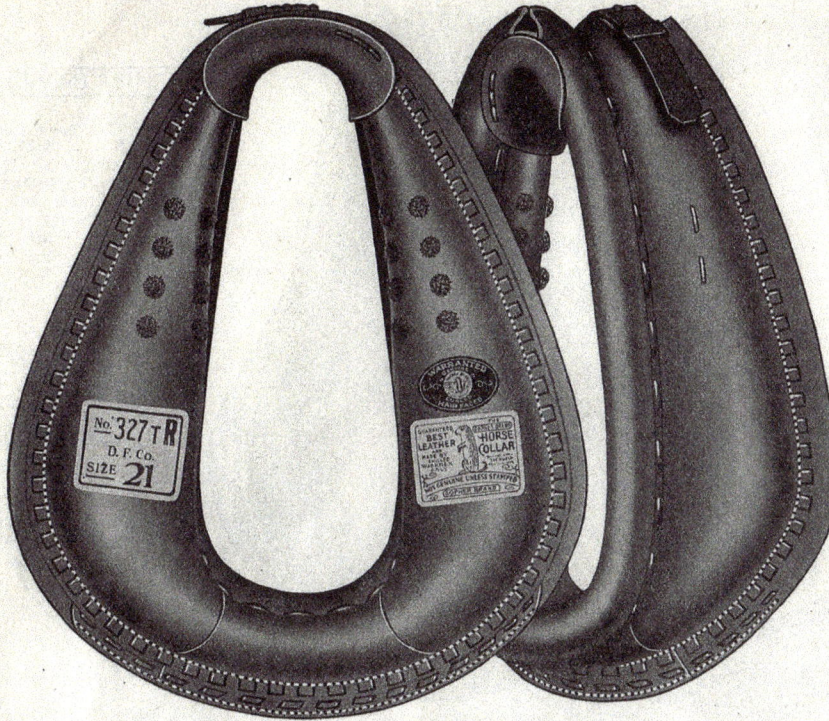

(GOPHER BRAND)

Per Dozen
No. 327TR $184.70

Norman shape tufted.
Russet collar leather back and face.

Heavy full shoulders.

Thong middle seam.

Wide ribbon outside seam.

Genuine curled hair face.

Pressed sole leather collar caps.

17 inch draft.

Buckle pieces and billets thonged on.

(GOPHER BRAND)

Per Dozen
No. 600T $197.90

Norman shape tufted.

Black collar leather back and russet face.

Heavy pieced shoulders.

Thong middle seam.

Wide ribbon outside seam, 2 rows.

Genuine curled hair face.

Pressed sole leather collar caps.

18 inch draft.

Buckle pieces and billets thonged on.

SATISFIED CUSTOMERS MEAN MORE SALES

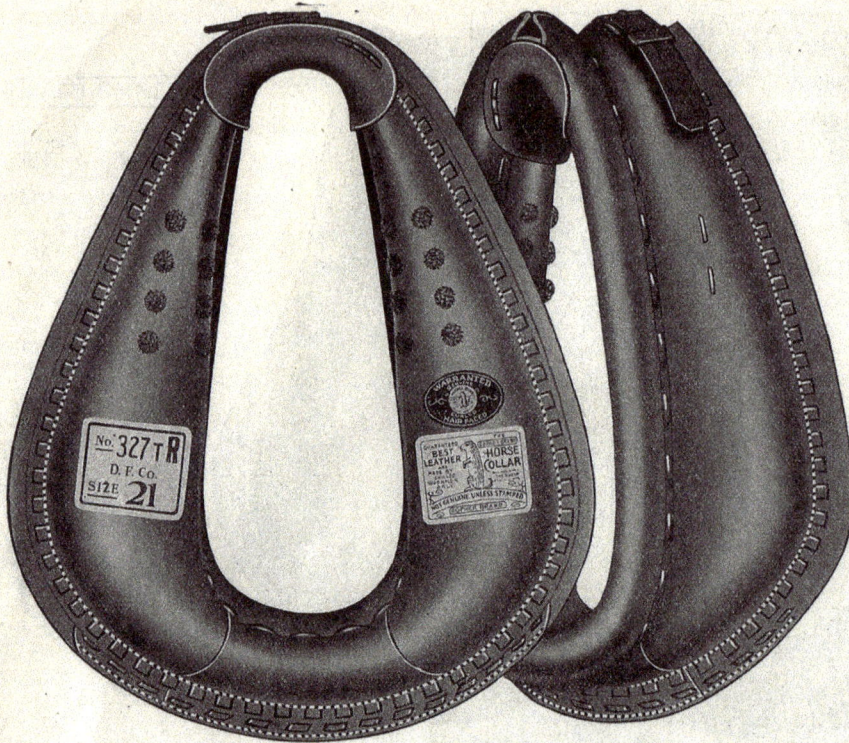

GOPHER BRAND

Per Dozen
No. 327TR $184.70

Norman shape tufted. Russet collar leather back and face.

Heavy full shoulders.

Thong middle seam.

Wide ribbon outside seam.

Genuine curled hair face.

Pressed sole leather collar caps.

17 inch draft.

Buckle pieces and billets thonged on.

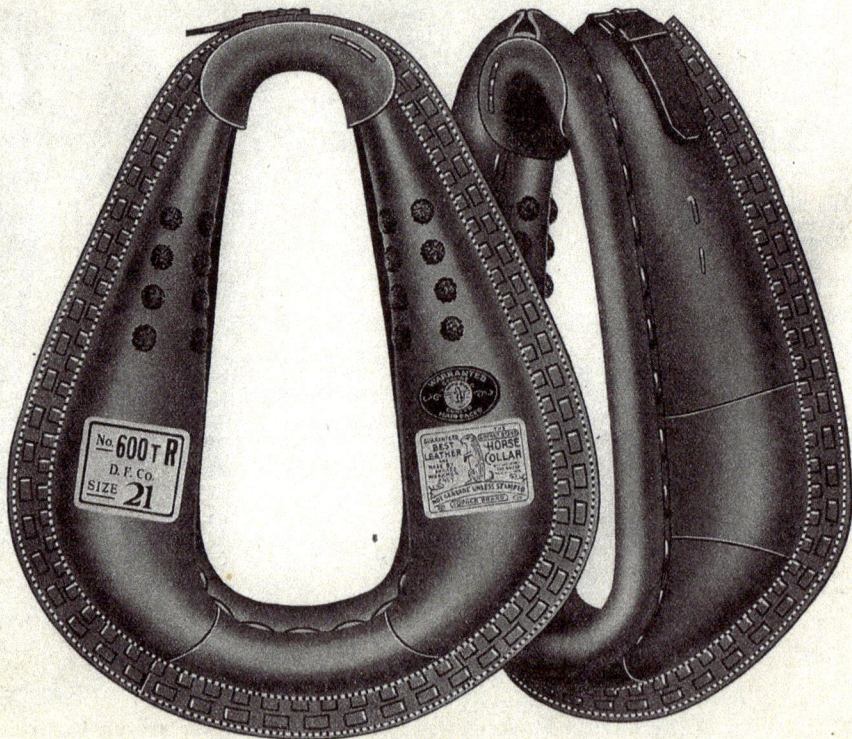

GOPHER BRAND

Per Dozen
No. 600T $197.90

Norman shape tufted.

Black collar leather back and russet face.

Heavy pieced shoulders.

Thong middle seam.

Wide ribbon outside seam, 2 rows.

Genuine curled hair face.

Pressed sole leather collar caps.

18 inch draft.

Buckle pieces and billets thonged on.

The information on these catalog pages, especially pricing, is of course outdated. We've left this here for what it tells us about advertising, construction, and cultural preferences.

WALLACE-SMITH HORSE COLLARS

No. 68.

No. 068.

No. 68—Heavy team, black kangaroo face, whole shoulder, russet rim and back, genuine curled hair face with inter liner, double thong stitched rim at throat, thong sewed rim, outside ribbon stitched, machine finished top.

No. 068—Heavy team, half Sweeney, black kangaroo face, whole shoulder, russet rim and back, genuine curled hair face with inter liner, double thong stitched rim at throat, thong sewed rim, outside ribbon stitched, machine finished top.

Even if in heavy, regular use, a well-constructed and well-cared for collar will last for generations. Keep them clean, removing salt build-ups occasionally wiping the surface with a rag soaked in neatsfoot oil compound.

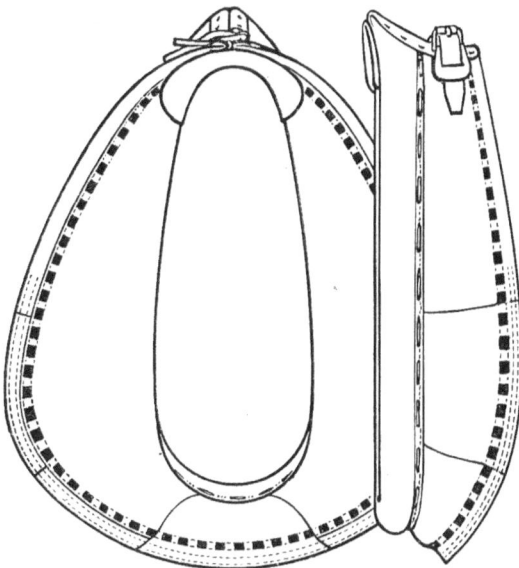

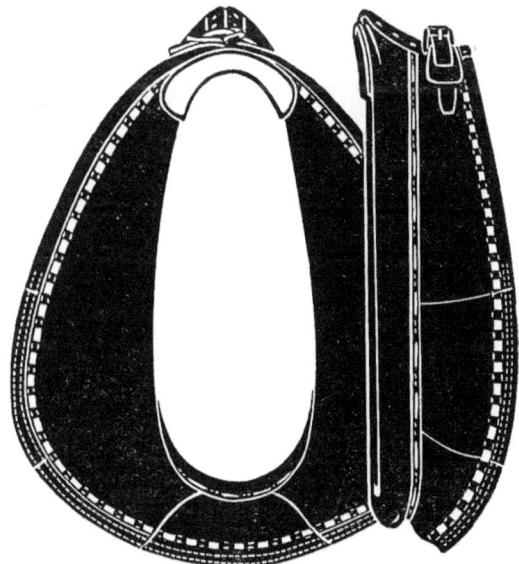

Heavy duty pulling or logging collar styles feature thicker wider construction with stronger stitching.

WALLACE-SMITH HORSE COLLARS

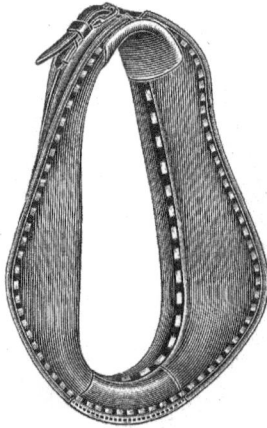

No. 158.

No. 190.

No. 158—Extra heavy team, quarter Sweeney, all russet, whole shoulder, genuine curled hair face with interliner, ribbon stitched throat, hand finished top.

Per Dozen

Sizes, 16, 17, 18 inch....................$......
　　　19, 20, 21 inch.................... ......
　　　　　　22 inch.................... ......
　　　　　　23 inch.................... ......
　　　　　　24 inch.................... ......

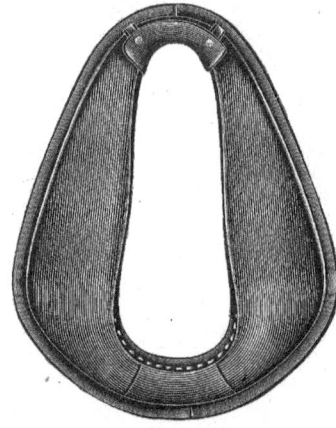

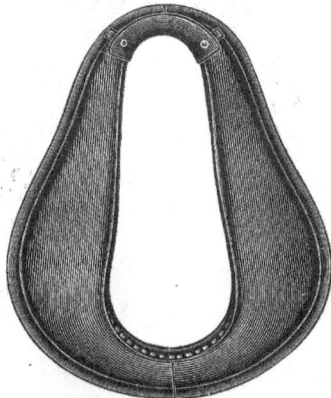

No. 190—Extra heavy blind seam, imitation case, russet face, whole shoulder. black rim and back, genuine curled hair face with inter liner, thong sewed, hand finished top.

Per Dozen

Sizes, 16, 17, 18 inch....................$......
　　　19, 20, 21 inch.................... ......
　　　　　　22 inch.................... ......
　　　　　　23 inch.................... ......
　　　　　　24 inch.................... ......

No. 195.

No. 210.

No. 195—Extra heavy blind seam, imitation case, half Sweeney style, russet face, whole shoulder, black rim and back, genuine curled hair faced with inter liner, thong sewed, hand finished top.

Per Dozen

Sizes, 16, 17, 18 inch....................$......
　　　19, 20, 21 inch.................... ......
　　　　　　22 inch.................... ......
　　　　　　23 inch.................... ......
　　　　　　24 inch.................... ......

No. 210—Extra heavy Norman style, imitation case, all black leather, whole shoulder, reinforced throat, thong sewed, hand tufted and hand finished top.

Per Dozen

Sizes, 16, 17, 18 inch....................$......
　　　19, 20, 21 inch.................... ......
　　　　　　22 inch.................... ......
　　　　　　23 inch.................... ......
　　　　　　24 inch.................... ......

STORING COLLARS: *The best way to store collars, short or long term, is to hang them upside down, preferably on a wide round surface. I have mounted empty one-pound coffee tins to the wall and they have worked fine for one collar each. A length of three to four inch diameter aluminum pipe can be made into a multi-collar hanger. If the collar is hung right side up, with time it will shift and break at the throat.*

"Big Jim"

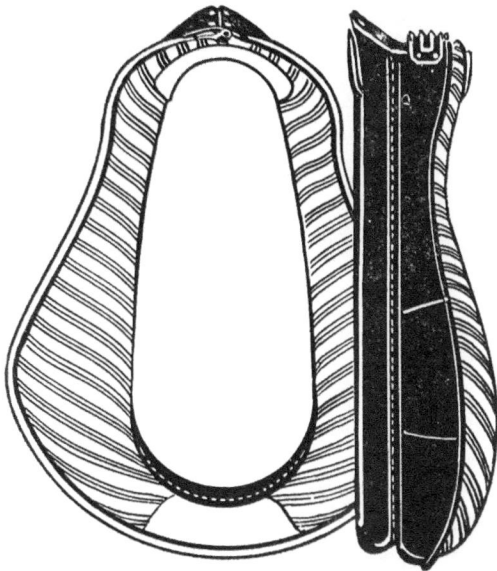

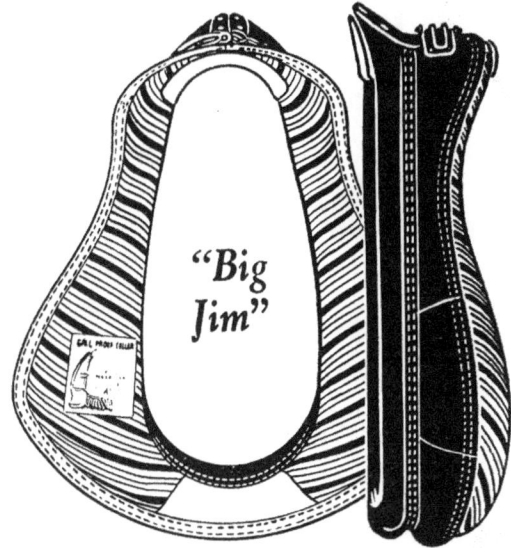

METAL PIPE THROATS

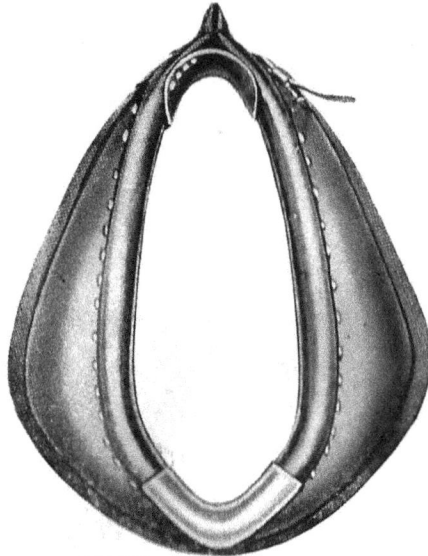

Pipe throat collars were the 'cat's meow' in the lineup. Far more expensive, they feature (see illustration to the left) a bronze shaped pipe which added a V to the throat area allowing a snugger fit while not obstructing breathing. If you find any, hang onto them. They can solve problems for animals that are otherwise difficult to fit.

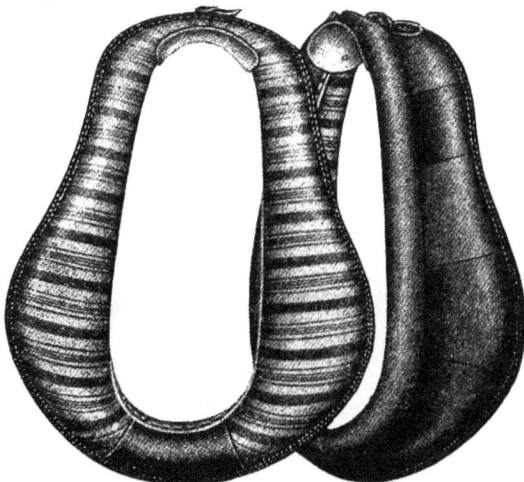

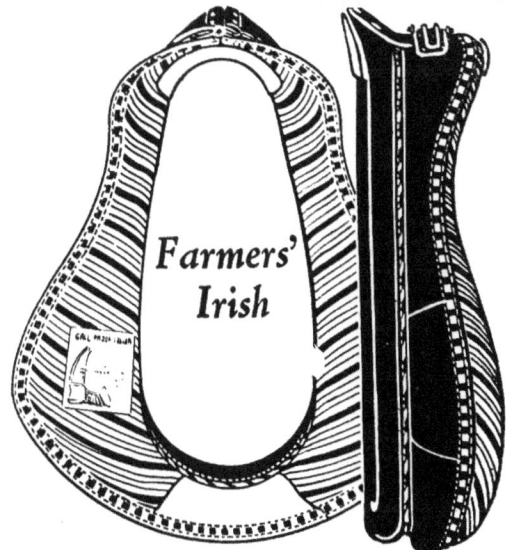

Farmers' Irish

BLUE EDGE

THINNER

AWNING

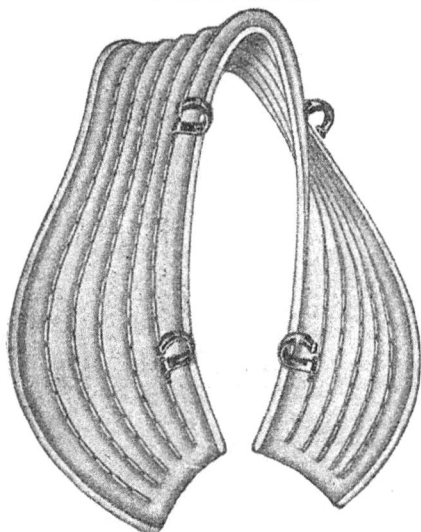
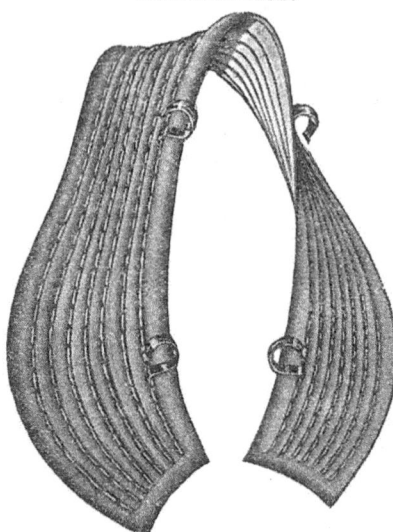
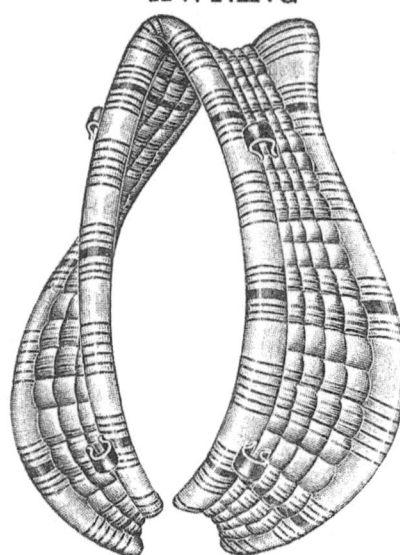

FLEXIBLE THROAT COLLARS.

Functioning somewhat like a pipe throat, the 'flat' or 'flexible' throat collar allows more room for the windpipe.

Why pads? Back in the early 70's when I was getting started, I asked Charley Jensen why he used collar pads. He answered, "Would you work in your leather boots without socks?" He didn't ask "could you?" He asked "would you?" Our equine workmates trust us to make such decisions. After a half century of working horses, my preference, for their sake, is to use collar pads, but ONLY if the combination of pad and collar results in a perfect fit.

PADS: Some teamsters never use collar pads, others would never be without them. And then there are those of us, forced to work with what we have, who mix it up and use pads on some horses while doing good work without pads on other animals.

Usually when we think of collar pads we mean the one-piece style you see illustrated on this page. But the category of 'collar pads' also includes the half pads you see on the bottom of the next page, and *top* or cap pads.

Collar pads may be stuffed, tufted and sewn construction such as those above. Or they can be made of single thickness felt as the one on the left. There are also modern synthetic construction pads that have come on the scene over the last 30 years.

Collar pads, or sweat pads, clip on the inside of the collar and offer a cushion to the horse's shoulder. While they are not absolutely essential to efficient draftwork, they do offer an additional level of comfort and expand the range of fit possibilities. Customarily sweat pads run two inches larger than collars. This allows for the pad to follow the concave curve of the inside of the collar just short of the throat. Depending on the thickness of the pad, the horse that wears a padless 24" collar will need a 26" collar with pad.

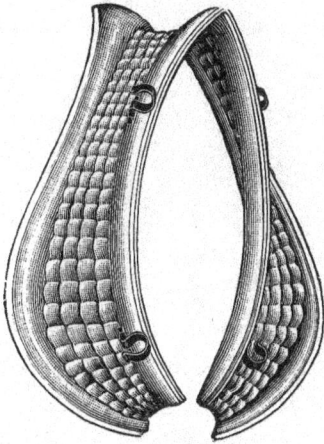

No. 212

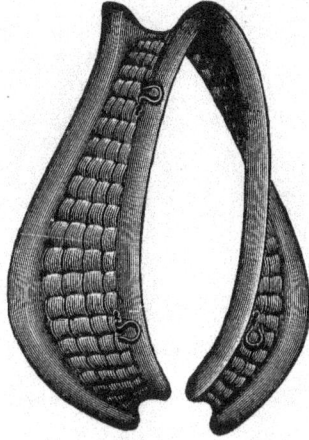

No. 2012

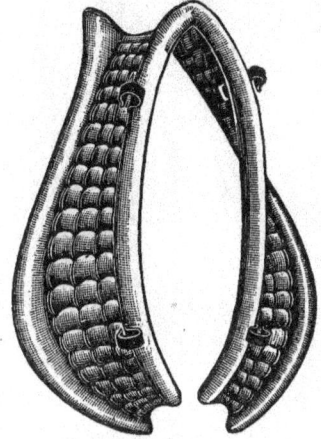

Nos. 512 and 120

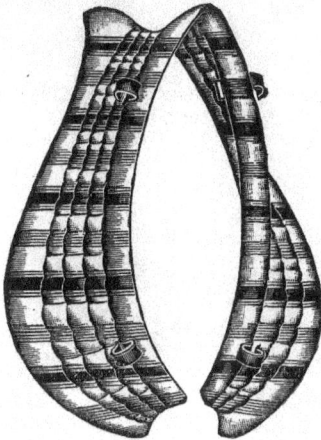

No. 50

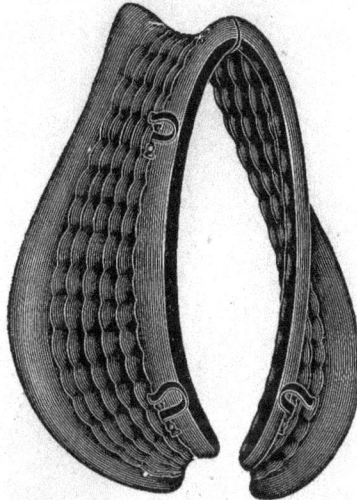

No. 711

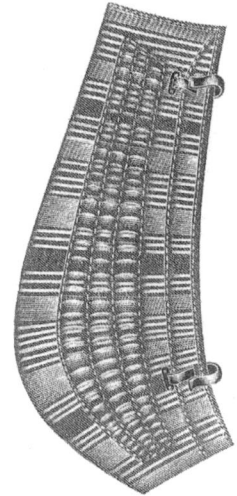

Nos. 7 and 8

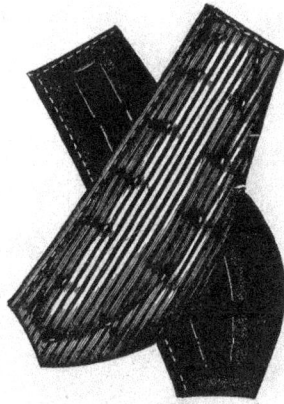

No. 100

TUFTED TOP SCOTCH CASE COLLARS.

YELLOW, RED EDGE

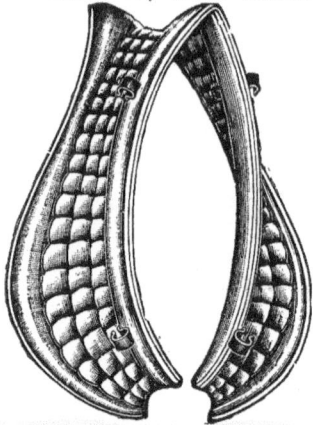

STRIPED DEER HAIR

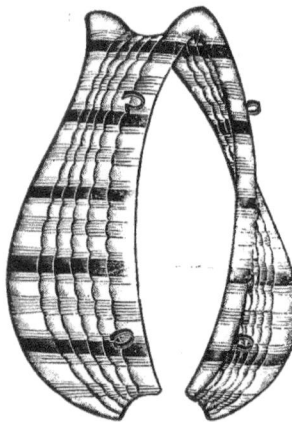

I have put into this book just about every attractive collar pad illustration I could find.

I do not use felt pads as I have found that sharp bits of hay and weed seeds can too easily embed themselves into the felt and irritate the surface of the animal's shoulder.

Protect and Ventiplex.

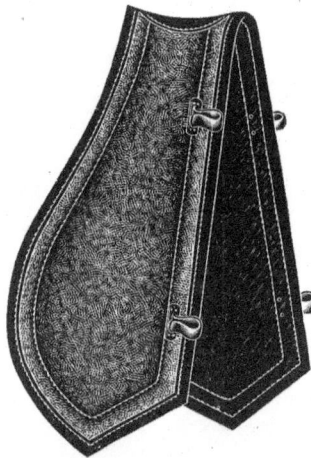

No. 315—Kurem.

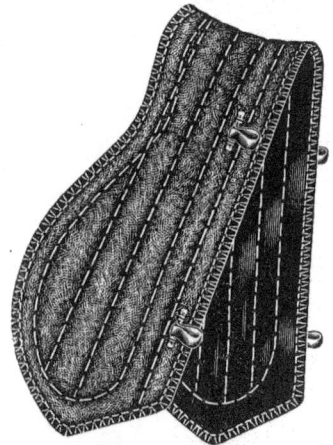

No. 320—King.

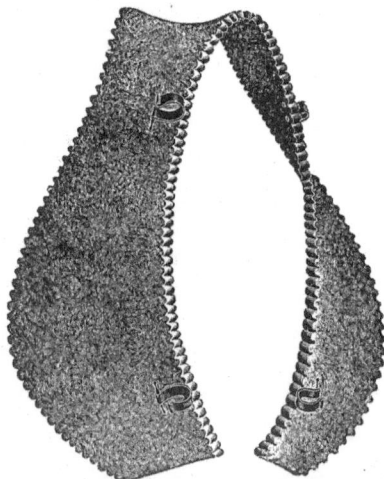

No. 78—Felt.

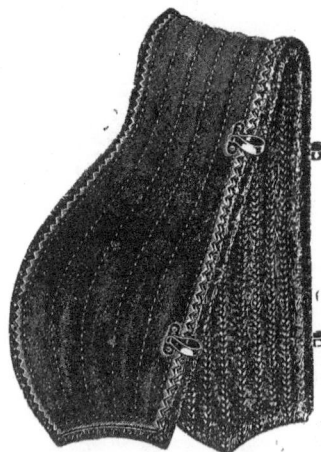

No. 610—Zero.

No. 710—Sanitary.

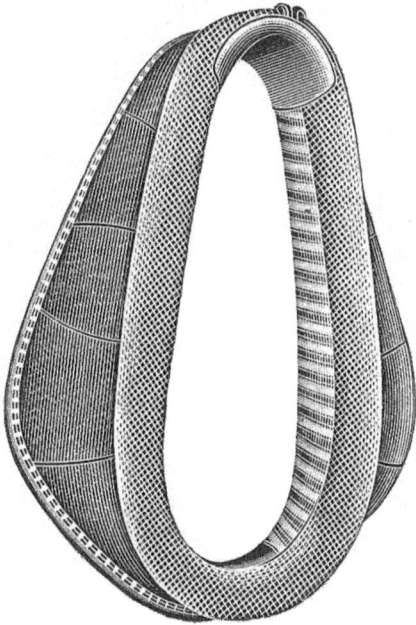

No. 16.

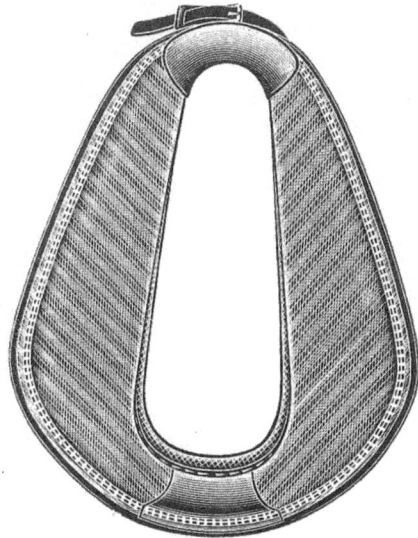

No. 17.

They have many names, a few include Irish, Scotch or Ticking Collars.

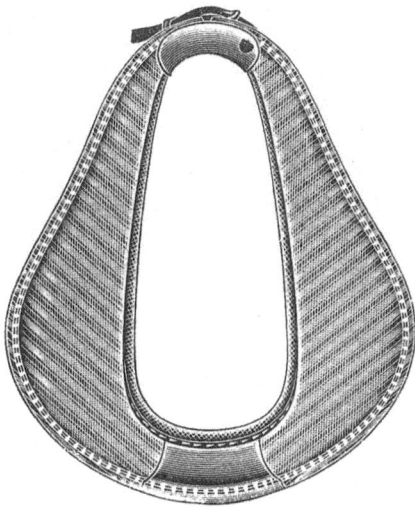

No. 016.

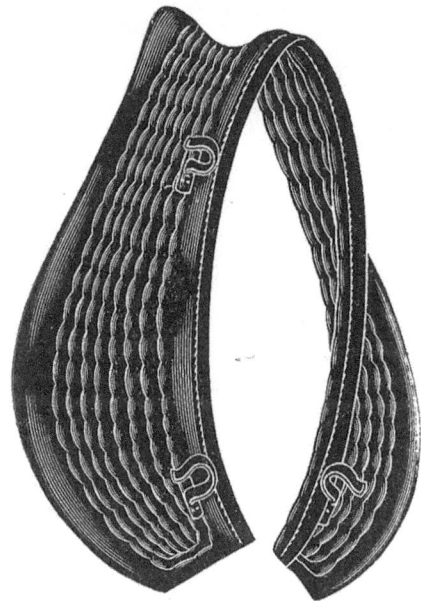

No. 100 WEAR CLIPS

JONES' HAME STRAP ATTACHMENT

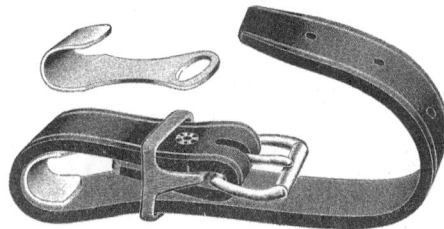

In both of these strap reinforcements, the thin metal protects strap leather as it rubs around rings or hame ends.

They may look tight to your eye, but these collars on Les Barden's team fit perfectly. Notice by his choice no pads.

This collar is way too big even with the pad.

This collar and felt pad are too big.

This collar is too small, notice how it pinches the windpipe on the bottom side.

The collars on these four appear to be the right length but they may be a tad bit too wide.

Buy the "DEPENDABLE" Line of Horse Collars

Per Dozen

No. 4—Heavy duck back, 8 oz. tick face, 5-inch web rim, leather hame lug and throat pieces, sole leather pad, draft 15 inches $19.00
No. 6—Same as above except draft 16 inches 21.70
No. 66—Same as No. 6 with split leather rim 22.70

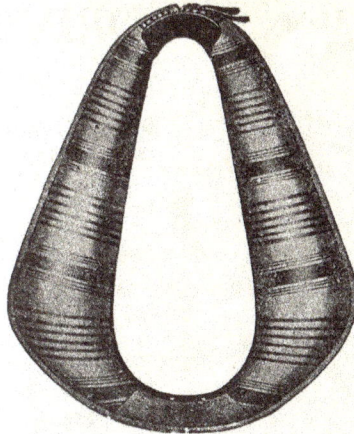

Per Dozen

No. 8—Chrome split leather back, 8 oz. tick face, 5-inch web rim, sole leather pad, draft 14 inches $24.00
No. 88—Same as No. 8 with bark split leather rim 25.00
No. 9—Same as No. 8 with web rim, draft 16 inches 27.00
No. 99—Same as No. 9 with bark split leather rim 28.00

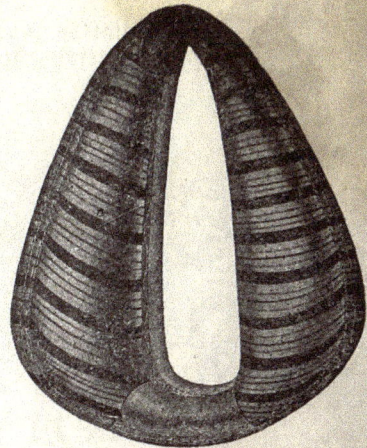

"TRIUMPH"

High grade heavy blue striped ticking face; heavy white duck back with leather re-inforcement; heavy 6-inch white web rim; special wide middle seam with leather welt; double thread outside seam; 17½-inch draft; pressed sole leather pads.
No. Triumph—Per Dozen $31.00

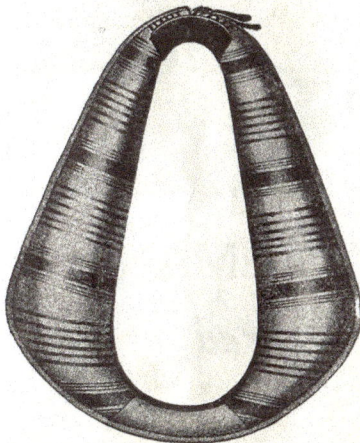

Per Dozen

No. 14—Bark split leather back and rim, 8 oz. tick face, sole leather pad, draft 14 inches $25.50
No. 16—Split leather back and rim, 8 oz. tick face, pressed sole leather pad, draft 16 inches 31.00
No. 17—Like No. 16 except 17-inch draft 34.00

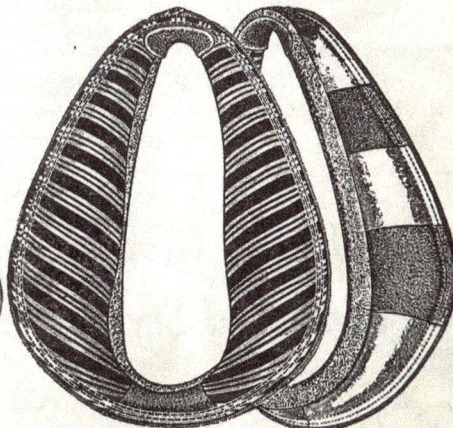

No. 780 heavy blue striped ticking face; hair pad white duck back with leather line ring and hame lugs; split leather rim; two rows thread sewed outer and inner seams; reinforced stuffed throat, 17½-inch draft; pressed sole leather pad. Per Doz. $31.50
No. 110—As above, without hair pad; 16-inch draft. Per Dozen 25.00

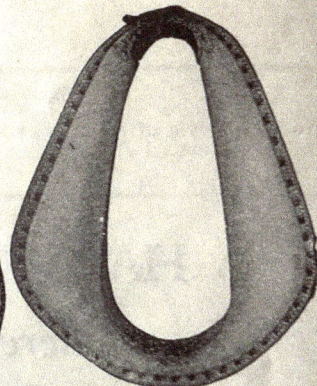

No.   **Per Dozen**

Senior—Split back and rim, blue chrome face, heavy 17-inch draft $48.00
Junior—As above, 16-inch draft 45.00
216—Is the Junior made in throatless style 45.00
217—Is the Senior made in throatless style 48.00
17K—Split back and rim, blue Chrome face, heavy 17-inch draft, machine sewed only ... 44.00
14K—As above, 14-inch draft, machine sewed only 35.00

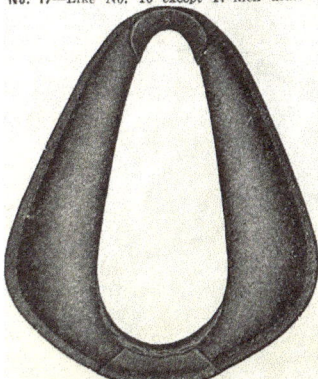

No. 321—Bark split leather back and rim, russet kip leather face, grain leather throat pieces and extended hame lug, hame lug laced along outside edge, sole leather top pad, draft 15 inches. Per Dozen $44.00

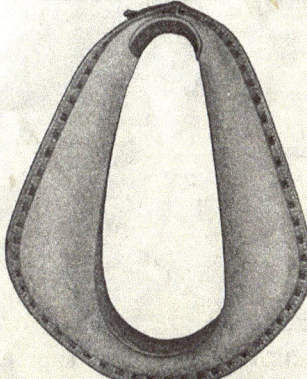

No. 145—Full russet kip, pieced back, ribbon laced outseam, whang laced middle seam; 15-inch draft. Per Dozen $73.00
No. 144—As above, 14½-inch draft. Per Dozen 69.00

No. 345—Full russet kip, pieced back, thong laced middle seam, ribbon laced outseam; stuffed throat, 15-inch draft. Some slightly scarred stock used in this collar. Specially priced. Per Dozen $69.00

No. 150 — As above, 16-inch draft. Per Dozen $74.00

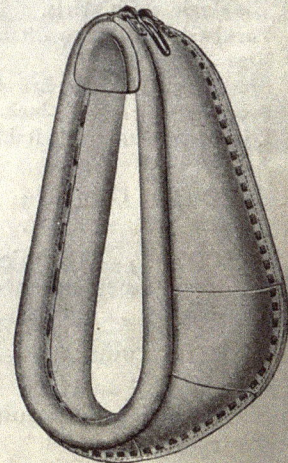

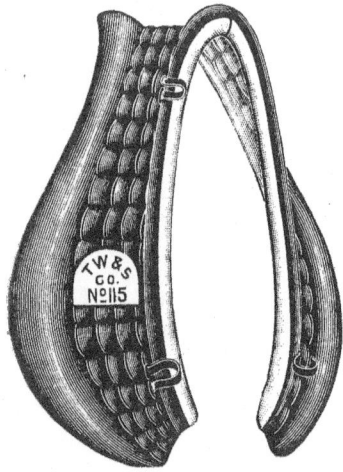

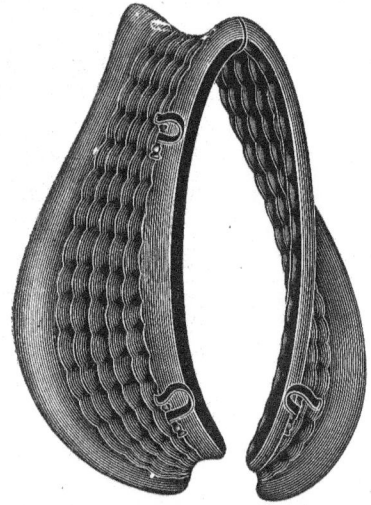

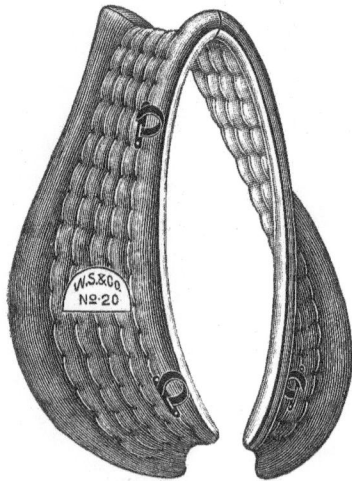

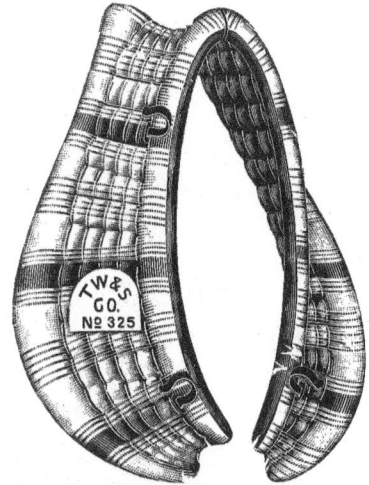

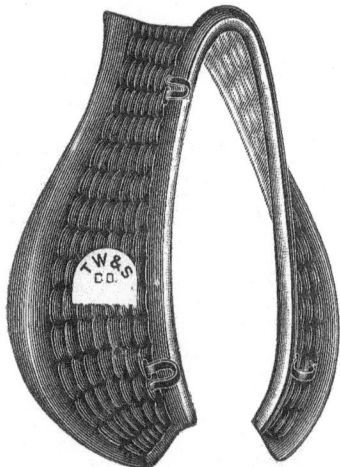

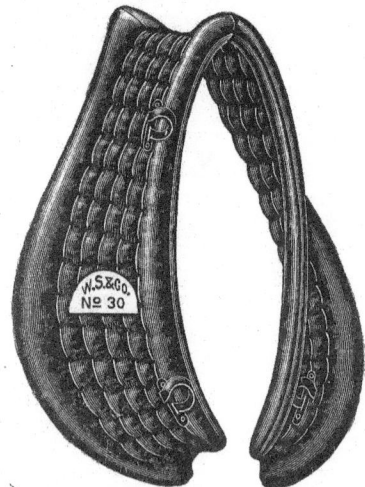

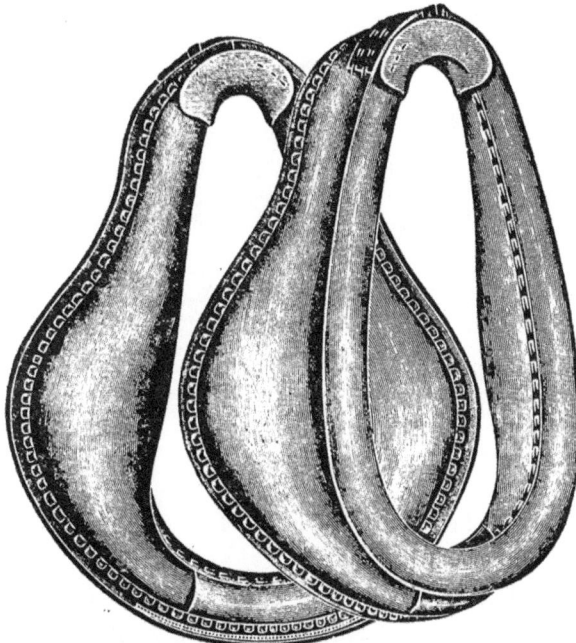

"UTAH"

Full Sweeney, all russet kip leather. Gall cure hair pad. Solid back, extended fender, reinforced throat, ribbon laced outseam. Thong laced inseam, 1½-inch sewed buckle and billet, large pressed sole leather pad thonged on. 18½-inch draft, regular sizes 15 to 22 inches.

No. Utah, price, per dozen $120.00

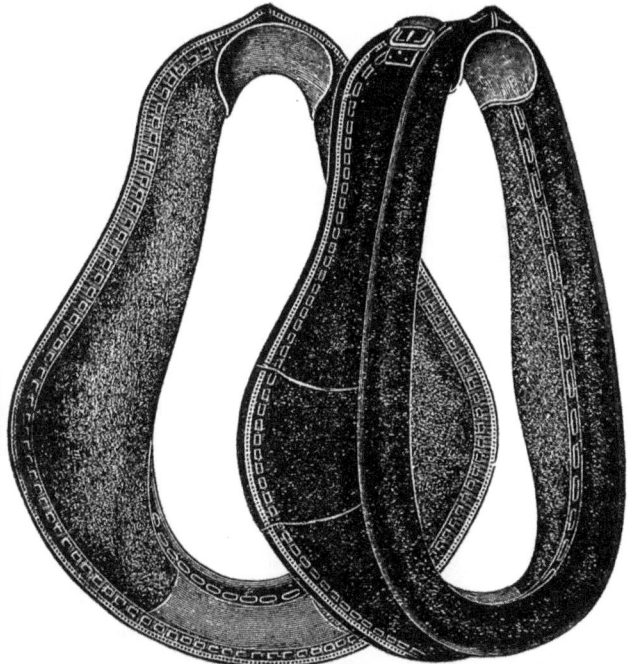

No. 550

Full Sweeney, black flexible leather wool face, black kip leather back and rim, reinforced throat, thong laced middle seam and outseam, pressed sole leather pad, two buckles and billets. Draft about 16 inches.
No. 550—Per dozen $86.00

GENUINE WOOL STUFFED
No. 560—As above, larger draft 17 inches, two buckles and billets. The body of this collar is filled with wool and then backed up with straw, making it 100% wool next to the shoulder. Per Dozen $100.00

A FINE LINE OF COLLARS

An Old-Fashioned Bench-Made Collar

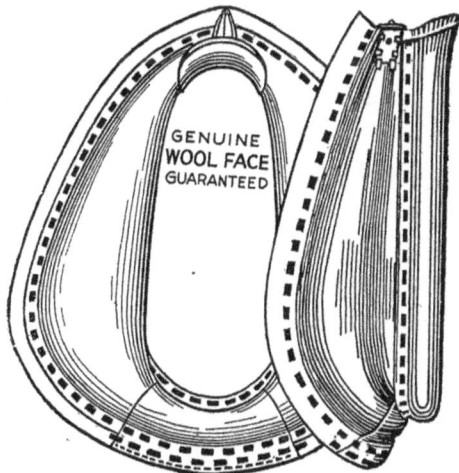

GENUINE
WOOL FACE
GUARANTEED

"DANDY"

Full russet kip, solid back with wide 1-inch extended fender on back; thong laced middle seam, whipped on cross stitch making wide hame bed, wide ribbon laced outer seam; large reinforced throat, two rows ribbon laced; sewed buckle and billet; pressed sole leather pad; about 18-inch draft. Genuine wool face.

DANDY—Per Dozen $108.00
No. 506—Same as above, only 17-inch draft. Per Dozen.. 96.00
No. 505—Same as 506, 16½ inch draft 90.00
No. 505L—As 505 with black flexible leather lugs. Price, per dozen 98.00

The Pride of the West

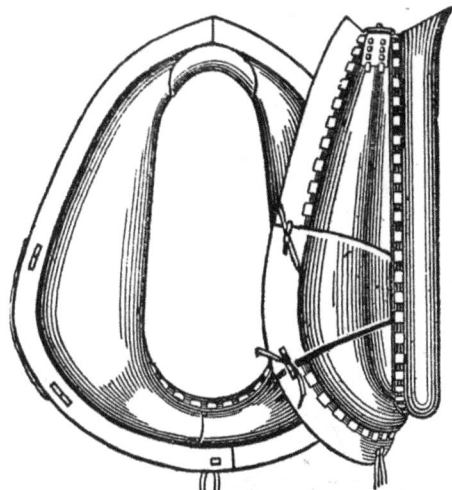

"OIL KING"

Genuine Scotch, full russet kip, largest and best collar made; it's a "Paramount." Extra large 20-inch draft. Extra big rim and large deep hame bed. Solid back with extra large hame lugs laced on side of fender. Special reinforcement throat. Wide Concord middle and outer lacing. Genuine curled "Hair face"; it will not gall. 1½-inch whang sewed buckle and billet. Leather strapped pad.
Per Dozen $144.00

Many new collars feature the latch shown above rather than a buckle. While it works well, it does take away an inch or so adjustment that is available with a two or three hole buckle strap.

Putting collars on over the horse's head: While 90+% of collars are designed to open at the top for ease of putting around the horse or mule's neck (see below left), occasionally there is efficiency to be gained in harnessing routines by having animals that will allow you to put the closed collar up over the horse's head. If you are working day in and day out, and using a sweat pad that has seated and formed to the collar, carefully pushing the collar up past the face of the animal, over the ears, and then down the neck to the shoulder, shaves minutes off the preparations. Once the horse has become accustomed to having his ears rubbed, the next piece of business is to make sure that the collar, with or without pad, is sufficiently wide to pass the ears and brows without force. (Most times the width of a well fit collar is wider than the face of that same animal.) Here's a trick that might work for you: rotate the collar so that you are passing the wider, lower portion over face and ears, then carefully at the throat, rotate the collar back to topside up (as in illustration below right.)

   It might take some time to get an animal willing to allow the contact and process, but I assure you that it takes but a few seconds of abusive pain and discomfort to *'train'* the horse to **refuse** the collar this way.

These collars appear to be the slightest bit too loose. And the blinders are tight up against the eyes. In both these cases, though I may not agree, it is apparent in the consistency we see that the teamster has made these choices. So I go back to the 'no only way' rule that says:

'what's right for you may not be right for someone else.'

STORING COLLARS

When storing collars it is best to hang them upside down, preferably over a round surface (like a one pound coffee can) that will hold the collar's shape.

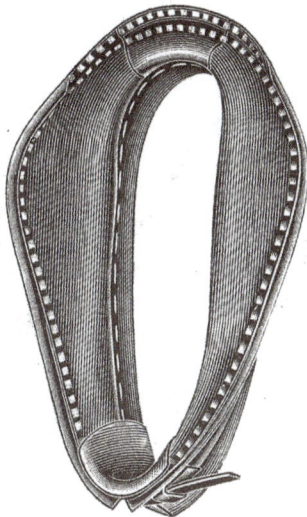

Of course the exception to the upside down rule applies with any specialty collar made to open at the bottom, such as the one on the right.

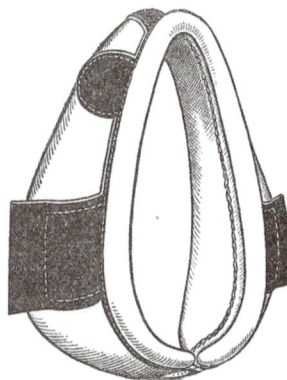

PERFECTION COTTON STUFFED

This collar is made of heavy sail duck, with heavy leather bearings for loggerheads and line rings. Is stuffed with specially treated cotton that will not pack and become hard.

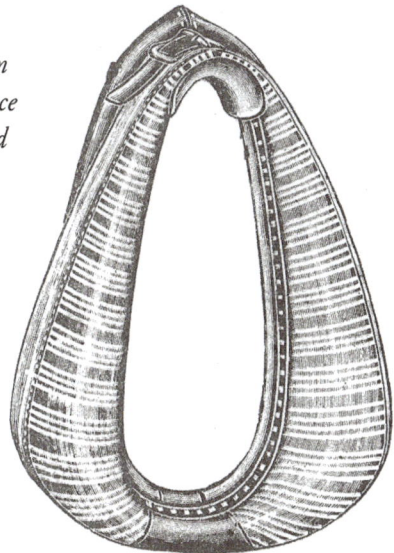

LEATHER AND CLOTH.

A stiff brush is best used on a dry cloth collar surface to clean debris.

adjustable collars

We have not found any examples in our set of circa 1900 harness catalogs of what are commonly called *adjustable collars.*

These pictures are of one of my adjustable collars, one I don't much care to use except when it's the only way to get through the day. As you can see it goes from 20" to 22" BUT when it goes long it goes too wide.

Also this collar was machine-made by stuffing chopped straw into the leather tube resulting in a lumpiness that 'rubs' the animal in the wrong way.

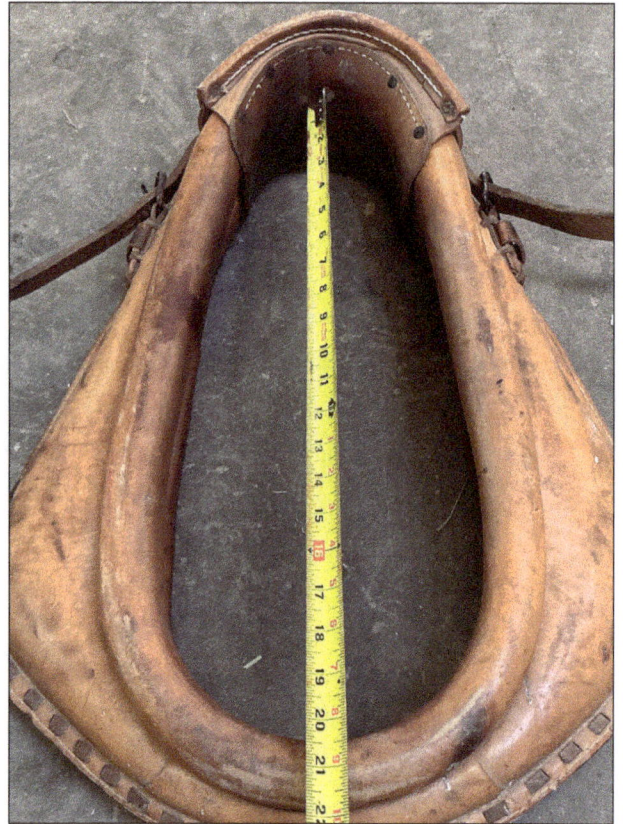

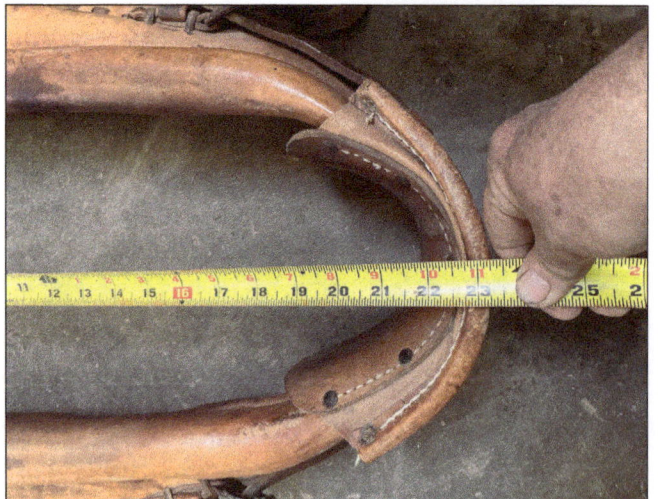

Above, the collar is lengthened out to 22" and you can see how this interrupts the curvature of the collar and makes it too wide, also making it difficult to get hames seated on.

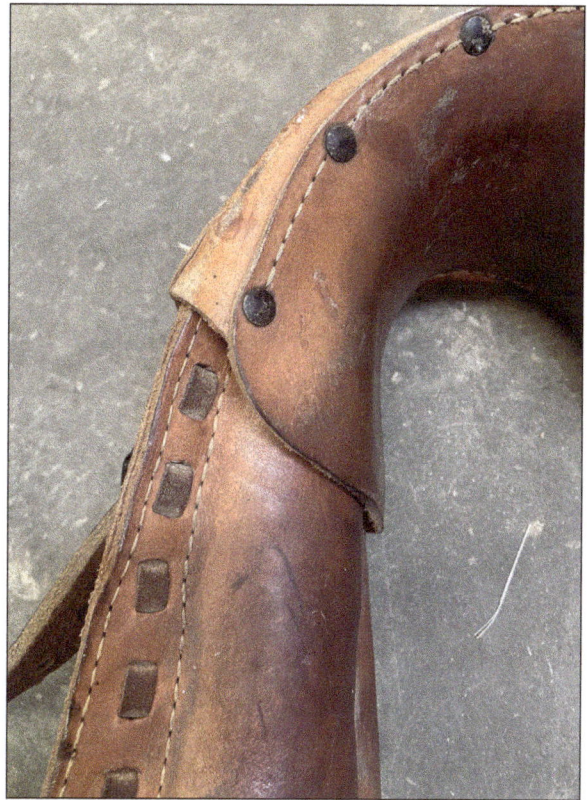

At the shortest position you can see how the collar sides fit into the cap snuggly. Bottom left is full length, right is shortened.

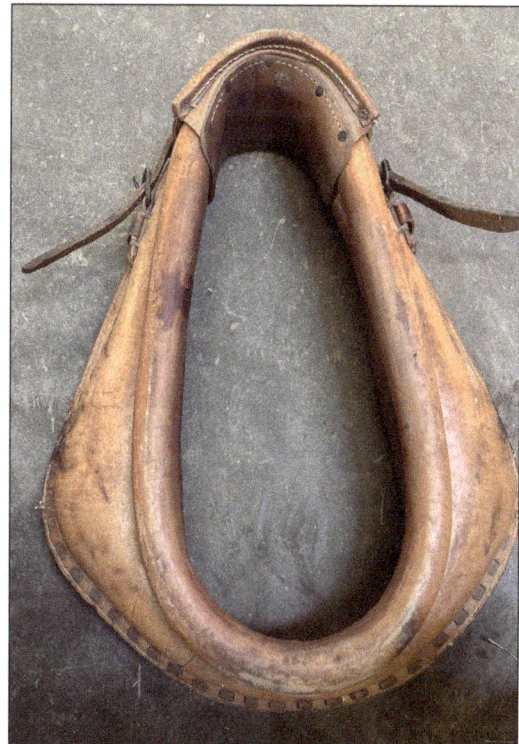

BE THOUGHT-FUL USE COLLAR PADS!

This illustration from an old collar pad advertisement makes an excellent case.

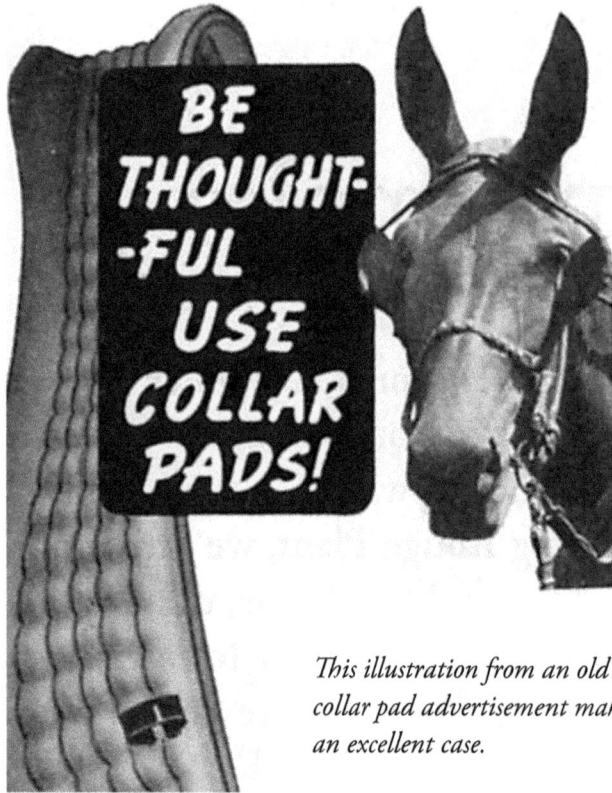

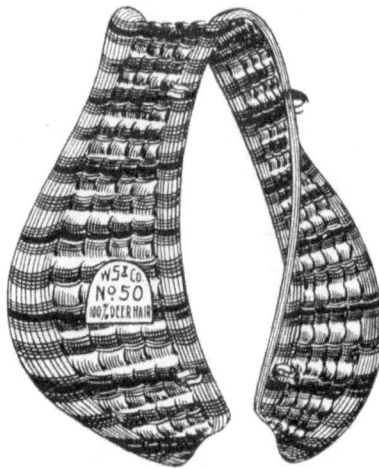

"SOLO" FABRIC COLLAR PAD

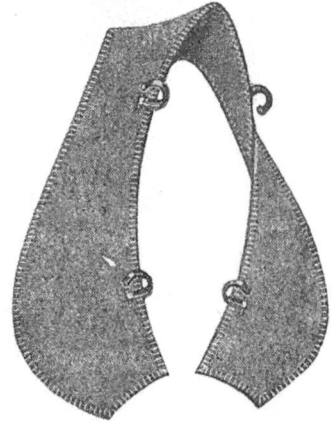

With time and practice you will learn to recognize the look of a comfortable horse working. These superb Belgians have that look, and notice they are all wearing collar pads.

I have this 100-year-old, oft-repaired and reinforced, high-end tufted collar. It shows its age and yet, were I to replace the hanger-abused top cap, it would still be serviceable on a good horse.

Before I owned this collar, which I have used as a demonstrator for students, someone hung it rightside up on a peg for a long time. This misshaped the top cap and cracked the throat. Then they patched the throat and tried to use the collar only to have the horse get sore withers from the top cap. So many lessons from one old collar.

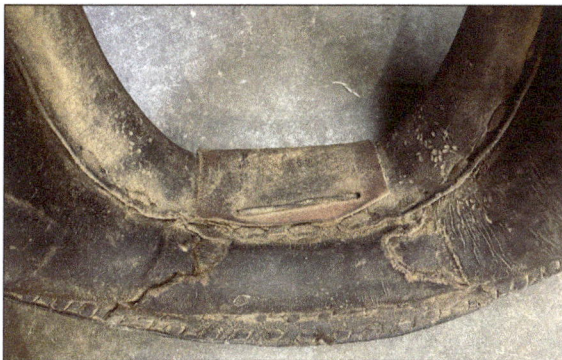

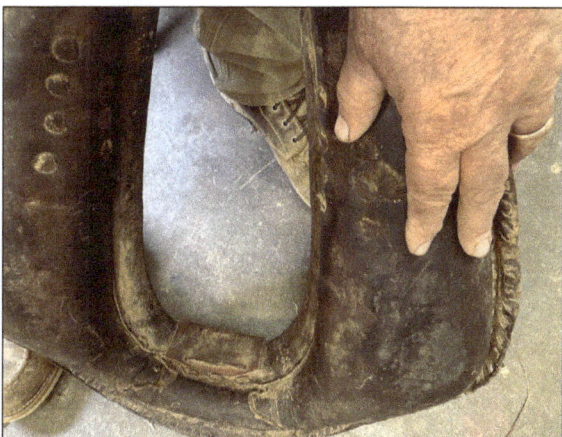

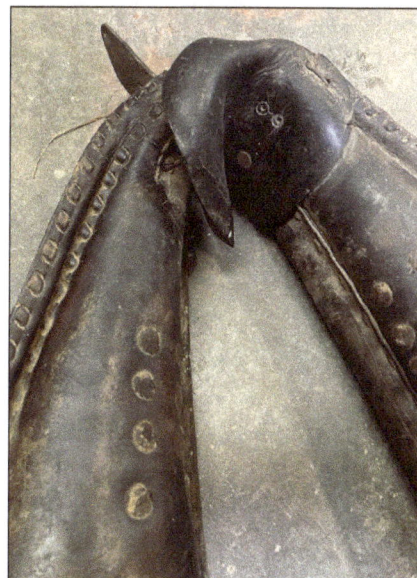

COLLAR PADS AND SWEAT PAD HOOKS

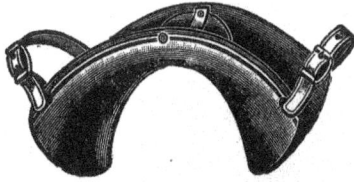

SOLE LEATHER COLLAR PADS
With Iron

	Per Dozen
No. 3	$ 7.50
No. 4	8.70
No. 5	11.60
No. 6	13.00

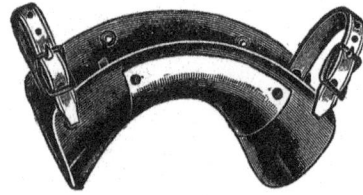

SOLE LEATHER COLLAR PADS
Without Irons

	Per Dozen
No. 3	$ 7.30
No. 4	8.50
No. 5	11.40
No. 6	12.80

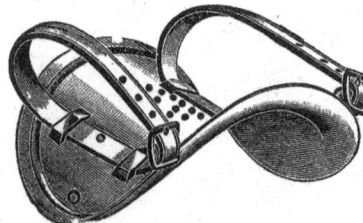

SOLE LEATHER COLLAR CAPS
Without Irons or Straps

	Per Dozen
No. 12	$3.80
No. 13	5.40
No. 14	6.60
No. 15	9.50

BOSS COLLAR PADS
Leather and Zinc

	Per Dozen
No. 3	$11.00
No. 4	12.30
No. 5	15.00
No. 6	16.50

ZINC COLLAR PADS
Perforated on Top

	Per Dozen
6½ inch	$8.90
7 inch	8.90
7½ inch	10.30
8 inch	10.30

2 Dozen in box.

SWEAT PAD HOOKS

	Per Dozen
Team size	$0.30

Made of Finely Japanned Spring Steel

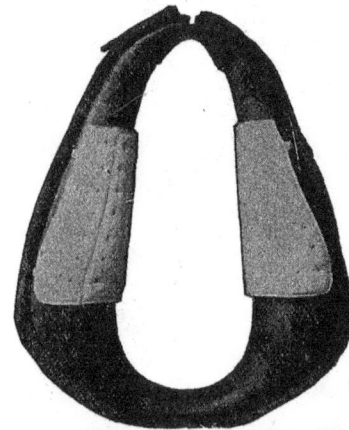

TRAUB COLLAR PADS

	Per Dozen
Small, 18 to 20 collars	$18.00
Medium, for 20 to 23 collars	18.00
Large, for 24-inch collars and larger	18.00

*Make sure collar size is increased one inch or more when you add top pads,
and two inches when you use full pads.*

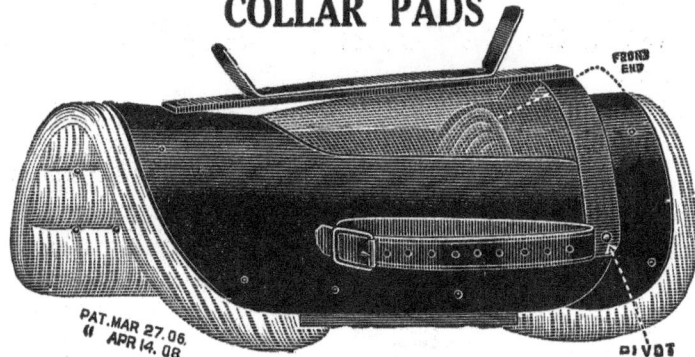

COLLAR PADS

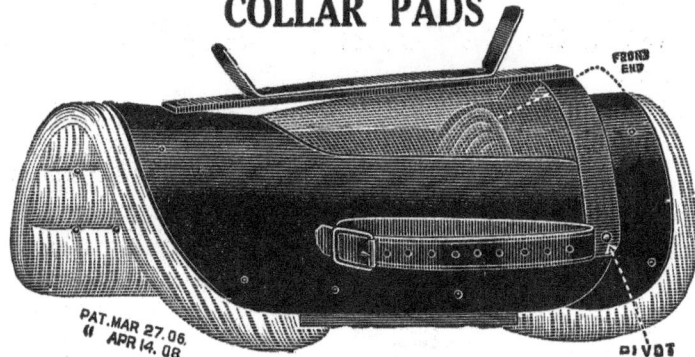

PAT. MAR 27.06.
APR 14. 08.

SACHSE SURE CURE NECK PADS

Per dozen $60.00

The SACHSE SURE CURE NECK PAD sells at sight as its merits can be seen at once.

The length of this pad is 15½ inches, and the free open space is 7 inches. It has a stationary bridge upon which the collar rests so it cannot come in contact with with the sore part.

The front part of this pad works on a pivot so the horse can move his head up or down with ease. The pad is made of sheet steel and black japanned. The open part is covered with screen so the flies cannot get at the sore.

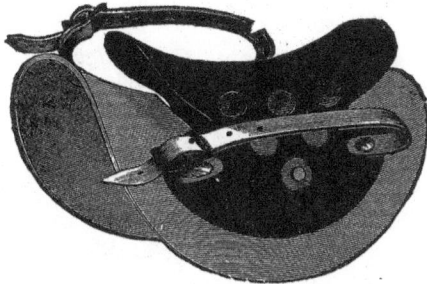

SWING BACK COLLAR PADS

Per Dozen

No. 2 Swing Back $14.00

This pad is made of high grade, carefully selected sole leather, so shaped as to adjust itself to the horse's neck. An open space in the center leaves the afflicted part free.

The metal plate is considerably elevated and is so secured on each side as to admit of a swinging motion.

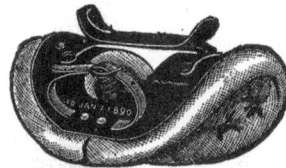

CYCLONE COLLAR PADS

Per Dozen

"Cyclone" $12.50

The Cyclone is the most effective remedy where the top of the horse's neck has been injured. Its constant use is a certain preventive of injury to the top of the neck.

The padded plate rests on the neck surrounding the afflicted part, while the elevated bar holds the collar clear of it.

Fistulous withers is a condition where injury at the withers results in a pus-filled pocket, or pockets, at the point of the neck where collars typically rest. This may be the result of horses or mules forced to wear ill-fitting collars and/or harness. It can be so painful that

WABASH BRIDGE COLLAR PAD

TRADE WABASH MARK

PATENTED
JUNE 3, 1913

the animal will fight the collar and/or refuse to move.

Over the last 150 years many top pads were offered to reduce the contact with the surface and allow that a healing animal might still be used in harness. On this page we offer a few examples.

DEER SKIN COLLAR PADS

Nos. 01 and 1.

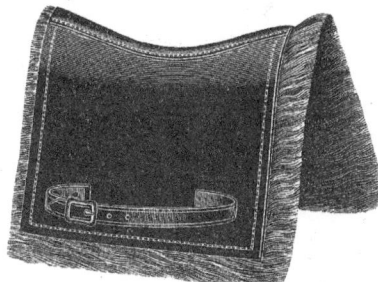

No. 3.

No. 10.

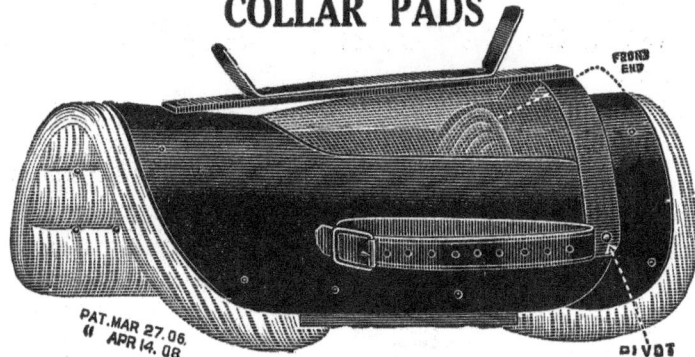

(a) Cloth and leather combined.   (b) Light duty remedial collar.   (c) Zinc salt mine collar.

A pile of consigned collars at the last SFJ Horsedrawn Auction in 2013.

RATCHETS FOR COMBINATION LOOP HAMES.

TOP LOOPS FOR COMBINATION LOOP HAMES.

With combination loop hames, the top hame strap stays flat and adjustable.

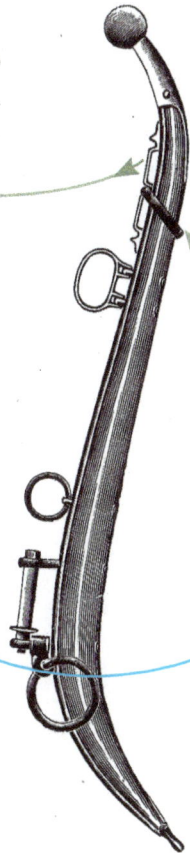

HAMES: The most popular style of work harness hame is of tube steel construction (with or without ornamental balls on top). Steel reinforced hardwood hames are also seen in use. As illustrated, there are two common styles of attachments for the top hame strap. One has it where the strap twists and slides through one of two slots in the hame top. The other features a sliding 'ratchet' square that rests in one of three notches in the receiver; the strap threads this on both hames. This is the more common and most preferable setup.

No. 155 Screw Bolt and
Extra Back Strap Ring.

A pair of hames, affixed top and bottom by hame straps and illustrating the relationship to the pair of tugs or traces.

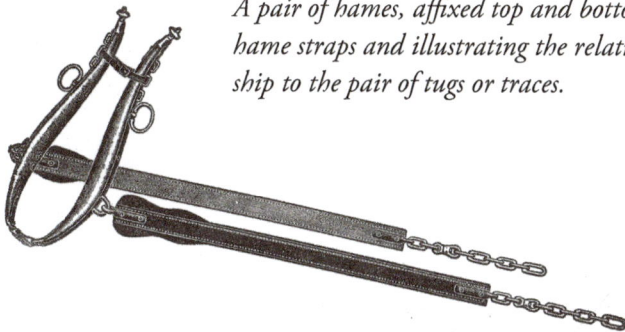

Without top loops, the hame straps must twist and pass, sideways through the slots at the top of the hame. This makes adjustment tedious.

Dandy Favorite Ball Top Screw Bolt.

BALL TOP IRON CLAD HAMES

 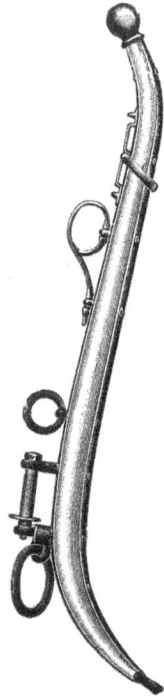 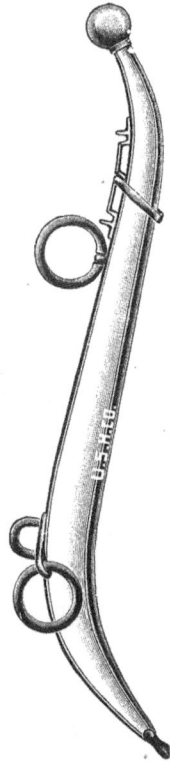 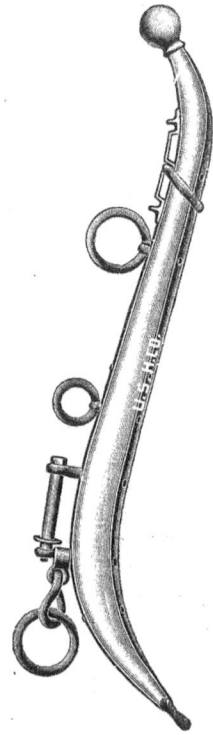

No. 40 Clip.   No. 40 Screw Bolt.   No. 92 Clip.   No. 92 Screw Bolt.

EBERHARD'S HOLLOW IRON HAMES.

No. 50.   No. 30.   No. 10.   No. 20.

HAME BALLS

Dandy and Favorite.

Globe.

No. 92.

Nos. 500, 710 and 720.

DANDY AND FAVORITE HAME BALLS

The Consolidated Hame Company was formed around 1896 when Baker, Carr & Company merged with Bartlett & Rowell. The Consolidated Hame Company purchased the Rome Hame Company and in 1902 were merged with various concerns which became known as the U.S. Hame Company.

Note: Early twentieth century commerce found it useful as advertising to dress up delivery horses through their harness as a means of 'branding' product and service. The most common example of this were Brewery Draft Hitches. Customizing hame designs proved a direct way to accomplish this.

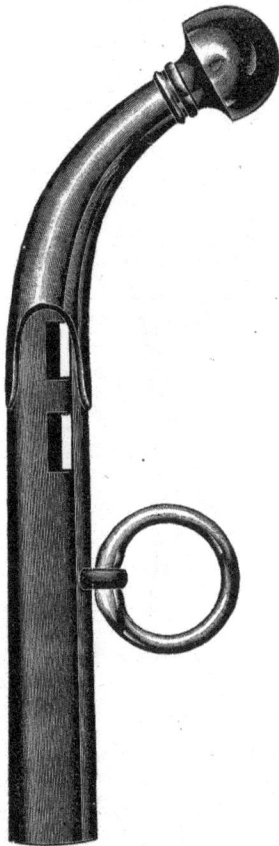

No. 93.

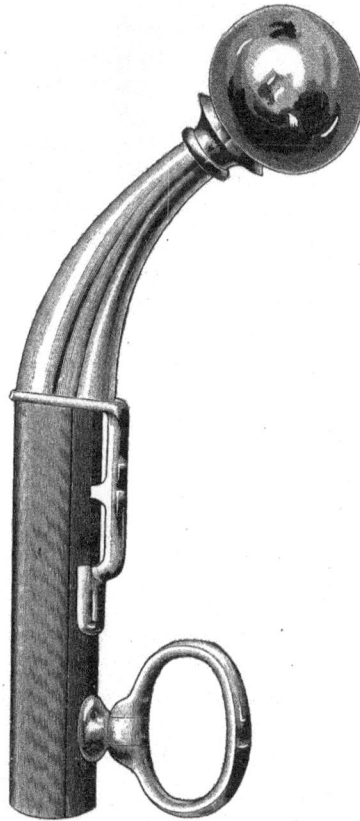

No. 94.

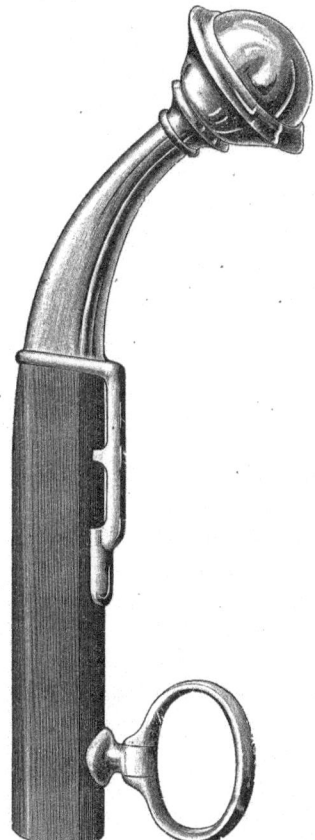

No. 95.

HAME TIPS

No. 2693 and 2702 HAME BALLS

		Per Dozen Pairs	
		Nickel	Brass
No. 2693	2 inch ball for No. 500 hames..	$17.30	$16.00
No. 2702	2 inch ball for No. 522 hames..	17.30	16.00

No. 2694 and 2699 HAME BALLS

		Per Dozen Pairs	
		Nickel	Brass
No. 2694	2 inch ball for No. 920 hames	$20.50	$18.70
No. 2699	2½ inch ball for No. 930 hames	24.00	22.00

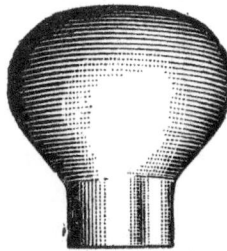

No. 700 HAME BALL

For Saco Hames.

		Per Dozen Pairs	
		Nickel	Brass
No. 700	2 inch ball....................	$17.30	$16.00

No. 2689 DANDY HAME TIPS

Size of Ball 1¾ Inch

		Per Dozen Pairs	
		Nickel	Brass
No. 2689	1⅛ inch hole.................	$25.50	$24.00
No. 2689	1¼ inch hole.................	29.30	28.00
	1⅛ inch hole, special........	20.40	19.20

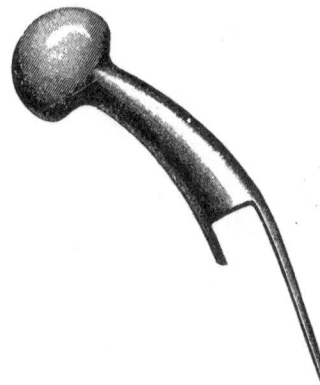

No. 2690 FAVORITE HAME TIPS

Size of Ball 2¼ Inch

		Per Dozen Pairs	
		Nickel	Brass
	1⅛ inch hole	$39.00	$37.30

PRICES SUBJECT TO CHANGE WITHOUT NOTICE

The information on these catalog pages, especially pricing, is of course outdated. We've left this here for what it tells us about advertising, construction, and cultural preferences.

HAMES

No. 61
HOOK HAMES

Per Dozen Pairs

No. 61   Polished Concord $28.00
        Packed 5 Dozen pairs in a case.

LONE STAR ADJUSTABLE DRAFT HAMES

Per Dozen Pairs

Polished Concord, for 17 to 20 inch collars, with back strap rings $30.00
        Packed 5 Dozen pairs in a case.

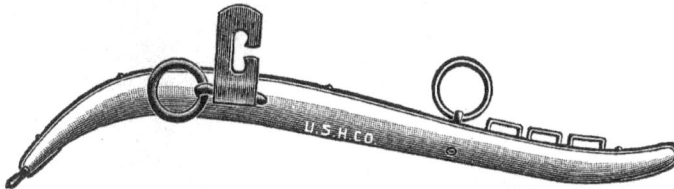

No. 55
HOOK HAMES

Per Dozen Pairs

No. 55   Varnished, hook on side... $17.00
        Packed 6 Dozen pairs in a case.

No. 150
WOOD HAMES—CLIP

Per Dozen Pairs

No. 150   Varnished clip with back strap rings... $25.00
        Packed 5 Dozen pairs in a case.

No. 460
WOOD HAMES—CLIP

Per Dozen Pairs

No. 460   X. C., long spot, black clip with back strap rings................................. $30.60
        Packed 5 Dozen pairs in a case.

PRICES SUBJECT TO CHANGE WITHOUT NOTICE

HAMES

No. 150

WOOD HAMES—BOLT

Per Dozen Pairs

No. 150   Damascus bolt $25.50
           Packed 5 Dozen pairs in a case.

Nos. 460 and 450

WOOD HAMES—BOLT

Per Dozen Pairs

No. 460   X. C., long spot, black bolt.. $32.00
No. 450   All black bolt 27.00

CONCORD HIGH TOP HAMES—CLIP

Per Dozen Pairs

No. 5   Damascus finish, with back strap ring.. $31.30
      Packed 5 Dozen pairs in a case.

CONCORD HIGH TOP HAMES—BOLT

	Per Dozen Pairs			
Nos.	5	8	10	12
Size of rollers, inches	2	2¼	2½	2¾
Damascus finish, mortise loop	$31.30	35.00	38.80	44.50
Damascus finish, extra long, mortise loop	....	42.00	....	....
Red, black trimmed, mortise loop	33.80	37.50	41.30	....
Quantity in case. Dozen pair	5	4	3	3

PRICES SUBJECT TO CHANGE WITHOUT NOTICE

*Mass produced hames came in cast iron, tube steel and steel-reinforced wood.
The tube steel hames are the most prevalent. Aside from weight and rot, the
choice is a matter of availability and preference.*

STEEL HAMES

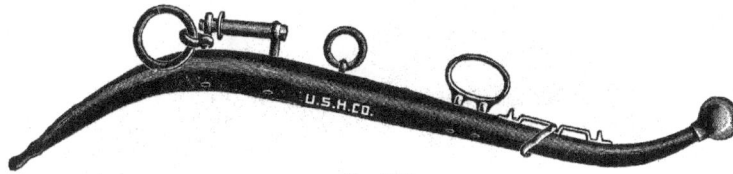

No. 500
TUBULAR STEEL HAMES—BOLT B. S. R.
2 Inch Balls
Made in two lengths only to fit 19-20 and 21-22 inch collars.

		Per Dozen Pairs
No. 500	Nickel ball only, japanned body.	$42.00
No. 500	Brass ball only, japanned body.	42.00
No. 1225	Nickel Chicago ball, japanned body, extra heavy, wood filled. Size 23/25.	78.00
No. 1225	Brass Chicago ball, japanned body, extra heavy, wood filled. Size 23/25.	78.00

No. 520
RED BAND STEEL BOLT HAMES B. S. R.
2 Inch Balls

		Per Dozen Pairs
No. 520	Nickel ball only. Fits collars 19 and 20 inches. Red Band	$49.00
No. 520	Brass ball only. Fits collars 19 and 20 inches. Red Band	49.00
No. 522	Nickel ball only. Fits collars 21 and 22 inches. Red Band	49.00
No. 522	Brass ball only. Fits collars 21 and 22 inches. Red Band	49.00
No. 524	Nickel ball only. Fits collars 23 and 24 inches. Red Band	54.00
No. 524	Brass ball only. Fits collars 23 and 24 inches. Red Band	54.00

BUFFALONIAN BLACK STEEL TUBE BOLT HAMES B. S. R.
2 Inch Balls

		Per Dozen Pairs
No. 622	Nickel Ball only, Fits Collar 21-22	$42.00
No. 622	Brass Ball only, Fits Collar 21-22	42.00
No. 624	Nickel Ball only, Fits Collar 23-24	47.00
No. 624	Brass Ball only, Fits Collar 23-24	47.00
No. 622	Nickel Long Spot and Ball, Fits Collar 21-22	74.70
No. 622	Brass Long Spot and Ball, Fits Collar 21-22	68.00

BUFFALONIAN RED STEEL TUBE BOLT HAMES B. S. R.

No. 622 R.	Nickel Ball only, Fits Collar 21-22	$47.00
No. 622 R.	Brass Ball only, Fits Collar 21-22	47.00

No. 522
RED BAND STEEL HAMES—CLIP B. S. R.
2 Inch Balls

		Per Dozen Pairs
No. 522	Brass ball only, Japanned body, size 21-22	$49.00
No. 522	Nickel ball only, Japanned body, size 21-22	49.00

No. 500-520-522-622 Hames Packed 3 Dozen pair in a case.
No. 524-624 Hames Packed 2 Dozen pairs in a case.

PRICES SUBJECT TO CHANGE WITHOUT NOTICE

The information on these catalog pages, especially pricing, is of course outdated. We've left this here for what it tells us about advertising, construction, and cultural preferences.

STEEL HAMES

No. 568 STEEL HAMES—BOLT

All black bolt, black enameled body; fits collars 19 to 22 inch ....

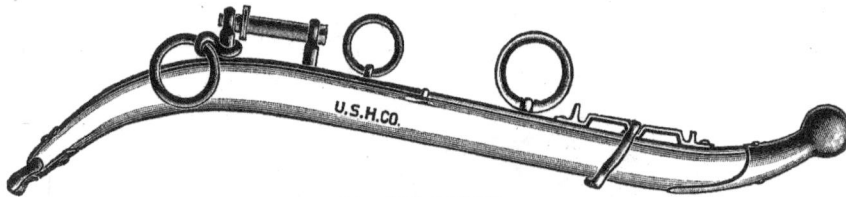

No. 568 STEEL
1¾ Inch Dandy Balls
2¼ Inch Bolt. Made in one size only

Nickel Dandy Ball only, red enameled body; fits collars 19 to 22 inch ....
Brass dandy ball only, red enameled body; fits collars 19 to 22 inch ....

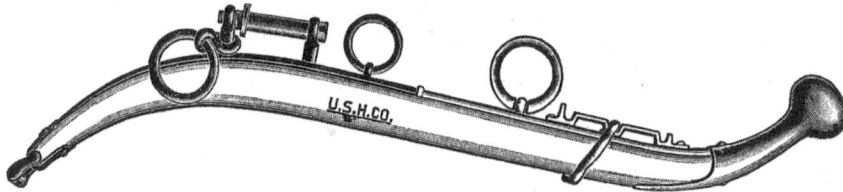

No. 568 STEEL
2¼ Inch Favorite Balls
2¼ Inch Bolt. Made in One Size Only

Nickel long spot, red enameled body; fits collars 19 to 22 inch ....
Nickel ball only, red enameled body; fits collars 19 to 22 inch ....
Brass long spot, red enameled body; fits collars 19 to 22 inch ....
Brass ball only, red enameled body; fits collars 19 to 22 inch ....
Packed 2 Dozen pairs in a case.

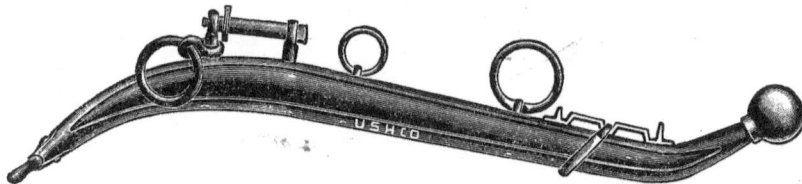

No. 920
STEEL HAMES—BOLT

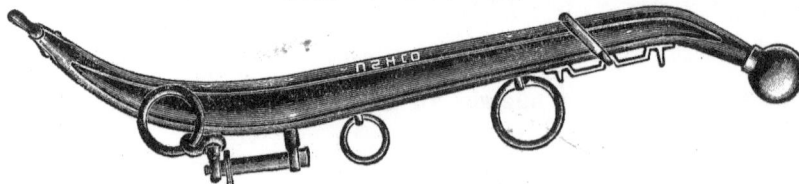

Nos. 920 AND 568 STEEL CONCORD HAMES

No. 920 Clip.

No. 920 Screw Bolt.

No. 568 Concord Screw
Bolt.

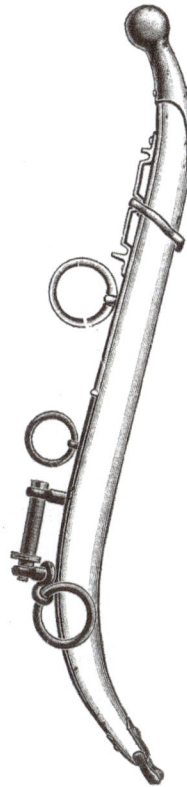

No. 568 Screw Bolt,
Dandy Ball.

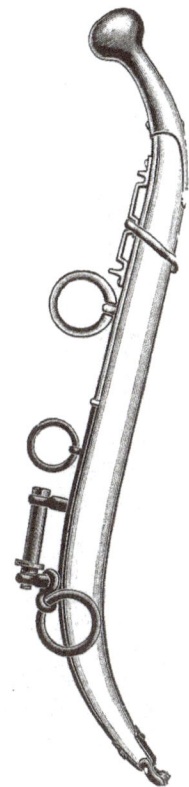

No. 568 Screw Bolt,
Favorite Ball.

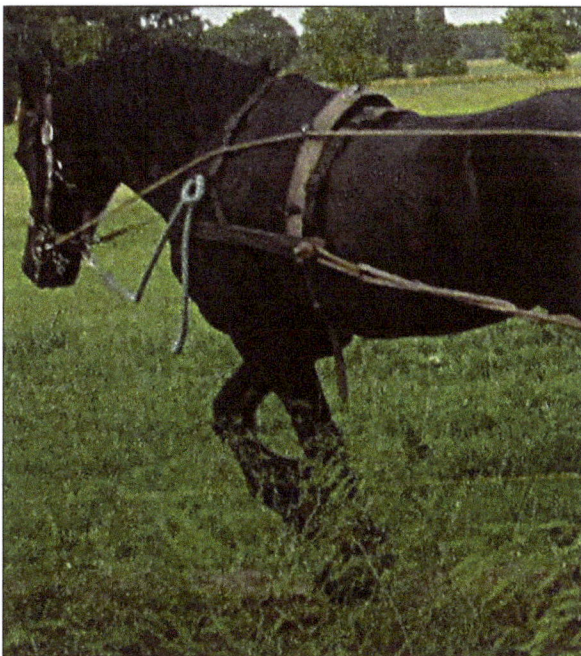

A working breast strap or breast collar.

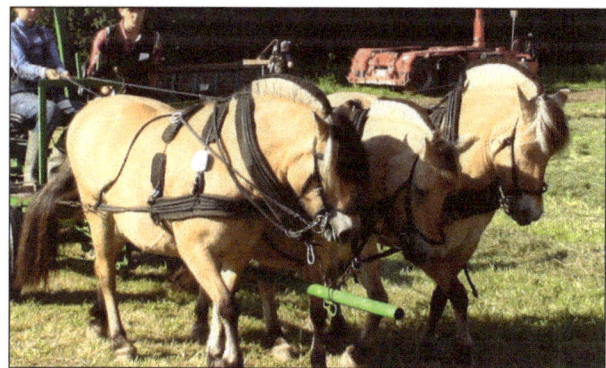

This detail of a William Castle photo taken in Germany shows a Fjord troika hitch with, to my eye, a most unusual combination of collar and breast strap harness. A clear example of just how various the world of harnessed equine has become, worldwide. We take risks by assuming that what we have in North America is the pinnacle when it is possible we don't know all that is out there, nor where we fit in.

HAMES

No. 5—CONCORD LOW TOP—BOLT
With Back Strap Rings

No. 156
WOOD TEAM—BOLT

No. 156
WOOD TEAM—CLIP

STEEL LOW TOP—BOLT

HAME START

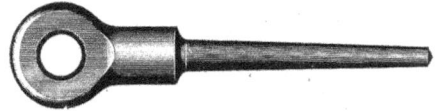

	Price, per Doz.
Concord Hame Starts, not threaded; polished	$1.70
Concord Hame Starts, threaded eye, polished	1.70

BOTTOM HAME LOOP

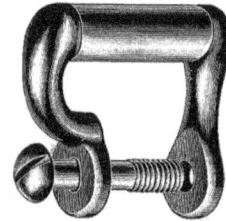

Concord—No. 5

		Price, per Doz.
No. 5	1 -inch;; polished, to screw	$1.30
	1⅛-nich; polished, to screw	1.70
	1¼-inch; polished, to screw	2.00

FLANGES

	Price. per Doz.
Hame Flanges	$0.90

BREAST STRAP LINK AND RING

	Price, per Doz.
Links and Rings	$3.50

"SACO" STEEL TUBULAR HAMES

I once had a set of these. They had a lovely nickel-plated top portion, the bottom was red. The finish did not last long. The nickel did not withstand all the rustle activity of putting on, taking off and hanging up.

No. 500 STEEL HAMES

Saco—Clip.   Saco—Screw Bolt.   No. 500 Clip.   No. 500 Screw Bolt.

World Champion plowman Mike Atkins.

hames curved or straight?

As you can see from the catalog illustrations in this volume, almost all work hames are straight along the side length. But the collars are not. Tightening the hame straps will tend to straighten the collar along the animal's neck. To illustrate a fine point that is indeed a FINE point here's a true story:

Mike Atkins, good friend and mule and Belgian man extraordinaire, tells the story of an exceptional red Belgian gelding that had an issue with his collar. To Mike's capable measure the collar size and shape seemed perfect. But when he was working in it Mike could tell the horse was struggling for air. The collar worried that gelding so much that, whenever Mike brought the collar to him to harness up, the horse would take a dump right there and then. So he had his brother watch the horse while he was pulling and he saw the collar seemed to tighten up on the sides quite a

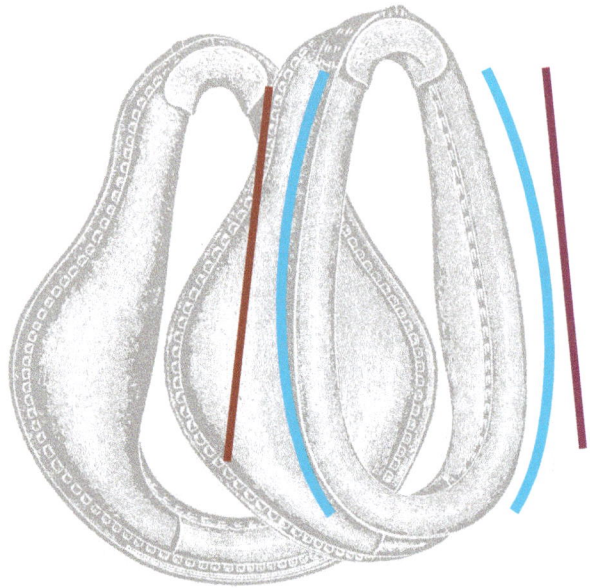

Straight

bit. So tight that he could not hold his hand between the side of the collar and the neck. Thinking this through Mike took the collar and hames to the Chubb Hame Company outside of Mt. Hope, Ohio, and asked if he could bend the hame enough to remove the pinching action. That done, Mike's gelding was comfortable from then on.

Mike told this story to me the year I met him at the US Plowing Championships in Ohio. I was there as an announcer. Later at one of our last horsedrawn auctions, I purchased a set of new CURVED steel-reinforced wood hames for four horses, two of which are pictured on the next pages. To this day I have not been able to find out who made these grand hames. But they illustrate handsomely and obviously how allowing the collar to more perfectly match the curvature of the horse or mule's neck can pay dividends.

Curved

Renee and Kenny Russell's lovely Percheron team.

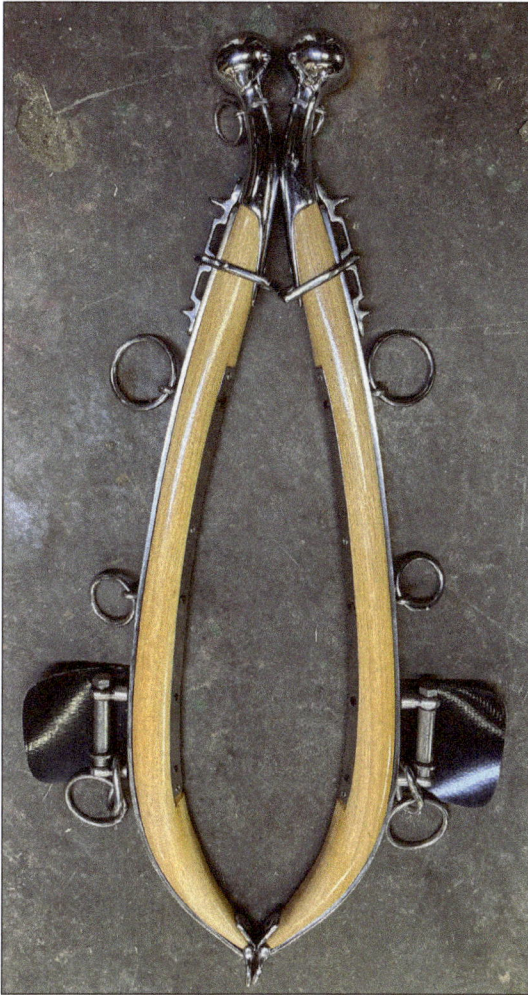

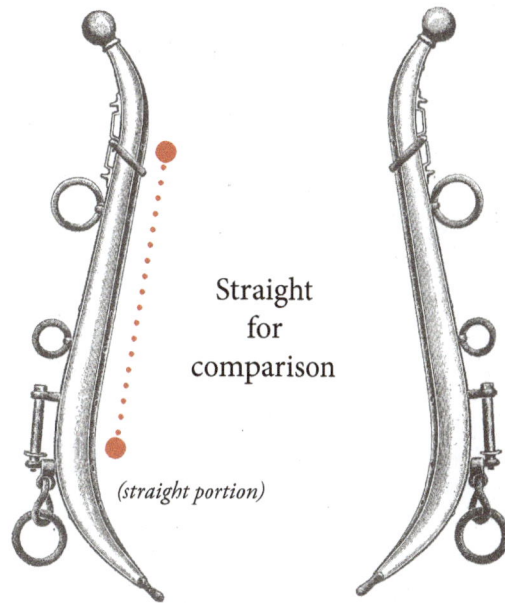

Straight
for
comparison

(straight portion)

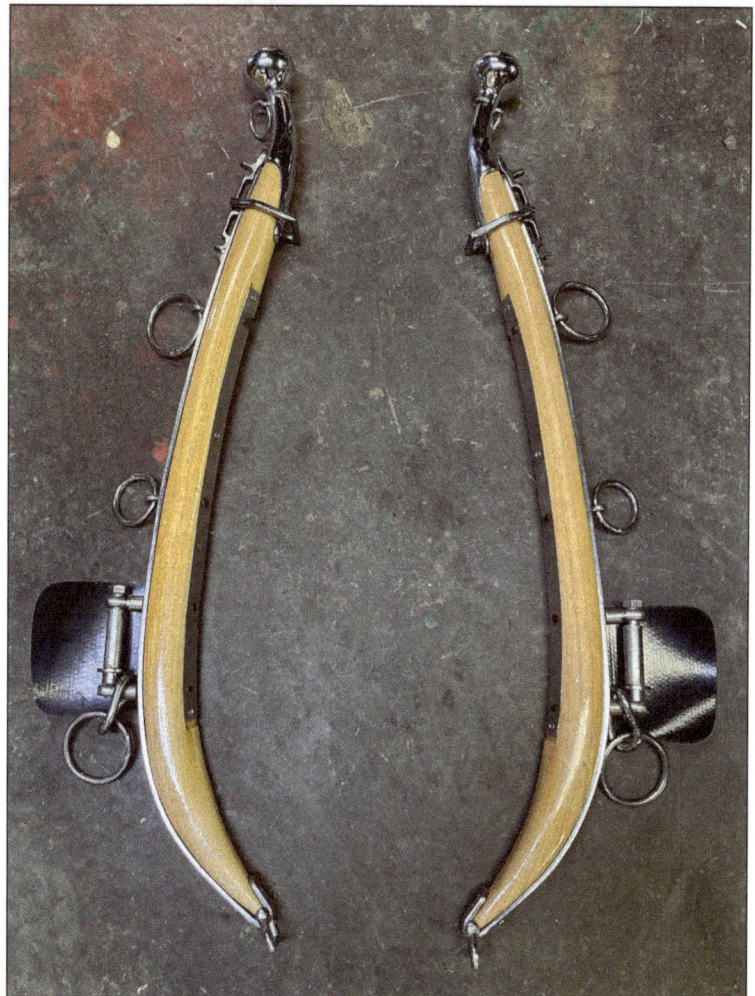

STILL LEARNING

I worked horses for forty years, wrote books about it and taught classes on the subject for 30+ years, and I never once asked why all the hames I had in my tack room were straight through the long middle. Then, as an announcer at the US Plow Championship, I met Mike Atkins and marveled at the comfort, intelligence and poise his mules and Belgians displayed. So in hunt and peck fashion I picked his brain for three days with questions and at one point he told me the story of how his Red gelding was so nervous about the hames – and what he did to remedy that. Such a small detail, but for me it spoke legions of the workaday genius of Mr. Atkins.

Now I am an old man remembering all the wondrous moments my grand equine friends gave me and sad to think I came so late to useful secrets. Life always seems to wait...

Collars fit perfectly. I only wish those blinders were wider and clear of those magnificent eyes.

HAME TRIMMINGS

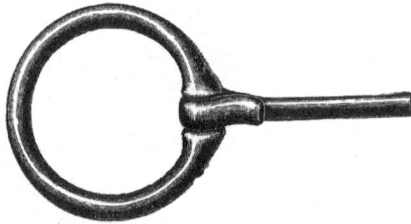

CONCORD LINE RINGS AND STUDS

		Per Dozen
No. 995	Polished	$1.10
No. 973	X. C.	2.10
No. 973	Nickel	4.20
No. 973	Brass	3.80
No. 973	Japanned, for No. 920 and No. 930 Steel Hames	1.80
No. 991	Studs only for Concord Line Rings	.24

2 Dozen in box.

TWO STUD LINE RINGS

		Per Dozen
No. 963	X. C.	$1.20
No. 966	Japanned	1.60
No. 966	Nickel	5.70
No. 966	Brass	5.30
No. 700	Japanned	1.60
No. 700	Nickel	5.70
No. 700	Brass	5.30

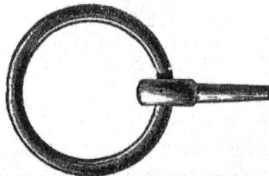

HAME BREAST RINGS AND STUDS

	Per Dozen
No. 930 Polished, 2 inch	$1.10

2 Dozen in box.

No. 964 HAME LINE RINGS AND STUDS

	Per Dozen
Polished, with burrs	$0.86

2 Dozen in box.

LOOSE BRACE TERRETS

	Per Dozen
No. 968 Japanned for steel low top hames	$2.40

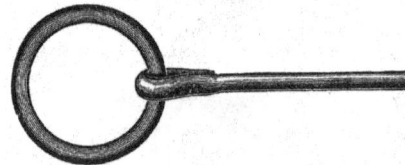

No. 975
BACK STRAP RINGS AND STUDS WITH BURRS

	Per Dozen
Team, polished, 1¼ inch	$0.68
Concord, polished, 1½ inch	.68

2 Dozen in box.

No. 990
HAME TOP LOOPS

	Per Dozen
Polished wire	$0.60

12 Dozen in box.
For common wood hames.

No. 978
SCREW HAME BOTTOM LOOPS

	Per Dozen	
Size, inch	⅞	1
Polished	$0.90	$0.90

3 Dozen in box.

No. 977
CONCORD SCREW HAME BOTTOM LOOPS

	Per Dozen	
Size, inch	1⅛	1¼
Polished	$1.34	$1.60

3 Dozen in box.

In my long lifetime I have worked hundreds of horses and two mules. Every single one of them taught me something. It feels like with that teaching I have been given a sacred trust – I know I was never completely worthy of that trust. I know now that each of the work horse books I have written have been my way of begging forgiveness for any second I might have spent making one of my working partners uncomfortable, frightened or unduly confused.

HAME TRIMMINGS

No. 931   LINK AND RING
Per Dozen
Polished, one link...........$2.10
1 Dozen Pair in box.

No. 932   LINK AND RING
Per Dozen
Two links $2.70
1 Dozen Pair in box.

No. 970 CLIP HOLD BACK PLATE AND RING
Per Dozen Pairs
Polished $3.30
X. C. 5.10
½ Dozen pair in box.

COMBINATION LOOPS

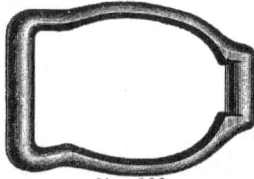

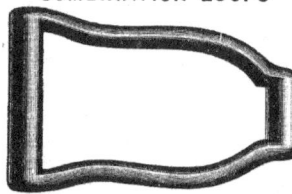

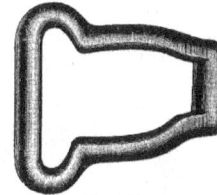

No. 936
For No. 450-460 Hames.
Per Dozen
Polished $0.96
X. C. 1.60
2 Dozen in box.

No. 937.
For No. 920 Hames.
Per Dozen
Japanned $1.30
2 Dozen in box.

No. 939
For No. 500 Hames.
Per Dozen
Japanned $1.30
2 Dozen in box.

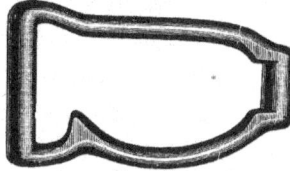

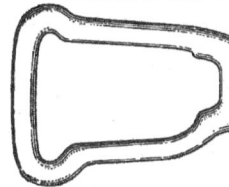

No. 940
For Concord and No. 568 Hames.
Per Dozen
Polished $0.96
2 Dozen in box.

No. 700
For Saco Steel Hames.
Per Dozen
Japanned $1.30
Brass 2.70
Nickel 2.90
2 Dozen in box.

No. 941 HAME RATCHET
For No. 450-460 Hames.
Per Dozen
Polished $0.96
X. C. 1.60

No. 944 HAME RATCHET
For No. 500 Hames.
Per Dozen
Japanned $1.30

No. 945 HAME RATCHET
For Concord and No. 568 Hames.
Per Dozen
Polished $0.96
Brass 5.30
Nickel 5.40

No. 946 HAME RATCHET
For No. 920 Hames.
Per Dozen
Japanned $1.30

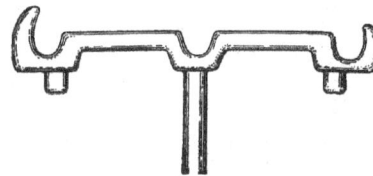

No. 700 HAME RATCHET
For Saco Hames.
Per Dozen
Japanned $1.30
Brass 5.30
Nickel 5.40

How tight should hame straps be? As tight as you can make them. In the beginning, when you first fit a collar to the horse, and a set of hames to that collar, it might seem like you have tightened the hame strap or tightener to the maximum, but a short bit of work may have hames work their way down into the collar groove and you might find the strap has loosened. This is NOT a good thing. To avoid having hames come free of the collar as the horse is working, you need to double check the tightness and be prepared to put a shoulder to the pull as you tighten that hame strap once again. With regularly worked horses, where the harness has found its 'fit,' it is less likely you will have to retighten.

HAME TRIMMINGS

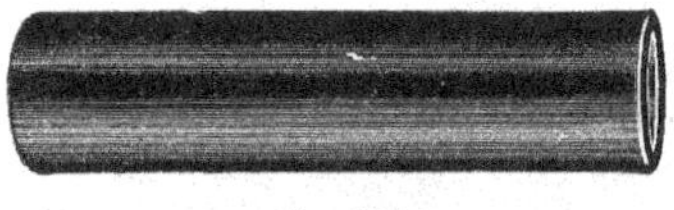

No. 994

CONCORD HAME ROLLERS—COMMON

		Per Dozen
2 inch.	Entire length	$1.06
2¼ inch.	Entire length	1.06
2½ inch.	Entire length	1.06
2¾ inch.	Entire length	1.06

2 Dozen in box.

No. 1000

CONCORD BOLT WASHERS OR COLLARS

Per Dozen

Polished $0.66

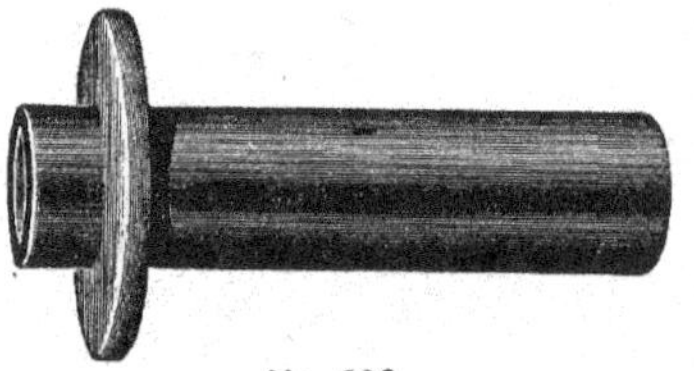

No. 993

CONCORD HAME ROLLERS—IMPROVED

		Per Dozen
Entire lgth. 2⅜ in.	For 2 in. trace.	$1.30
Entire lgth. 2⅝ in.	For 2¼ in. trace.	1.30
Entire lgth. 2⅞ in.	For 2½ in. trace.	1.30
Entire lgth. 3⅛ in.	For 2¾ in. trace.	1.30

2 Dozen in box.

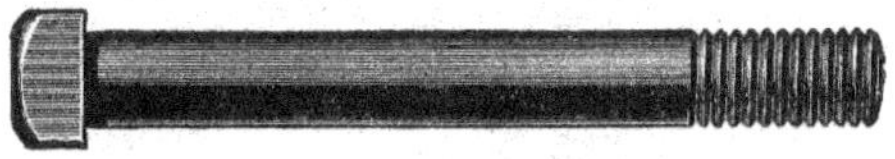

No. 996 CONCORD HAME BOLTS

			Per Dozen
2 inch.	Polished.	Entire length 3½ inches	$1.06
2¼ inch.	Polished.	Entire length 3¾ inches	1.06
2½ inch.	Polished.	Entire length 4 inches	1.06
2¾ inch.	Polished.	Entire length 4¼ inches	1.06

2 Dozen in box.

SACO STEEL HAME BOLTS

Per Dozen

No. 701 Entire length 3½ inches................$1.06
2 Dozen in box.

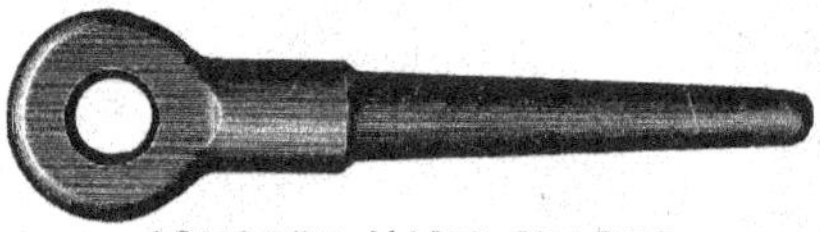

CONCORD HAME STARTS

		Per Dozen
No. 997½	Small, plain, for Nos. 5 and 8 hame	$1.60
No. 997	Small, threaded, for Nos. 5 and 8 hame	1.60
No. 998½	Large, plain, for Nos. 10 and 12 hame.	1.60
No. 998	Large, threaded, for Nos. 10 & 12 hame	1.60

2 Dozen in box.

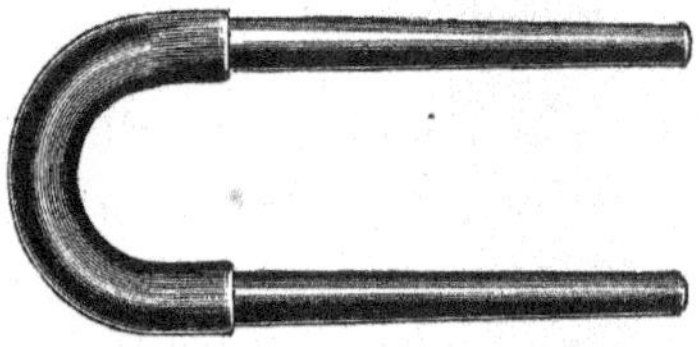

TEAM HAME STAPLES

Per Dozen

No. 960 Regular polished, ⅜ inch, with burrs, length shank to shoulder, 2¼ inches, width of shank at shoulder, 1 inch......$0.64

No. 959 Polished, ⅜ inch, "extra wide," with burrs, length shank to shoulder, 2⅜ inches; width of shank at shoulder, 1 1/16 inches .... .64

2 Dozen in box.

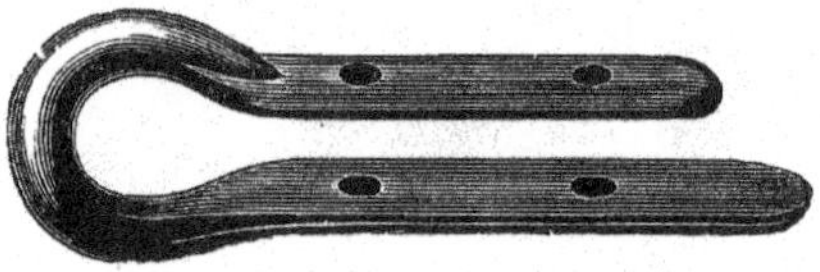

No. 982 TEAM HAME CLIPS
2 Hole

		Per Dozen
No. 982	Polished, ⅜ inch	$1.40
No. 982	Japanned, ⅜ inch	1.90
No. 983	Japanned, 7/16 inch, extra heavy	2.00

2 Dozen in box.

No. 981 COACH HAME CLIPS
2 Hole

Per Dozen

No. 981 Polished, 5/16 inch.......................$1.10
2 Dozen in box.

SQUARE HAME STAPLES

		Per Dozen
Concord, 1¾ inch,	polished	$1.60
Concord, 2 inch,	polished	1.60
Concord, 2¼ inch,	polished	1.60

2 Dozen in box.

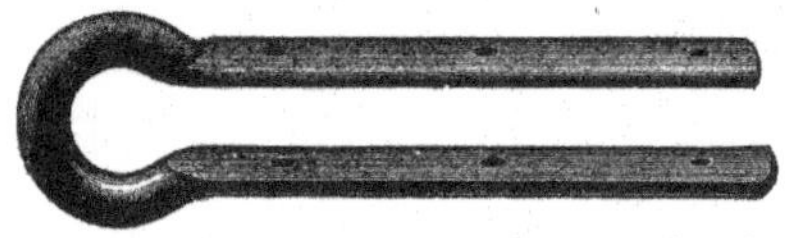

Nos. 984 and 985 TEAM HAME CLIPS
3 Hole

		Per Dozen
No. 984	Japanned, ⅜ inch, extra long	$2.30
No. 985	Japanned, 7/16 inch, extra long	2.60

2 Dozen in box.

Why it matters that hames are the right length? The relative position of the draft on the hames, that point where the tug or trace is connected, is most important to the comfort of the horse as it leans into a pull. If the hames are too long for the collar, that position might end up being too low and irritating the horse or mule, perhaps even to the point of causing it to refuse to move ahead (baulk) as it tries to avoid the discomfort. If, on the other hand, the hames are too short it could change the shape of the collar and pinch the animal. Within a two to three inch length range, hames should fit the collar and its shape.

HAME HOOKS AND REPAIR CLIPS

HAME WASHERS
Polished

		Per Pound
No. 954	Hame staple washers	$1.00
No. 953	Hame line ring washers	1.00
No. 955	Hame rivet washers	1.00

Quantity in box, 1 pound.

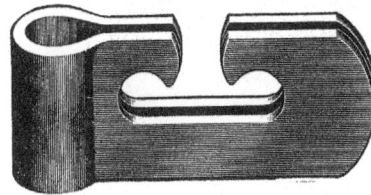

No. 948
DOUBLE HAME HOOKS

Per Dozen Pairs

Polished, double wrought steel...................$2.60
½ Dozen pairs in box.

No. 900
LONESTAR HAME HOOKS

Per Dozen Pairs

Polished $1.90
1 Dozen in box.

No. 950
CONCORD HAME HOOKS

Per Dozen Pairs

No. 6   Polished forged steel $5.00
½ Dozen pairs in box.

REPAIR CLIPS AND LOOPS

Per Dozen

No. 1968	Japanned matchless, for wood hames, 1 inch	$1.00
No. 1969	Clips only, for wood hames	.56
No. 1983	Clips only, for steel hames	.56

REPAIR CLIPS AND LOOPS
For Steel Hames

No. 1981	Japanned, for U. S. Steel hames	$1.10
No. 700	Japanned, for No. 700 steel hames	1.10

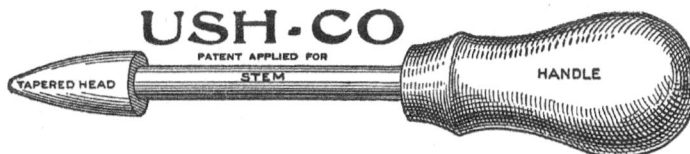

USH·CO
PATENT APPLIED FOR
TAPERED HEAD   STEM   HANDLE

BOLT HAME ROLLER INJECTOR
For Inserting Rollers in Hame Tugs

Each $1.40
1 in a box.

Directions for Its Use

Remove stem from handle. Slip the roller over the stem with the plain end (if a flange roller) near tapered head. Replace handle on stem. Insert tapered head in opening of trace. Drive stem through opening in trace until the bolt roller is correctly located. Remove handle and stem. The roller is now in the trace and USH-CO ready for more work. Furnished in one finish only, i. e., with a russet colored handle and nickeled stem.

PRICES SUBJECT TO CHANGE WITHOUT NOTICE

COOPER'S CLIPS

No. 20—Jointed
Price, per Dozen

	1½	1¾	2	2¼	2½
No. 21 Jointed	$3.50	$4.00	$4.40	$5.00	$5.70

COOPER'S CLIPS

No. 40—Jointed

	2x2	2¼x2½	2½x2½
No. 40 Jointed	$10.40	$11.40	$12.20

COOPER'S CLIPS

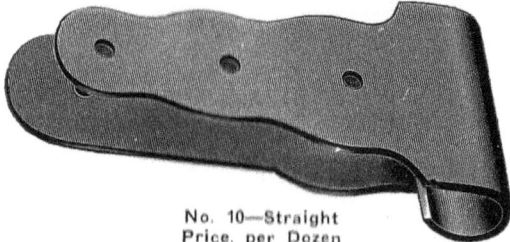

No. 10—Straight
Price, per Dozen

	1½	1¾	2	2¼	2½
No. 10 Straight	$1.40	$1.60	$1.70	$1.90	$2.10

BACK STRAP RING

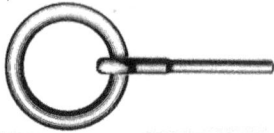

HAME TRIMMINGS

BOLT ROLLERS

	Price, per Doz.
Flange Rollers	$1.80
Plain Rollers	1.50

WROUGHT IRON HAME CLIPS—FOR WOOD HAMES

	Price, per Doz.
No. 982 ⅜-inch	$1.30

CONCORD HAME LINE RING AND STUD

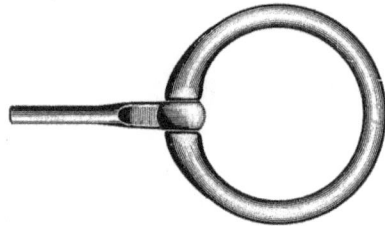

	Price, per Doz.
Polished	$1.20
Nickel	3.50
Brass	3.50

HAME BOLTS

O. L. HAME CLIPS.

	Per Doz.
No. O. L.—Japanned, one leg iron hame clips......................	$ 88
No. O. L.—Nickel, one leg iron hame clips.........................	1 32

"H. B." REPAIR CLIPS AND LOOPS.

	Per Gross.
Japanned, H. B. repair clips and loops..........................	$ 6 66
Japanned, H. B. repair clips, without loops......................	6 00

NO. 37 PATENT COMBINATION CLIPS.

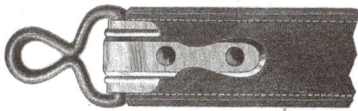

	Per Doz.	
	1½	1¾-Inch.
No. 37—Japanned	$ 77	$ 88

PHILLPOTT'S PATENT WEAR CLIPS.

	Per Gross.
Polished.	$ 2 76

Buffalo Line Rings.

HAME STAPLES.

No. 950.   No. 960.   Nos. 957, 958 and 959.

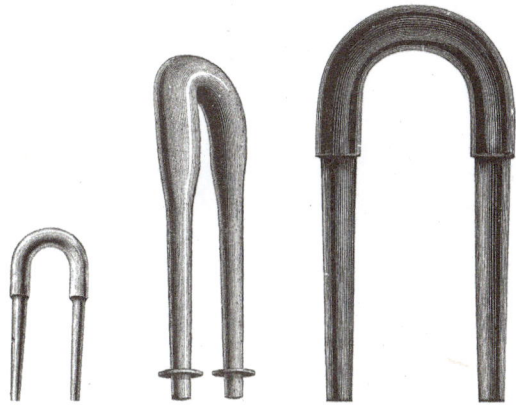

	Per Doz.
No. 950—Polished, 5/16 iron, for oval iron wood coach hames........	$ 33
No. 950—Japanned, 5/16 iron, for oval iron wood coach hames........	60
No. 960—Polished team, ⅜ iron, with washers.....................	28
No. 962—Polished team, long and wide, ⅜ iron, with washers........	28
No. 958—Polished, extra long and extra wide, ⅜ iron, with washers...	33
No. 959—Polished, extra long head, ⅜ iron, with washers, for Hayden hold back plates	42
No. 957—Polished, 7/16 iron, for heavy Concord hames.............	52

SQUARE HAME STAPLES.

	Per Doz.
No. 450—Polished, 1¾-inch	$ 44
No. 460—XC plate, 1¾-inch	52
No. 5—Polished Concord, 1¾-inch	44
No. 8—XC plate, oval iron wood coach, 1½-inch..................	70

Some purists believe that the rings on the hames, through which the lines pass, should be oblong to keep the flat of the line at the right axis. Whether I agree or not is not so important, but I do enjoy the way this craft draws people in and suggests they might insist on their own tunings.

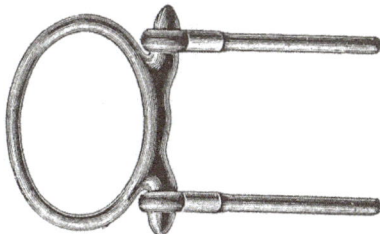

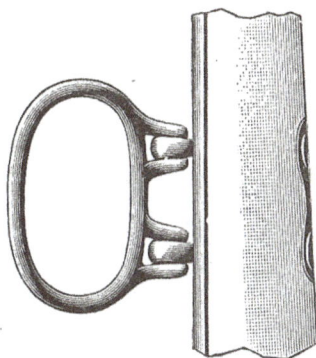

ADVANCE HAME ATTACHMENTS

EUREKA PATENT HAME CLIPS

No. 1—Advance Clip.   No. 1—Advance Clip in Use.   No. 2—Eureka Clip.

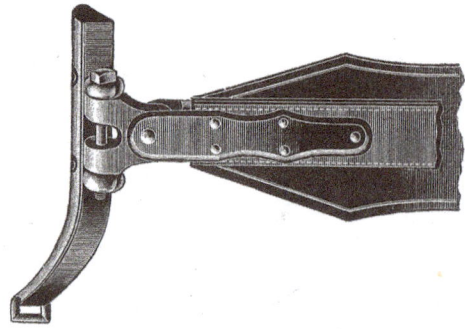

HOLLOW IRON HAME TRIMMINGS

No. 922 Clip.   No. 923 Screw Bolt.   No. 926 Terrets.

The author plowing with Lana and Cali.

Easy to adjust hames to fit perfectly all sizes of collars.
Fig. B.

Side cut away to show how end of chain is held. Easy to adjust to all size collars. No chance to lose hame chain, as end of chain cannot pass through opening. Each adjustment one inch.

No hame straps to buy.

Extra ½-inch adjustment for real close fit.

Once closed, the hames stay closed until opened by pulling down on lever—Jack knife spring acts only as extra safety.

Lever open ready to receive hame chain.

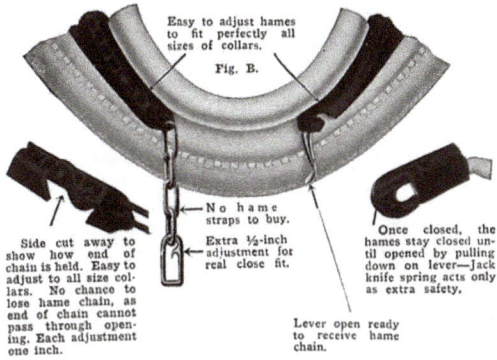

Walsh Hames The Neatest, Best Looking, Best Fitting, Strongest Hame Ever Made

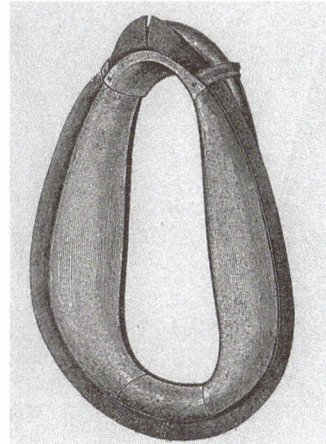

Fig. A.

See how hames fit under the roll of the collar.

Hames are made extra wide where most strain comes.

Heavy coat of zinc galvanizing under this coat of baked on red enamel.

To open pull down on hame lever—which throws hame chain off center allowing hame to open.

Note that link rests against the hame—not on the lever.

Notice nice smooth job — nothing dangling loose to catch or annoy—four times stronger than best hame strap.

Walsh Harness Advertising for their patented hame hookup.

Genuine Scotch Collar of old, different than Scotch Top Collars usually seen on show hitch harness.

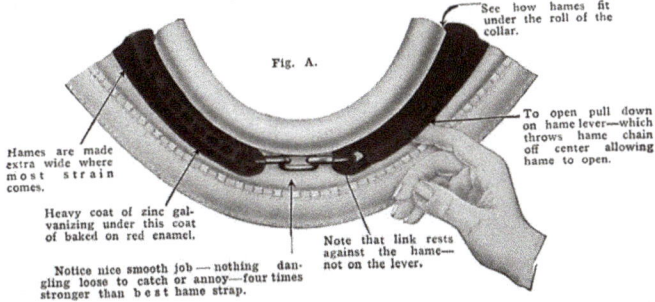

PAT'D FEB.18,06

G. G. G.—Open View.

Hame housings to cover and protect collars from weather and used as parade or hitch decoration.

Hame housings are typically big flaps of heavy leather stitched together to a ridge cap. They are meant to ride on the collar. When used with long ball-topped hames, slits are cut to allow them to pass through.

"LONE STAR" HAMES

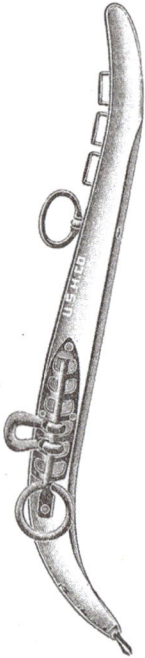

No. 3

With "Lone Star" Adjustable Draft Hook

Painted black, red top, Japanned trimmings. Three loops, steel over tops, oval hame back steel. Grooved woods. Fitted with "Lone Star" attachment. Spring is made of finely tempered clock spring steel, fastened securely to body of hame and acts as a lock, preventing draft hook from jerking out. Adjustment in length from 17 to 21 inches.

Per Dozen Pairs................ $9.00

No. 61

With "Lone Star" Adjustable Draft Hook

Wood part varnished; metal parts Damascus or polished finish. Three 1⅛-inch mortised top loops. Grooved white ash woods. ¾ x ¼ inch back steel. Fitted with "Lone Star" attachment. Spring is made of finely tempered clock spring steel, fastened securely to body of hame and acts as a lock, preventing draft hook from jerking out. Adjustment in length from 17 to 21 inches. Weight 80 pounds per dozen pairs.

Per Dozen Pairs................ $12.00

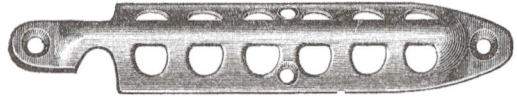

LONE STAR HAME RATCHETS

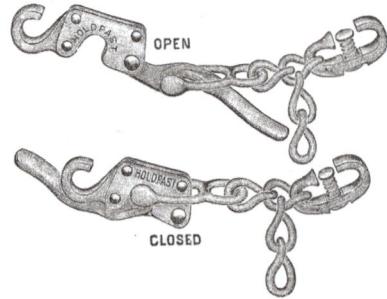

OPEN

CLOSED

LONE STAR HAME HOOKS

William Livingston and team, photograph by Ida Livingston.

har·ness

noun
a set of straps and fittings by which
a horse or other draft animal is
fastened to a cart, plow, etc. and is
controlled by its driver.

verb
1. put a harness on (a horse or other
draft animal).
"How to groom a horse and harness it."

2. control and make use of (natural
resources), especially to produce energy.
"Attempts to harness solar energy."

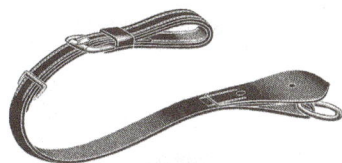

Chapter Three
Bridles, Bits
& Check Reins

Bridles

For centuries, workhorse and mule bridles have been designed to allow the teamster on the lines to turn, back and stop the employed animals. If properly constructed, adjusted and fit they will be comfortable. If they are the wrong size, poorly constructed and rough-edged they can be pure misery for our working partners. The 'honorable' thing to do, as my old friend Les Barden might have said, is to use some common sense and get the right gear for the horse's sake.

Here are a few of the less obvious fit issues: 1. The **crown strap**, which runs behind the ears and over the poll, or top of the head, should be adjustable as it needs to fit each individual animal properly. Its length should fit down to the top of the jaw on both sides so that the adjoining

2. The **throat latch** drops naturally along and inside the jawline. Some bridles, while seeming to be large enough and fully adjustable, actually crowd and pinch the ears quite a bit. When a horse tosses his head with the bridle on, it should be taken as an indication that something doesn't feel right for him. Check to see how the fit is around the ear base and back side. If the brow band is too long, the throat latch will drop ahead of the jawline and make it impossible to adjust (see diagram page 93). The throat latch must be snug enough, and well behind the curve of the jaw, so that it is hard for the animal to rub the bridle off entirely. Something, it should be noted, which does not happen when an (optional) over-check is employed properly.

3. The **cheek straps** of the bridle also need to be of proper length to allow enough fine tuning to the fit of the bit. Too short and it is possible you will not be able to adjust the **bit straps** long enough for a proper and comfortable position for the bit. Too long and

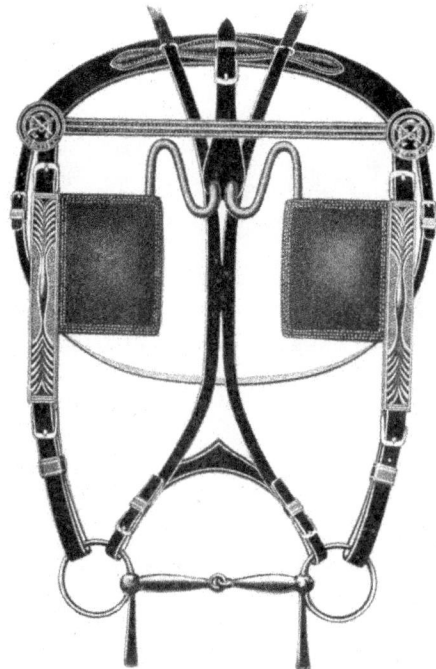

there is the risk that a loose bit will cause injury to the horse's teeth or fall out of the mouth altogether. The bit straps typically allow two or three hole adjustment, but this may be insufficient. Check to see if the cheek strap length is too short for the size and proportions of the head.

4. The **nose band** of the work bridle can, in certain situations, become an added discomfort for the horse. Pulling hard on the lines may put pressure on the bridge of the nose. While this could help stop the animal in a panic, with heavy-handed teamsters it could cause a baulky horse to result.

5. The **brow band**, running across the forehead, needs to be loose and not gather the crown and cheek

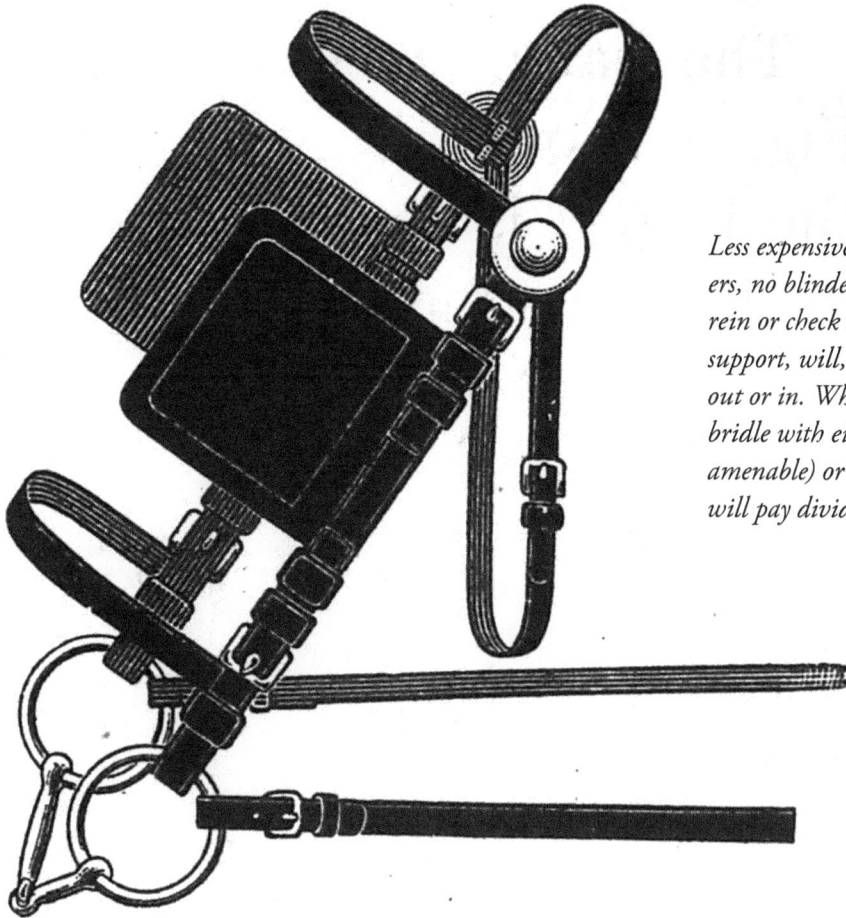

Less expensive plain work bridle. Flat blinders, no blinder stays, no face piece, no check rein or check loops. These blinders, without support, will, over time, bend and could flop out or in. While serviceable, replacing this bridle with either an open bridle (if horse amenable) or cupped and supported blinders will pay dividends in horse comfort.

straps forward to irritate the eyes or ears. It may or may not feature an attachment or a slit for the blinder stays.

6. The **blinders** restrict the side and back vision of the animals and should, in this author's opinion, be cupped and well away from the eyes.

7. The **check reins,** or the **overcheck**, serves to hold the horse's head up, but hopefully, in a comfortable position. The check reins run through guides at the throat latch union, and either fasten to the top of the collar or go round one or both hame tops. The overcheck, ideally, runs from the bit sides across the face, between the ears and back to where it fastens at the collar or back pad.

8. **Blinder** or **winker stays** are commonly fastened to the brow band center, or hung from the crown strap center.

NOTE: Itchy horses will develop a habit of rubbing their bridles and could end up pulling them off. Most times itchy horses are made because care has not been taken to have bridles with no sharp edges, scratchy stitching, sharp

edged rivets, or otherwise to fit comfortably. It is common in some parts, and especially at public events, to have the animals wear their halters under the bridles. In these cases similar care should be taken to see that halters are comfortable.

blinders versus no blinders

There are folk who believe that work horses and mules should be worked without blinders. They say to do otherwise is to disrespect the animals. This is not the place to invite prolonged arguments. Obviously the individual teamster is in a position to pick and choose. Suffice it to say that there are very few reasons why the properly trained animal cannot work well and safely without a blindered bridle.

However, know that the animal's eyesight is quite different from ours and that even the best behaved and trained horse can be surprised by moving objects that appear other than what they are. We have tried working in open bridles with older well-trained horses accustomed to blinders and their nervousness is real. Please use caution, common sense and compassion when presuming your notions on the working routine.

Walsh Bridles

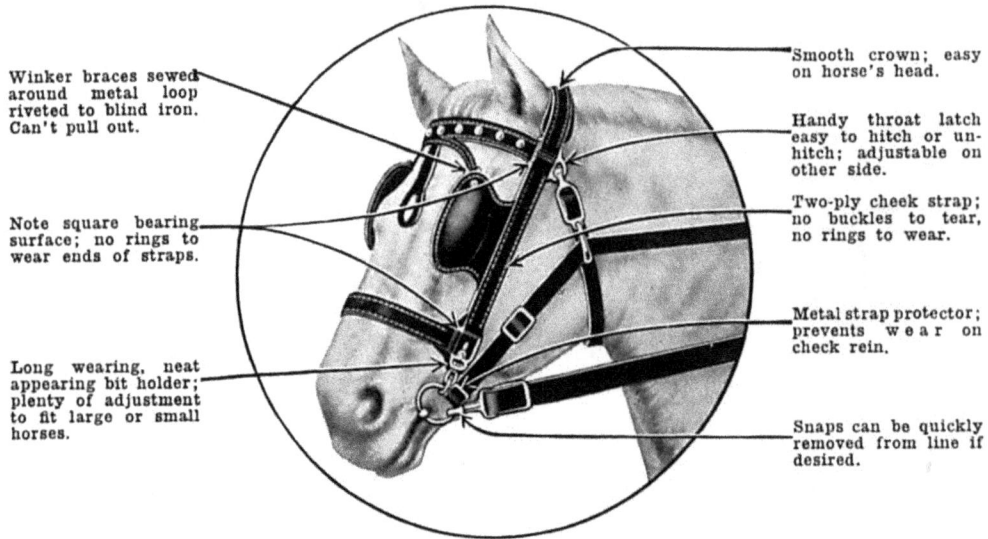

Winker braces sewed around metal loop riveted to blind iron. Can't pull out.

Note square bearing surface; no rings to wear ends of straps.

Long wearing, neat appearing bit holder; plenty of adjustment to fit large or small horses.

Smooth crown; easy on horse's head.

Handy throat latch easy to hitch or un-hitch; adjustable on other side.

Two-ply cheek strap; no buckles to tear, no rings to wear.

Metal strap protector; prevents wear on check rein.

Snaps can be quickly removed from line if desired.

This diagram was part of an advertising campaign for a brand of harness that used no con-ventional buckles. It serves a purpose here to illustrate a few things. The artist's rendering of the horse's head is poor and out of proportion which may be the reason that the throat latch appears too loose, the bit too loose in the mouth, and the crown strap tight against the back of the ears. Later we'll talk about line snaps versus other ways to attach lines.

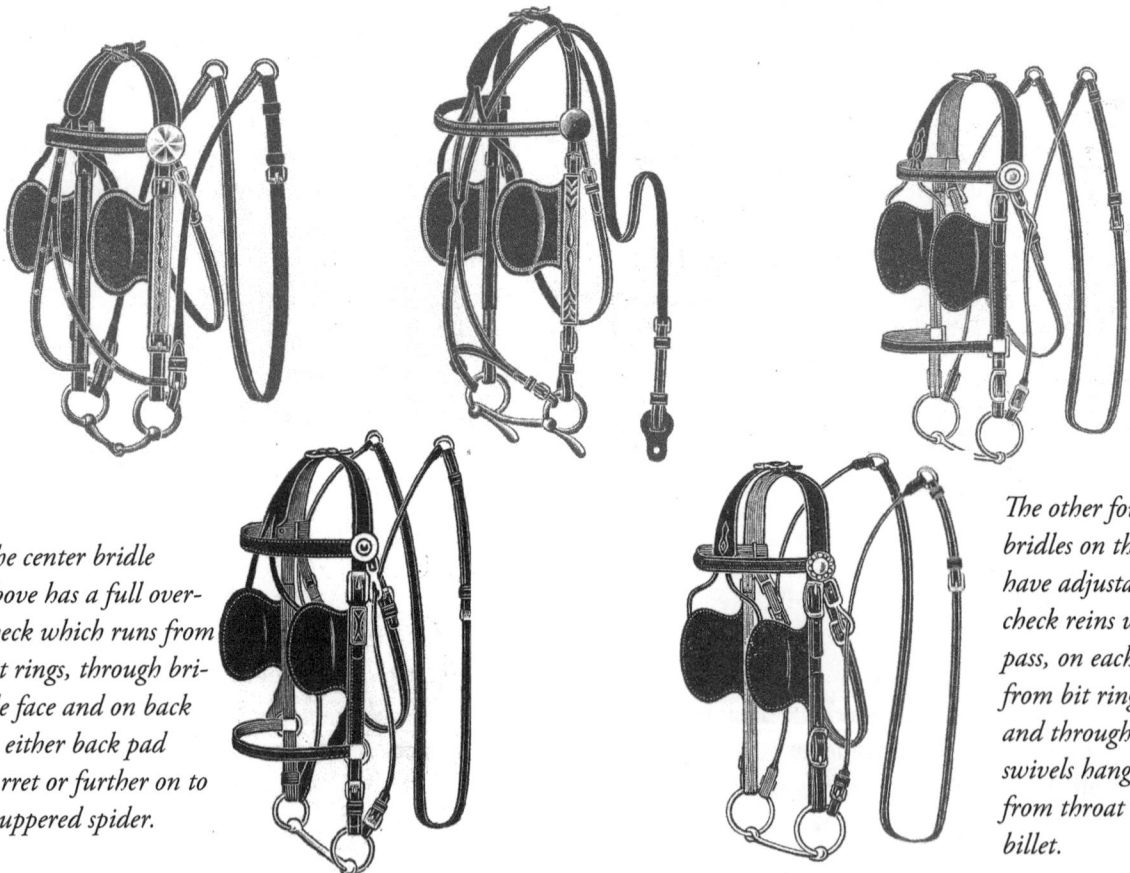

The center bridle above has a full over-check which runs from bit rings, through bri-dle face and on back to either back pad turret or further on to cruppered spider.

The other four bridles on this page have adjustable check reins which pass, on each side, from bit ring up and through gag swivels hanging from throat latch billet.

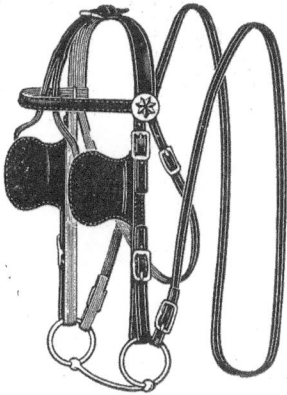 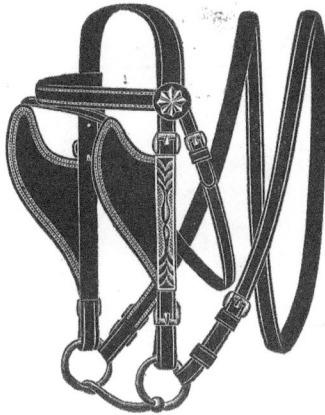 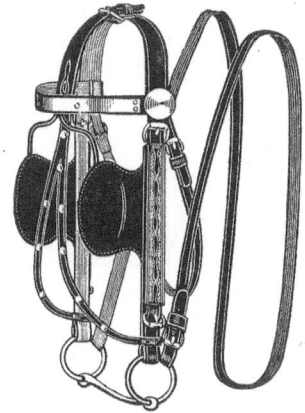

These bridles have 'relaxed' check reins which run straight back to hang over a hame top.

ROUND WINKER STAYS.

FACE PIECE

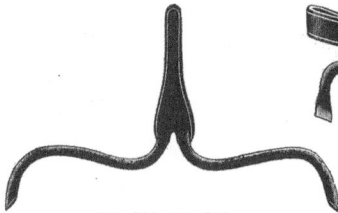 

No. 318.

No. 314, 315, 316.

Some bridles have winker stays to hold blinds from folding back, some don't.

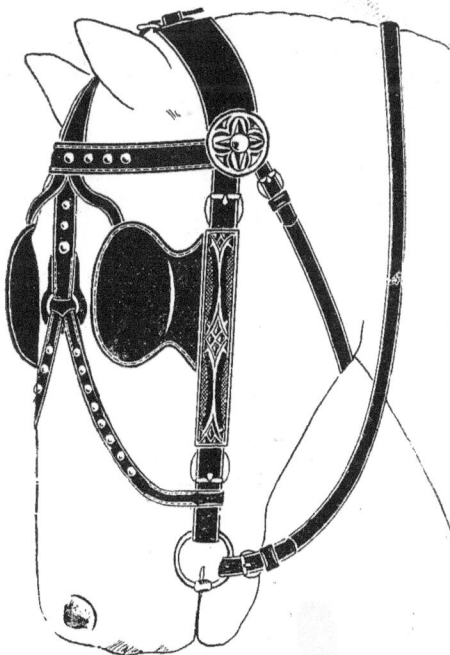

Gag Swivels

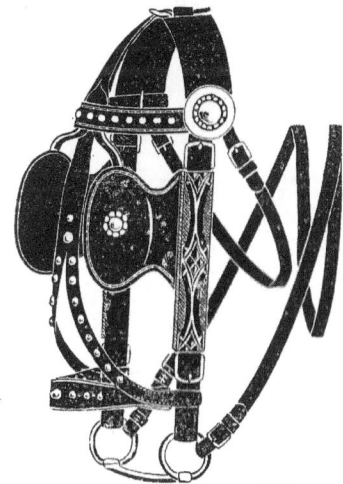

Bridles, early on, became the first place that farmers looked to individualize and make a 'fancy' statement with spots, stamped cheek patterns, rosettes over rings and rounded rather than flat winker stays.

Some bridles have nose bands while some have split faces and then there are those who have neither.

DODSON-FISHER CO.

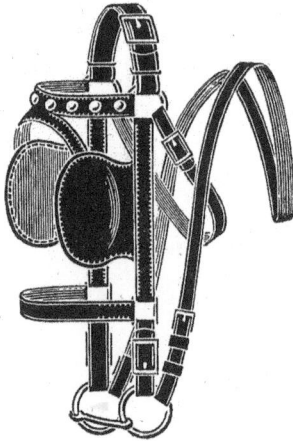

Nos. 3854-2768
Creased

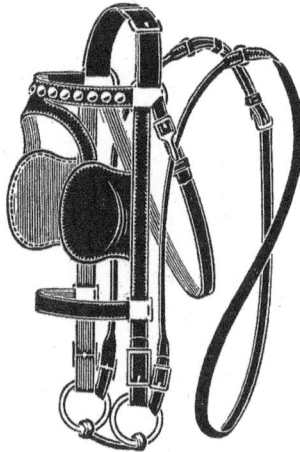

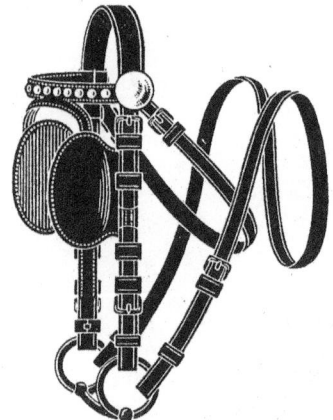

No. 2854
Creased

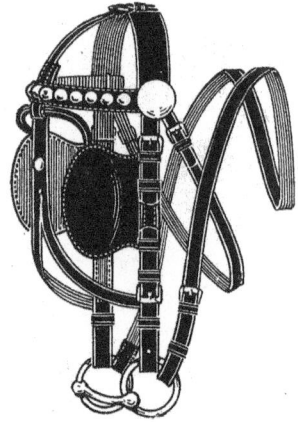

No. 2781
Creased

Per Dozen

No. 3854 ⅞-inch cheeks, ring crown, Concord harness, leather blinds, spotted combination fronts and plain winker braces, short flat side checks, ring bits. Japanned $63.20

No. 2768 1-inch short cheeks, ring crown, Concord harness leather blinds, spotted combination fronts and plain winker braces, short flat side checks, ring bits. X C folded crown........................ 78.20

No. 2854 ⅞-inch short cheeks, ring crown, Concord harness leather blinds, spotted combination fronts and plain winker braces, long round side reins, ring bits, japanned. 78.90

No. 2769 1-inch short cheeks, ring crown, Concord harness leather blinds, spotted combination fronts and plain winker braces, long round side checks, ring bits, X C folded crown.................... 99.00

No. 2781 ⅞-inch cheeks, narrow loops, Concord harness leather blinds, round Concord winker braces, ⅞-inch short flat side checks, spotted front, one spot at crotch on face piece, X C trimmed............ 78.40

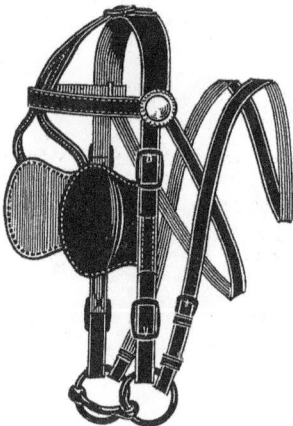

No. 3908
Creased

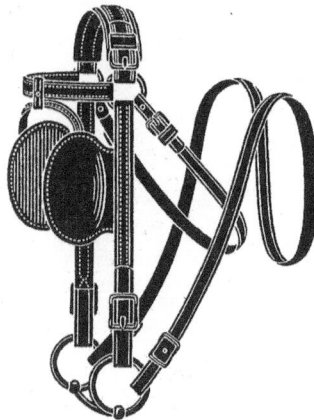

No. 3852
Creased

No. 2224
Gopher Grade, Smooth

Per Dozen

No. 3908 ⅞-inch cheeks, Japan bar buckles, split double and stitched winker braces, short flat side check, harness leather Concord blinds, ring bits..$56.80

No. 3852 ⅞-inch cheeks, ring crown with 2-inch wear leather and 1-inch full length layer, ⅞-inch short flat check, combination front and winker brace, double and stitched, Concord blinds. Japanned trimmed 50.00

No. 3853 ⅞-inch. Same as 3852, only has long round side check 69.50

No. 2224—⅞-inch cheeks, narrow loops, Concord harness leather blinds, combination fronts and winker braces, short flat side checks, Japan or X C trimmed.. 81.30

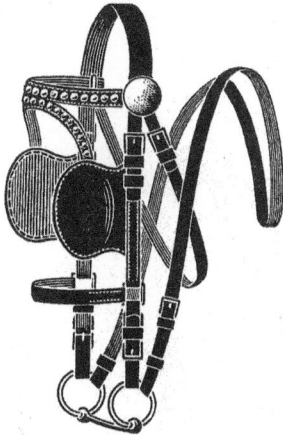

No. 1270

Gopher Grade, Smooth

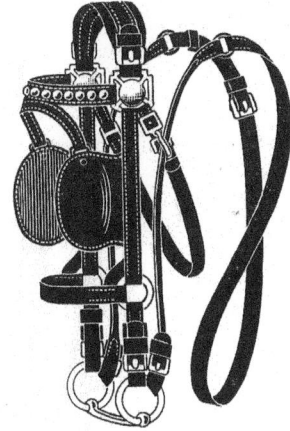

No. 2814

Gopher Grade, Smooth

No. 2530

Gopher Grade, Smooth

Per Dozen

No. 1270 ⅞-inch short cheeks, narrow loops, Concord harness leather blinds, spotted combination fronts and winker braces, short flat reins, X C trimmed..$87.40

No. 2814 ⅞-inch short cheeks, sewed into ring at crown, folded crown with 1-inch billets, Concord harness leather blinds, combination fronts and winker braces, spotted front, short flat reins. X C trimmed   87.30

No. 2530 Same as No. 2814, only has long round side checks 111.50

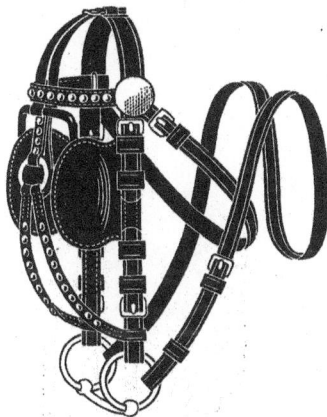

No. 3806

Gopher Grade, Smooth

⅞-inch cheeks, narrow loops, Concord harness leather blinds, round Concord winker braces, short flat checks, nickel spotted face piece with celluloid ring, nickel spotted front, X. C. trimmed.

Per Dozen

No. 3806 Bridles...$95.80

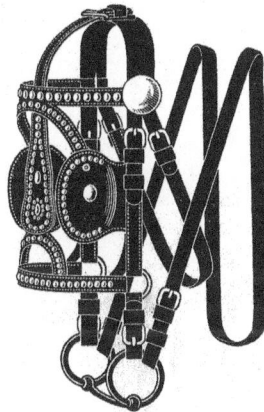

No. 2985

Gopher Grade, Smooth

⅞-inch short cheeks, narrow loops, short side checks, fancy spotted Concord blinds, closely spotted combination fronts and winker braces, doubled and stitched face pieces very fancy spotted.

Per Dozen

No. 2985 X. C. trimmed $121.40

No. 2985 Nickel or brass 134.40

No. 3620
STALLION

Flat cheeks, solid crown, flat throat latches, extra long lead rein.

Per Dozen

1-inch, black, nickel only $103.70

1¼-in., black, nickel only 123.40

STALLION LEAD REINS

No. 3620 12 foot long with buckle and billet and button.

Per Dozen

1-inch, black, no chain $43.10

1¼-inch, black, no chain 48.40

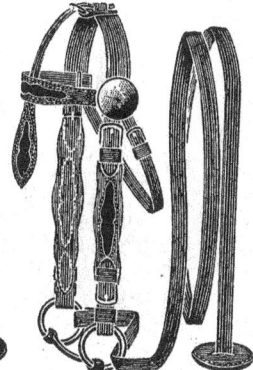

No. 3649

STALLION

Purple inlaid, swell cheeks with ⅞-inch buckles and billets, purple inlaid fronts, and face drops, heavy nickel bits, fancy rosettes, 1-inch lead reins, nickel trimmed.

Each

Russet leather......$14.70

The information on these catalog pages, especially pricing, is of course outdated. We've left this here for what it tells us about advertising, construction, and cultural preferences.

Heavy-duty open-face stallion bridles are a popular option for people who prefer to drive horses without blinds.

I prefer a soft blinder which sets wide and away from the eyes.

Drawing by the author.

DODSON-FISHER CO.

BLINDS

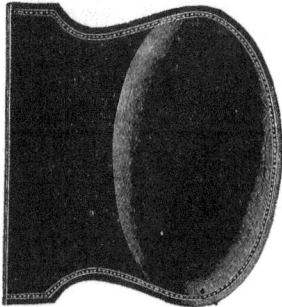

Nos. 358, 359

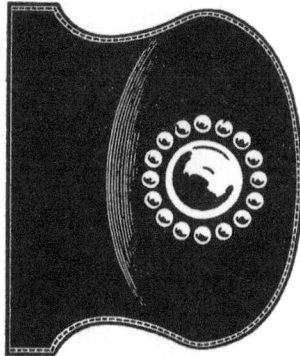

No. 656

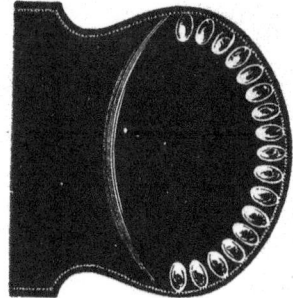

No. 657

SENSIBLE

		Per Dozen Pairs	
	5⁄8	3⁄4	7⁄8
Size of cheek, inch..			
Size of blind, length and width, inches..............................	L5 W3¼	L5⅜ W3½	L6½ W4¼
No. 358 Harness leather, one row extra heavy stitching....................	$8.70	$8.70	$10.80

SENSIBLE, SPOTTED AND PLAIN

		Per Dozen Pairs	
		3⁄4	7⁄8
Size of cheek, inch..			
No. 359 Harness leather, one row extra heavy stitching........................	.....	$9.70	......
No. 656 Harness leather, spotted, one row extra heavy stitching..............	.....	14.70	$15.70
No. 657 Harness leather, spotted, one row extra heavy stitching..............	.....	15.70	17.70

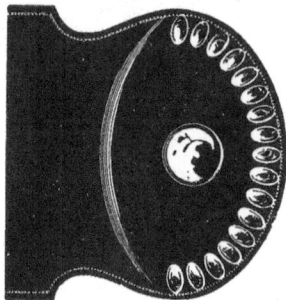

No. 658

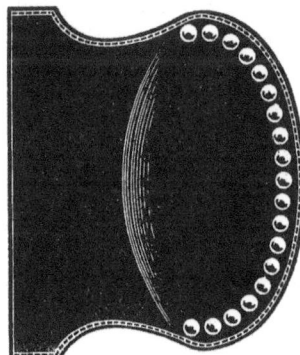

No. 659

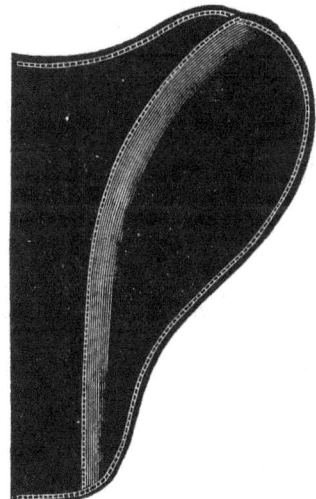

No. 488

SENSIBLE SPOTTED

	Per Dozen Pairs	
	3⁄4	7⁄8
Size of cheek, inch..		
Size of blind, length and width, inches..............................	L5¾ W4	L6½ W4¼
No. 658 Harness leather, spotted, one row extra heavy stitching	$18.00	$19.50
No. 659 Harness leather, spotted, one row extra heavy stitching	13.10	14.50

WESTERN BLINDS

Per Dozen Pairs

No. 488 Western, solid harness leather, without irons, one row extra heavy stitching, length 8 inches, width, 4½ inches, for 7⁄8-inch bridles..$14.60

DODSON-FISHER CO.

CHEEKED BLINDS AND OPEN CHEEKS

Nos. 210, 281

No. 253

No. 287

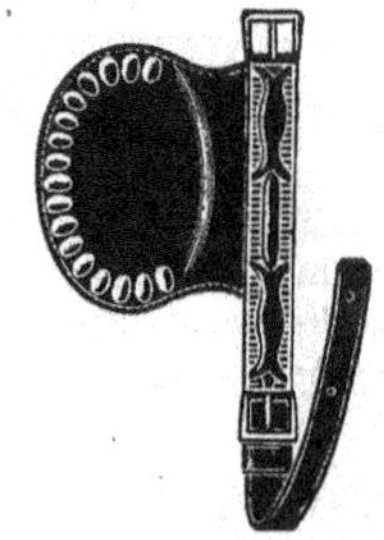

No. 657

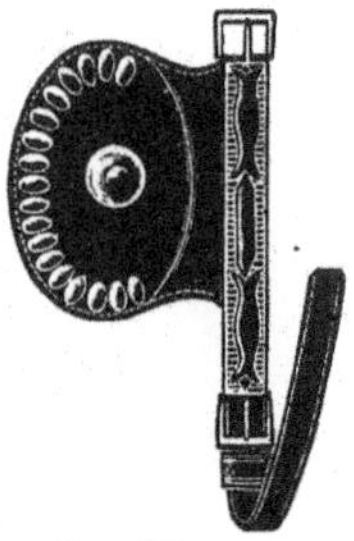

No. 658

CHEEKED BLINDS—WAGON AND TEAM

		Per Dozen Pairs ¾ in.	⅞ in.
No. 210	Japanned or X. C. plate	$27.90	......
No. 253	Japanned or X. C. plate	30.00	......
No. 253	Brass	31.50	......
No. 281	Japanned or X. C. plate	.....	$31.00
No. 287	Japanned or X. C. plate	.....	35.50
No. 287	Brass	.....	38.60

CHECKED BLINDS, SPOTTED

		¾ in.	⅞ in.
No. 657	Japanned or X. C. plate	$36.80	42.40
	Nickel or brass	38.40	45.80
No. 658	Japanned or X. C. plate	39.20	44.70
	Nickel or brass	41.00	48.10

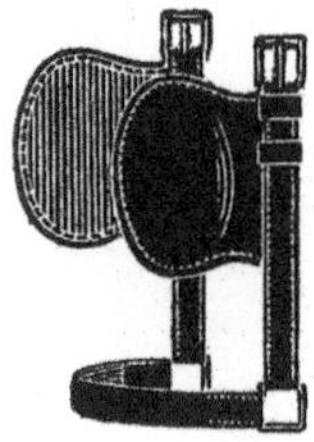

No. 3946

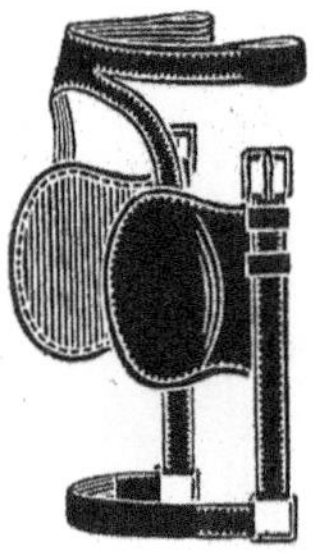

No. 3956

No. 3957

No. 2770

SHORT CHEEKED BLINDS

		Per Dozen Pairs ¾ in.	⅞ in.
No. 3946	Short cheeks, narrow loops, Concord harness leather blinds, X. C. or japanned buckles	$23.20	$25.00
No. 3956	Short cheeks, narrow loops, Concord harness leather blinds, combination fronts and winker braces, X. C. or japanned buckles	36.50	38.60
No. 3957	Short cheeks, narrow loop, Concord harness, leather blinds, combination fronts and winker braces, nickel or brass spots, X. C. or japanned buckles	43.10	44.20
No. 2770	⅞ cheeked blinds, for ring crown, ⅞ cheek, ⅞ nose band, spotted front and plain winker brace, less throat latches	....	41.00

OPEN CHEEKS

		Per Dozen Pairs
No. 244	⅞-inch, with wear leather, X. C. or japanned roller buckles	$22.10

The information on these catalog pages, especially pricing, is of course outdated. We've left this here for what it tells us about advertising, construction, and cultural preferences.

SENSIBLE TEAM BLINDS

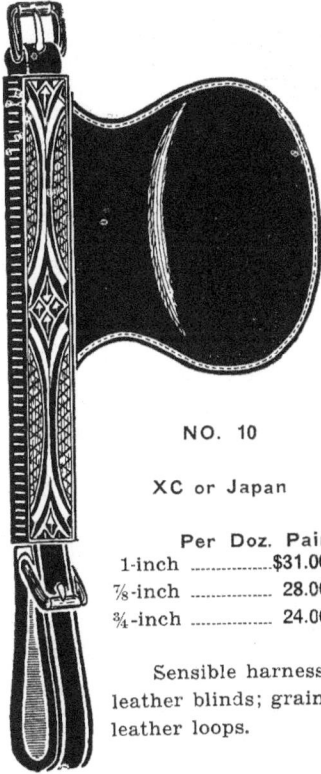

The mass produced blinders for work horse bridles circa 1900 featured stiffeners. Oft times these were shaped and sometimes cupped sheets of thin metal that were sandwiched between two layers of leather. Blinders were also made without stiffeners. In 1900 when millions of horses were depended upon day in and day out, finesse seldom found its way into the gear, its fit and its function. The few exceptional horsemen of the day took pains to make their workmates as comfortable as possible. Doubtless they knew that in the end, the performance of the animals reflected on their own reputation. But I like to imagine they took such care because they loved their animals.

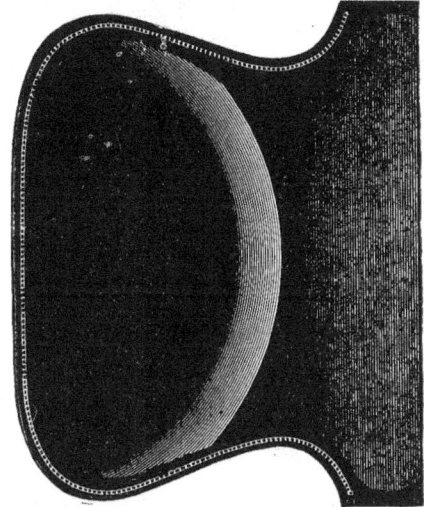

NO. 10

XC or Japan

Per Doz. Pair
1-inch $31.00
⅞-inch 28.00
¾-inch 24.00

Sensible harness leather blinds; grain leather loops.

WALLACE, SMITH & CO., MILWAUKEE, WIS.   [173]

BLINDS.

EXPRESS AND COACH BLINDS.

Nos. 127 and 128.   Nos. 136 and 1136.

	Per Doz. Pair.
No. 127—Coach, ⅝-inch extra quality, 3 rows silk stitched, new surrey shape	$14 98
No. 128—Coach, ¾-inch extra quality, 3 rows silk stitched, new surrey shape	16 10
No. 136—Chariotee, ¾-inch best grain, 3 rows silk stitched	11 10
No. 1136—¾-inch best main skirting leather, 3 rows black silk stitched	19 98

SENSIBLE CUP SHAPE BLINDS.

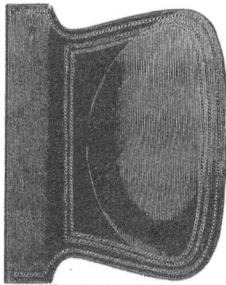

Nos. 200 to 207.

	Per Doz. Pair.
No. 200—⅝-inch patent leather, 3 rows stitched, cup shape	$ 6 22
No. 201—¾-inch patent leather, 3 rows stitched, cup shape	7 55
No. 202—⅞-inch patent leather, 3 rows stitched, cup shape	8 65
No. 205—⅝-inch harness leather, 3 rows wax thread stitched, cup shape	6 10
No. 206—¾-inch harness leather, 3 rows wax thread stitched, cup shape	7 22
No. 207—⅞-inch harness leather, 3 rows wax thread stitched, cup shape	8 32

SENSIBLE CUP SHAPE SPOTTED BLINDS.

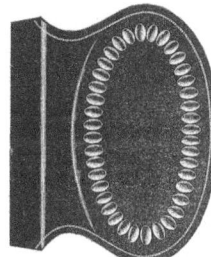

Nos. 1206 and 1207.   Nos. 2206 and 2207.

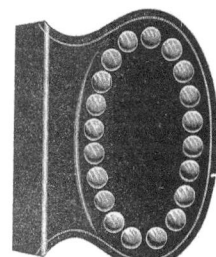

	Per Doz. Pair.
No. 1206—¾-inch harness leather, wax thread stitched	
800 Brass spotted	11 32
801 German silver spotted	12 22
No. 1207—⅞-inch harness leather, wax thread stitched	
800 Brass spotted	12 22
801 German silver spotted	13 32
No. 2206—¾-inch harness leather, wax thread stitched	
Standard brass spotted	11 32
Standard German silver spotted	12 22
No. 2207—⅞-inch harness leather, wax thread stitched	
Standard brass spotted	12 22
Standard German silver spotted	13 32

METAL BOUND BLINDS.

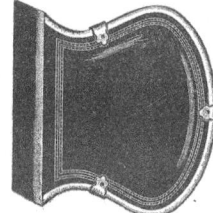

Nos. 155 and 156.

	Per Doz. Pair.
No. 155—Express, ⅝-inch nickel or brass bound, with clasps, 3 rows stitched	$12 54
No. 156—Express, ¾-inch nickel or brass bound, with clasps, 3 rows stitched	14 10

FRONTS

Per Dozen

No. 82   ⅝-inch, harness leather, solid diamond raised center, doubled and stitched $4.20
No. 82   ¾-inch, harness leather, solid diamond raised center, doubled and stitched............ 4.50
No. 20   ⅞-inch, harness leather, solid raised center, doubled and stitched 5.00
No. 20   1 -inch, harness leather, solid raised center, doubled and stitched 5.50

Inch	⅞	Per Dozen 1
No. 276 Nickel or brass, harness leather, doubled and stitched, Standard spots	$7.40	$7.90

Inch	¾	⅞	Per Dozen 1
No. 278 Nickel or brass, harness leather, doubled and stitched, Standard spots	$6.90	$7.40	$7.90

COMBINATION FRONTS AND WINKER BRACES

Per Dozen

No. 270   All harness leather, doubled and stitched, ornamented with nickel or brass spots, fronts are ⅞-inch
and winker braces are ¾-inch. Nickel or brass $12.80
No. 271   Same as No. 270 except without spots.. 11.40

Per Dozen

No. 280   All harness leather, doubled and stitched, oval spots, ⅞-inch fronts, ¾-inch winker braces. Nickel
or brass $21.60
No. 281   All harness leather, doubled and stitched, standard spots, ⅞-inch fronts, ¾-inch winker braces.
Nickel or brass 16.40
No. 283   Same as No. 281 except without spots.. 10.90

Per Dozen

No. 282   All harness leather, doubled and stitched, spotted, ⅞-inch fronts, ¾-inch winker braces. Nickel or
brass $20.90

GOPHER BRAND GOODS WILL INCREASE YOUR BUSINESS

Notice different headset on the two horses on the right.

Overcheck versus (Side) check rein.

A well-harnessed horse has a bridle path cut through the mane right behind the ears. This allows for comfort and better bridle fit.

Notice: Blinder well away from eye.

Notice: These horses have their bits buckled to the lines, not snapped.

PIGEON WING TEAM BLINDS

The pigeon wing bridle, sometimes referred to as mule bridle, allowed a longer purchase at the bridle cheek which supported the unstiffened blinder.

Throat latch should not run on this path.

Throat latch runs here.

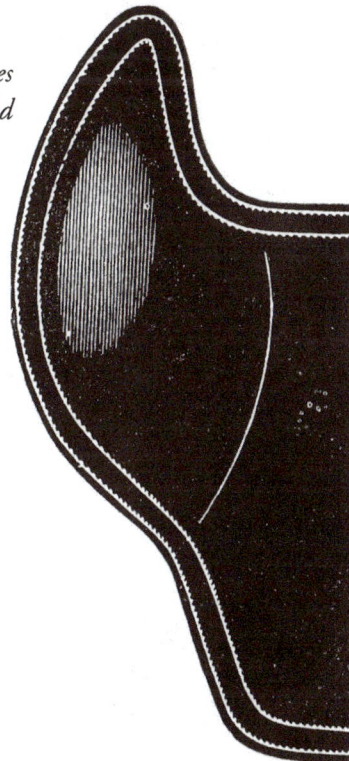

On the three-abreast pictured above (photo by Lynn Miller) please notice there are no nose bands on these bridles, and the blinders are all well away from the eyes. The far horse is in an overcheck and the other two are with side checks through guides off the bridles. This allows us to compare the 'headset' the two styles of check might result in. The overchecked horse cannot lower his nose without additional force applied to the bit.

I found this magnificent old photo of a hard working southern horselogger and his white mule. To my eye a fine pair, mule and man, comfortable and sure in their postures. Aside from the power of its composition and message, the image is a perfect candidate for some harness sleuthing. Notice the throat latch appears to be loose and up over behind the ears. This might likely be 'cuz the gentleman was interrupted from taking the bridle off. Otherwize the mule could easily rub the bridle off. Note that the collar is a perfect fit and that the harness employs a hooked chain trace. Also, there's no nose band to the bridle and the bit fits just right.

FACE PIECES

No. 1604   No. 2870   Nos. 3725 and 3726

Understanding Where It Hurts, *or employing harness variables to help give your workmates comfort.*

Some horses and mules are sensitive about any pressure on the bridge of the nose. It may have developed from abusive handling or an accident, or it just may be the way they are 'built.' If you find that your equine flinches, stops and/or backs up immdiately when any pressure is applied on the front of the face and above the nostrils, consider using bridles with facepieces such as those above, instead of nosebands. A sensitive nose bridge aggravated by pressure from a halter or noseband can create a baulky horse or mule.

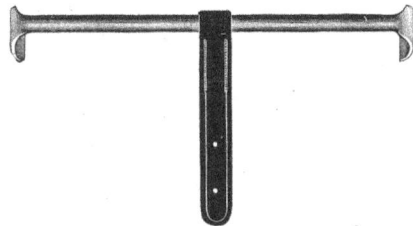

No. 1319—Round Concord ¾-inch billets..........................

NOSE BANDS.

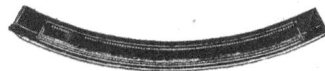

Per Doz.

½-inch, plain leather..

check reins and overchecks

On page 93, the far horse is wearing a full overcheck, which pulls up on the bit rings. If the horse lowers his head he feels a corresponding pinch and tug at his cruppered tail. Effective and uncomfortable all at the same time.

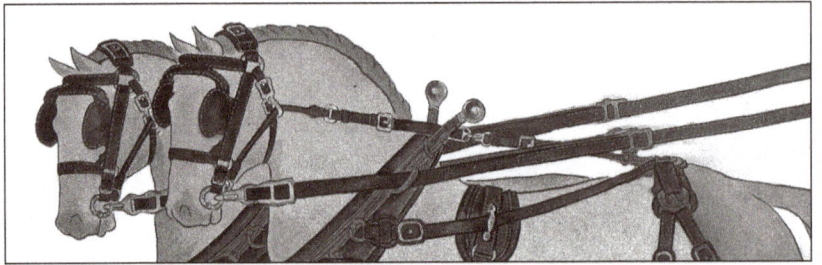

Mike Atkin's mules are wearing relaxed check reins that pass around hame tops. This is far and away the most prevalent setup in North America.

Sometimes big teams are rigged with lead lines passing through wider-stance gag runners (no. 374 on right). This allows room for both the check rein and lead lines to work freely.

Most of the old time overchecked harness styles portrayed in this book hook over a terret on the back pad. Option is another short strap that runs on back to the crupper as is seen in the diagram below.

Above (see also page 104), these check reins pass from the bit ring, up through the gag swivels at the bridle, then back between hame tops to a short strap that connects to the breeching spider. Nose down or head down this rigging comes tight. As effective as an overcheck and not nearly as uncomfortable if properly adjusted.

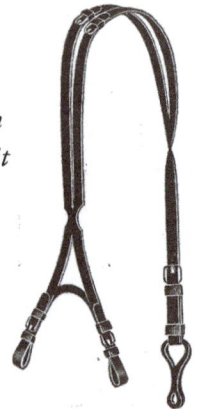

Full overcheck straps which can be fastened at bit rings or lower cheek straps if width allows.

SWIVELS OR GAG RUNNERS

No. 374.   No. 384.   No. 380.

A relaxed, gag swiveled check rein, laid around the top of the outside hame, will, when adjusted properly, prevent most horses from lowering their head to rub or graze.

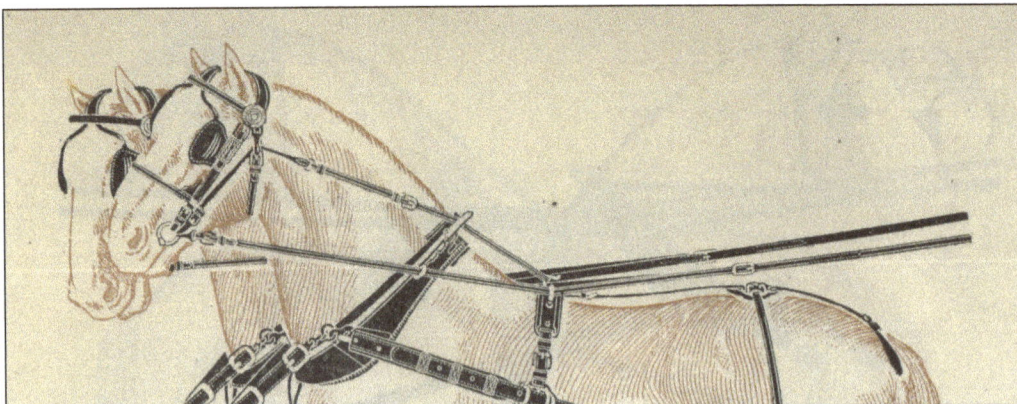

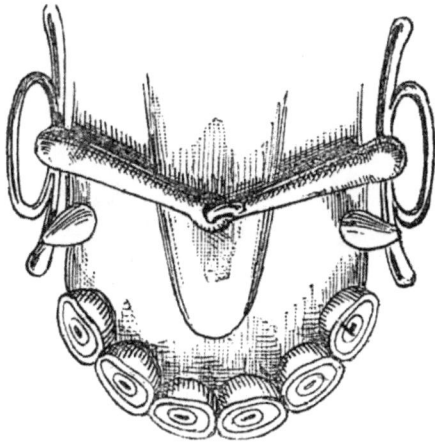

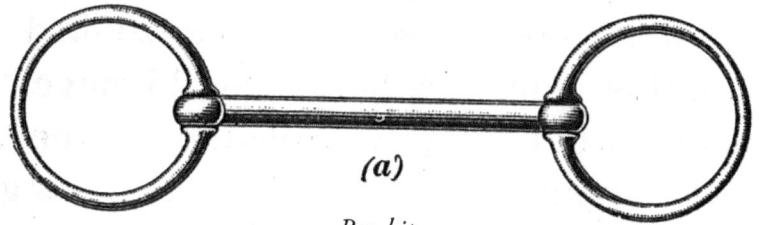

(a)

Bar bit

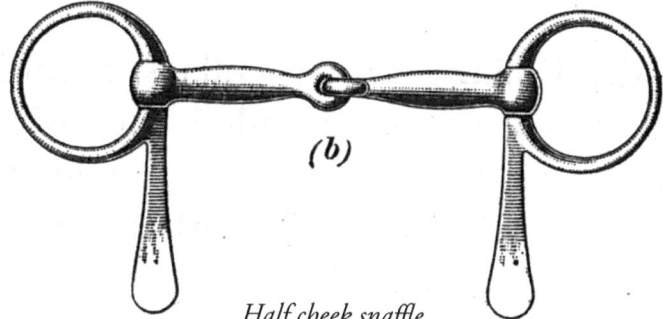

(b)

Half cheek snaffle

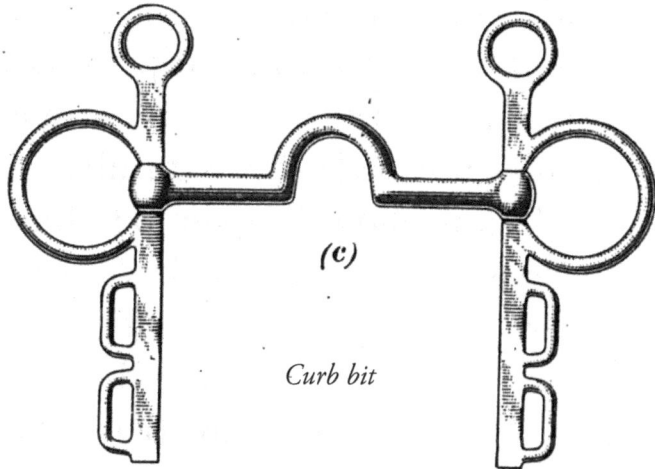

(c)

Curb bit

The bit is that instrument which goes into the mouth of the equine, hanging from the bridle, and fastening to the driving lines, check rein and sometimes curb chains. The bit and biting process is that arena where control argues with true horsemanship. Light and sure hands on the line result in partnership. Heavy hands gripping and pulling and worrying separate man and beast into a perpetual contest.

Bits commonly used with western harness encompass a wide array of styles. There are the straight bar bits with ring and bar diameters that vary. There are the snaffle, or jointed bits – ring-style, half cheek and full cheek, and an even wider array of levered or curb bits with straight bars, jointed snaffles, and some combined with raised port. These levered curb bits feature a design to gain leverage by fastening the lines lower on the shank. (See next page.)

Snaffle

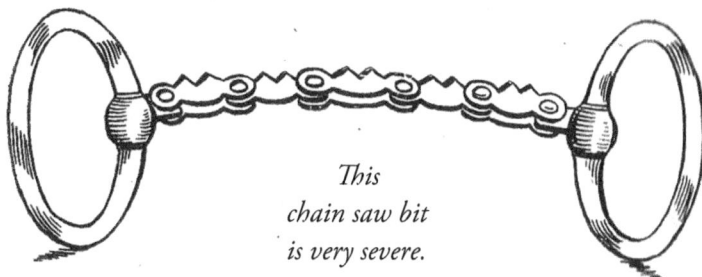

This chain saw bit is very severe.

Fitting the Bit

When you have determined what bit you wish to use with which animal, you need to determine the length required. This amounts to common sense. A bit that is too short will either pinch the sides of the mouth or not fit at all. While one that is too long will hold the bridle away from the cheeks and tend to cause the mouthpiece to saw back and forth when in use. Matching the horse's mouth width, from outside cheek to outside cheek, is the key to comfort.

BITS

No. 2371 Steadman.

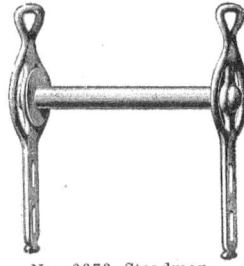

No. 2373 Steadman.

No. 2 Tubular.

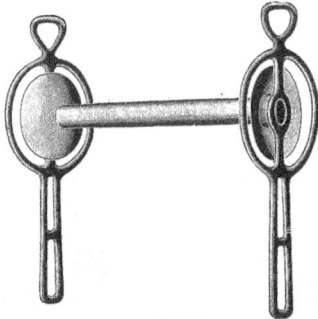

No. 1 Tubular.

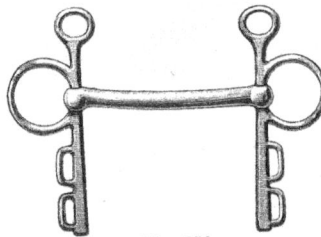

No. 759.

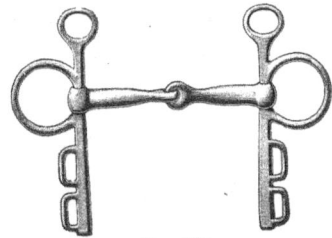

No. 762.

No. 34.

No. 160.

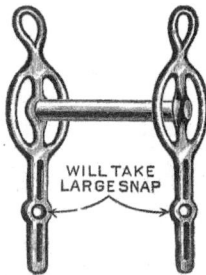

No. 0335.

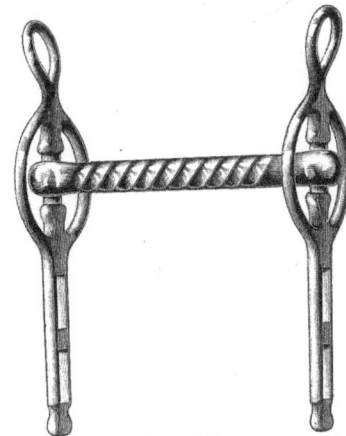

No. 738.
Loops are Large Enough to Take Snap.

No. 78.

Nos. 46 and 56.

Twisted bar

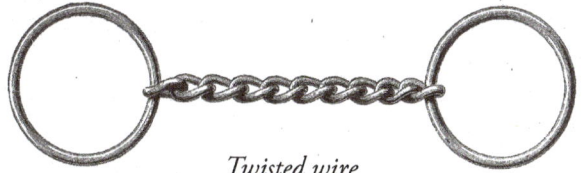
Twisted wire

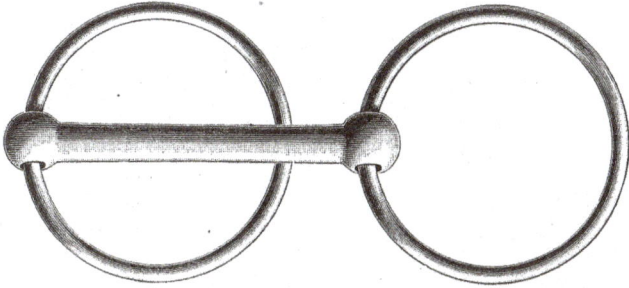

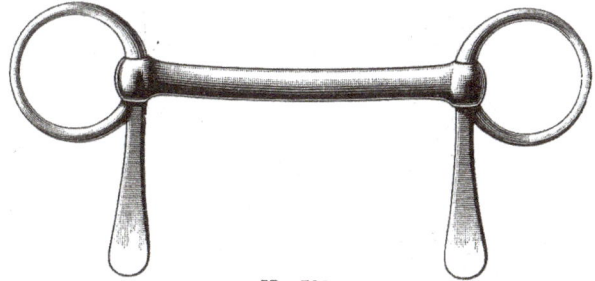
No. 502.

FULL CHEEK BITS.

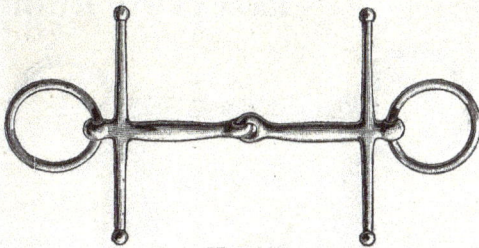
No. 892.

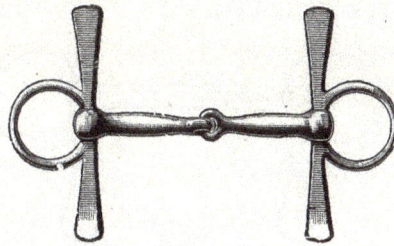
No. 504.

RUBBER MOUTH

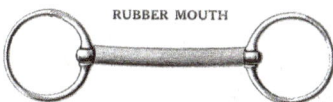

RUBBER MOUTH

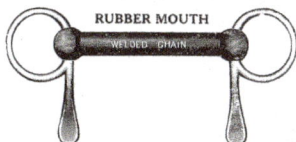

These horses are walking and you can see that the bits are not tight. The line pressure is perfect. The left side bit is a standard snaffle and the right is a half cheek snaffle. Occasionally an unruly or frightened horse might get the bit ring lodged inside of the mouth. The half cheek or full cheek prevents this.

J.I.C. DRIVING BIT

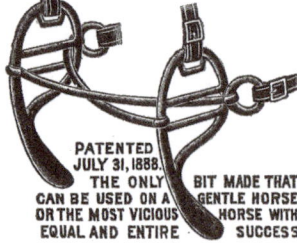

BERRY BIT.

PATENTED
JULY 31, 1888.
THE ONLY
CAN BE USED ON A   BIT MADE THAT
OR THE MOST VICIOUS   GENTLE HORSE
     HORSE WITH
EQUAL AND ENTIRE   SUCCESS

PAT. APPL D. FOR

Success Patent.

Pinching Bits

The three bits across the top of this page are classified as **training bits.** They allow that the teamster has great leverage and can pinch the horse's cheeks together with less pull on the bars of the mouth. While effective in the hands of one who understands what's involved and at stake, they are a mistake and a hazard for the novice.

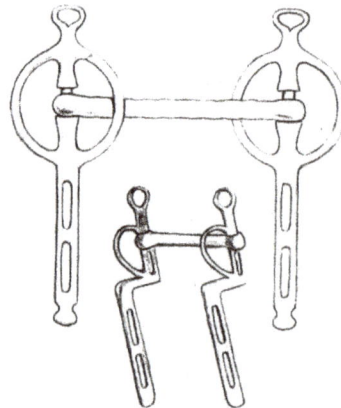

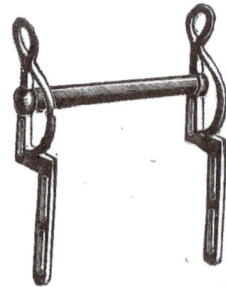

When the lines are pulled back, the bit pivots and the top of the bridle puts downward pressure on the poll or crown of the horse, simultaneous with a pinching pressure at the mouth and correspondingly at the chin.

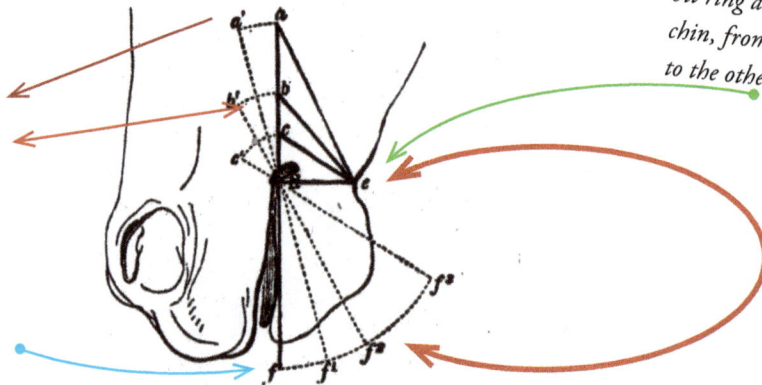

Obviously the curb bit could easily become a torture device in the wrong, or heavy, hands.

When finessed with light hands, by a teamster who is always concerned for the comfort of the animals, it can be a useful tool.

With the curb bit, the teamster may customize the relative pressure for the individual animal. The higher on the shank the less the pressure. If the curb chain or strap is lengthened you can do away with most all of the chin pressure.

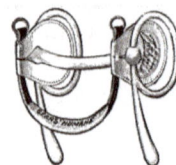

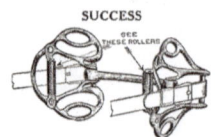

The curb strap or chain customarily attaches to the bit ring and runs under the chin, from one side of the bit to the other.

Diagram to show action of curb bit.

At the lower end, or shank, of the curb bit there are usually two or three places where the lines may be attached. As the driving line is pulled back the bit pivots in the mouth applying pressure to the curb chain and the bridle cheek straps and crown.

SUCCESS

SEE THESE ROLLERS

The line on this mule is buckled direct, with little or no leverage.

The line on this mule is at the first position allowing a leverage action that will tighten the curb chain beneath the jaw.

Notice how open these pigeon wing blinders are, basically only obstructing the view behind.

These are old original US army bridles. The bits are modern generation.

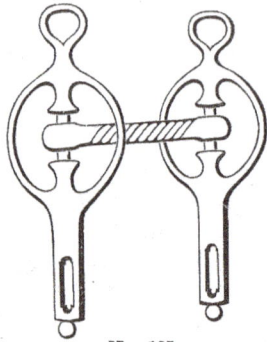

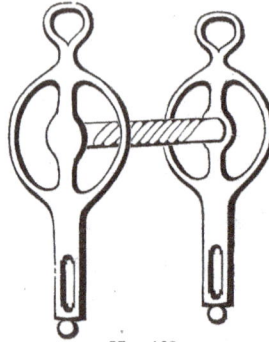

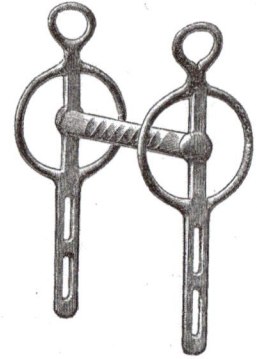

No. 175.   No. 185.   No. 190.   Nos. 200 and 210.

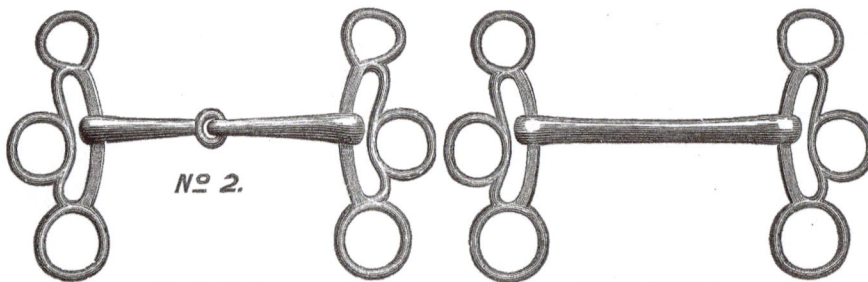

Nº 2.

No. 2. Buckeye.   No. 3. Buckeye.

At public events and contests many teamsters find that leaving the halter on beneath the bridle lends an additional level of security. (If the bridle should break or come off, the halter is still there.) I would like to see a halter designed and constructed for this purpose where the halter hardware and strap attachments do not compete with the bridle fit and function.

Please note: in this picture on the right you can see the leather driving line 'snapped' into the bit ring (the bit of chain coming down from above is part of the check rein). Some cautious teamsters always turn those snaps towards the horse to avoid the horse turning its head and hanging the snap up on the breast strap or other harness hardware, rendering zero control to the teamster.

In the photo above, of this superb three-abreast of Belgians portrayed here at the World Plowing Championship in Ohio a few years back, it is easy to see the comfortable posture of all three. The collars fit perfectly. The bridles do not crowd the ears. Note that these horses are wearing well-fitted nylon halters beneath their bridles.

Halters beneath the bridles or no?

Fit makes all the difference. If it doesn't fit, don't do it. The halter needs to ride above and clear of the bit, bit strap and lines, such as in the photo above. On the previous page the halters are interfering with the bridle and bit comfort and aggravating what is an already too tight fit.

Beyond the issue of fit, all things being equal, teamsters agree to disagree on the question of leaving halters on beneath the bridle. Some old timers feel that it's sloppy and denotes a less than experienced teamster. I do it when the halter is a necessary element in the hitch, such as for tying in buck rake horses, or when bucking back and tying in big hitches.

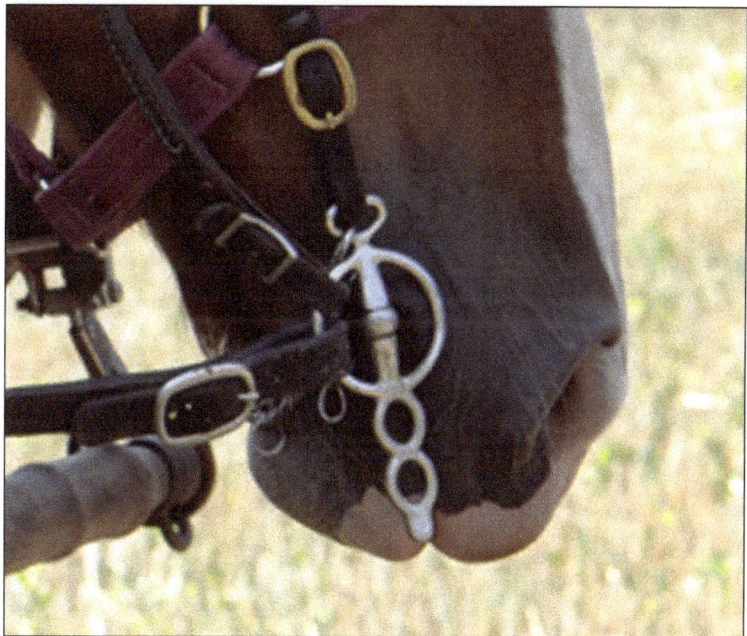

The photo directly above shows a levered bit perfectly set in the mouth and buckled straight away (no leverage). The red halter is properly fitted and completely free of the bit and bridle action.

Big field hitches of yore sometimes drove with halters and no bridles. The bits were buckled or snapped into the halter side ring.

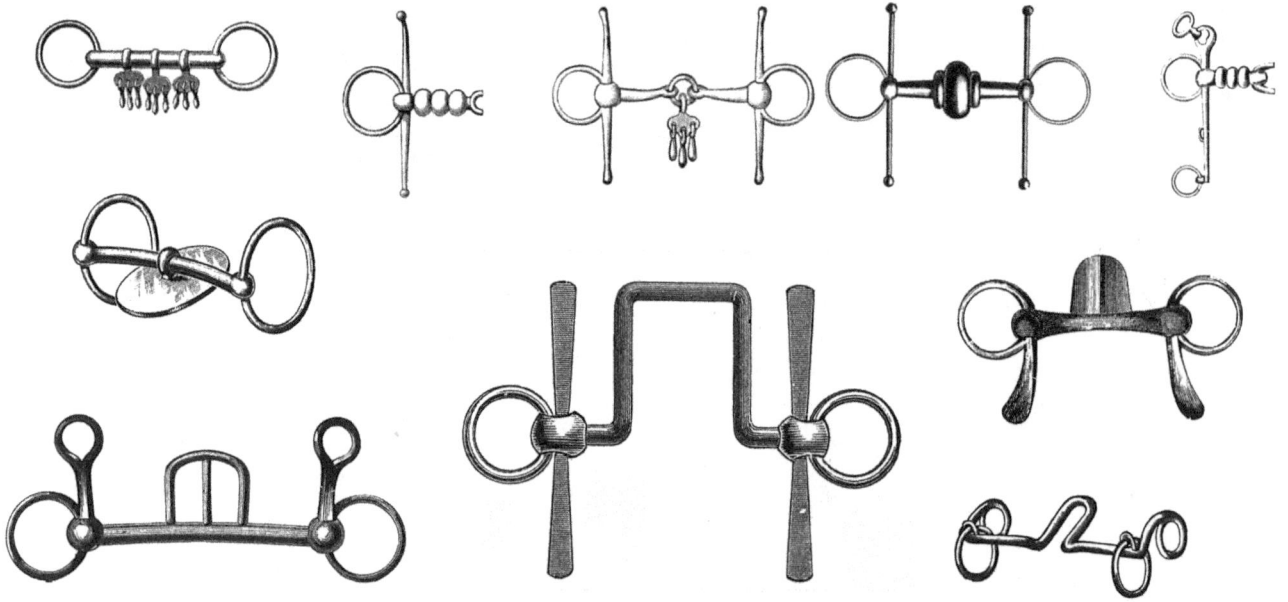

See how the check rein is snapped back to the brichen spider.

Look to the photo below of Les Barden's team and note that with his bridle design and fit it appears almost as though these are blinderless bridles – but they aren't. The point being that it is possible to keep some distractions at a minimum for the animals while still 'honoring' them with great freedom and comfort.

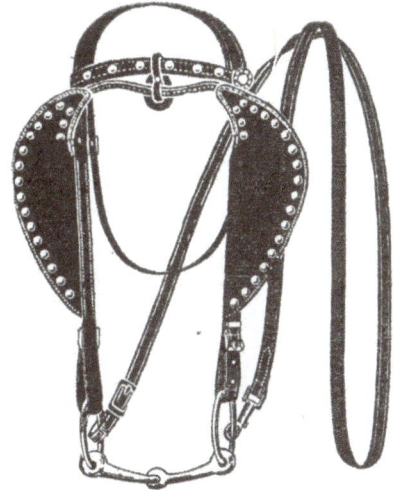

William Castle's Percheron mare works in Shropshire, England, without blinders.

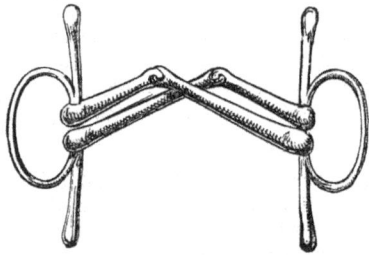

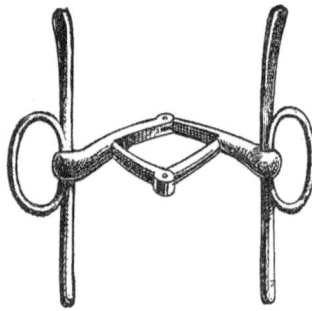

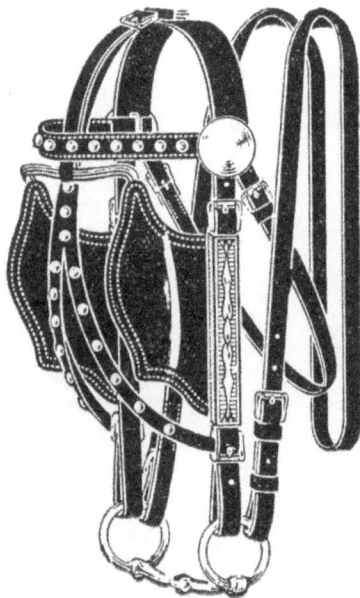

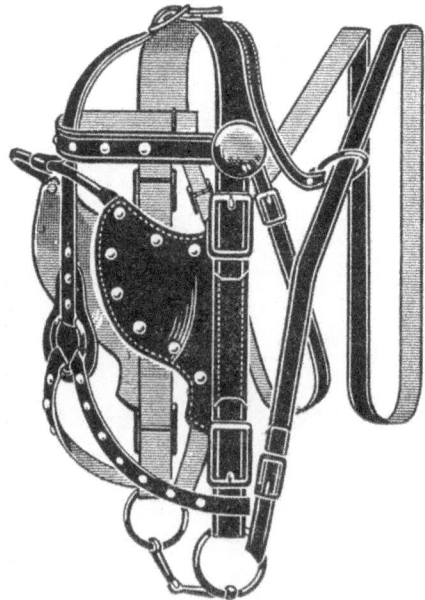

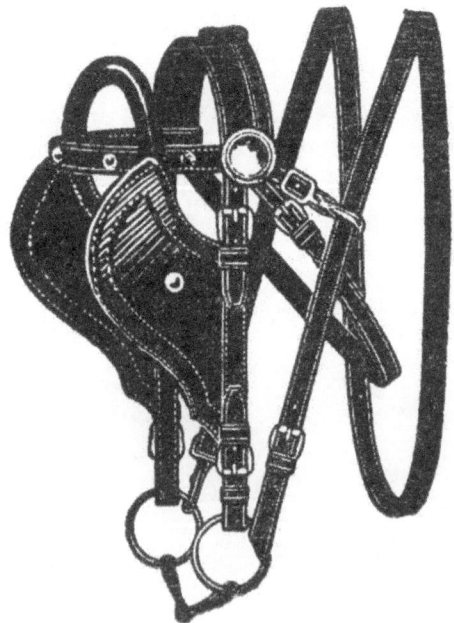

Notice that there are no nosebands on this page and that the 'Jack' cartoon above features a small, possibly twisted wire, snaffle bit leveraged by a rein chain running through bit rings and under the jaw for a pinching action.

Chapter Four

Tugs, Traces, Belly Bands & Back Pads

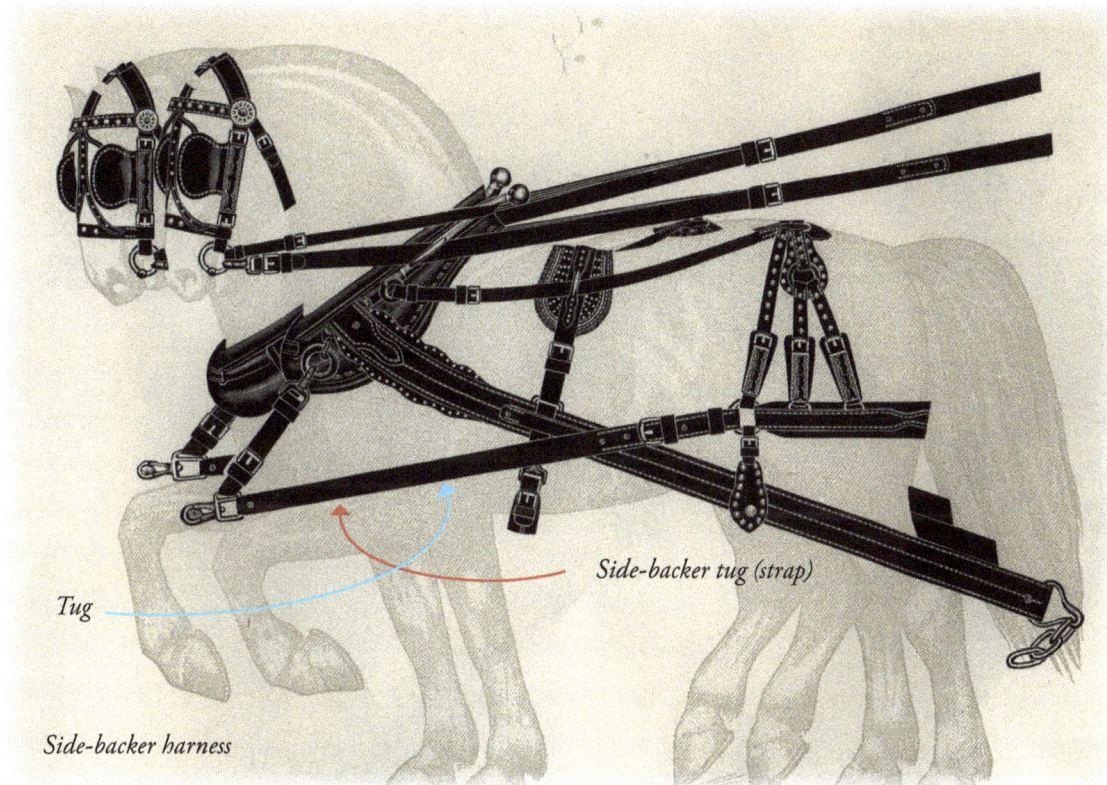

Tug

Side-backer tug (strap)

Side-backer harness

In the western hame and collar style harness structure, what translates the forward motion of the equine to a pull? It is, for each animal, two parallel tugs moving back from their attachment at the collar-held hames along the animal's sides clear back to where the tugs attach to the single tree.

For the purposes of this book, our primary focus has been and continues to be the western-style bas-ket-brichen harness. Within that outline there are a few variables. Above is an old harness catalog illustration of a heavy duty, parade spotted, western three-strap brichen harness with side-backers instead of pole strap assembly. On the next page is the more conventional two-strap brichen, unspotted, heavy work harness with pole strap backing and braking assembly. Note how the pole strap passes between the front legs.

On page 110 (top) there is another version of a side-backer harness, this one without back pads.

But back to the basics: as the animal steps forward into the collar, the hames, affixed into the groove along the outside circumference of the collar, provides a frame on which two tugs or traces attach, one on each side of the animal. These

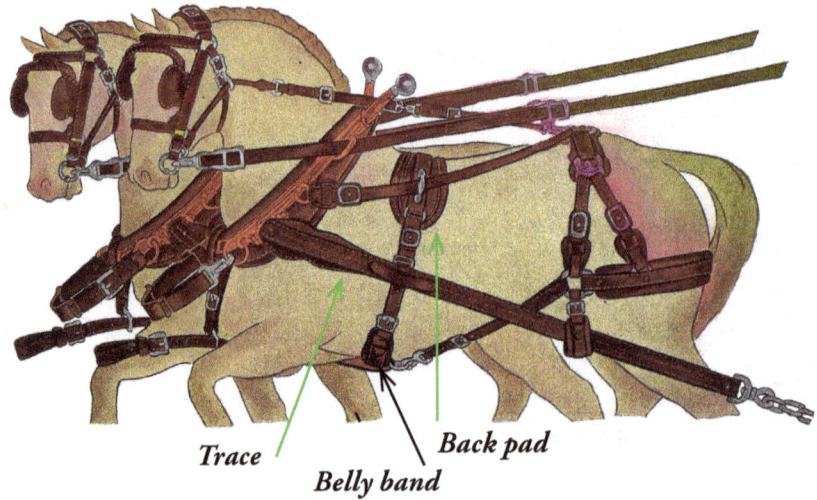

Trace   Back pad   Belly band

Pastern angle   Hame seat angle

The correct hitch for a plow. The tugs are parallel to the line of draft from the center of resistance of the plow through the clevis pin to the point on the hames where the tugs fasten the hitch. (Courtesy Oliver Chilled Plow Works.)

pivot on the hames at a spot which coincides with the point of the shoulder. This beginning spot for the traces or tugs also aligns with the widest portion of the collar.

Forgive me for the simplification but let's just use the more *common* term, *tug*, for the purposes of the rest of this discussion. That is, after all, what they do, they *tug* the load. (*Trace* is the more formal and long-lived term for this harness part.)

The tugs, from their attachment to the hames usually traveling back along the animal, best perform at an angle of 80 to 100 degrees to the line of the hames (see diagram above). If the angle is

significantly less than 80 degrees the tugs *may* pull up and back on the collar, but only if the belly band is not adjusted properly. When it is, this will 'interrupt' an aggravated angle preventing the horse from being choked by a forward rocking collar.

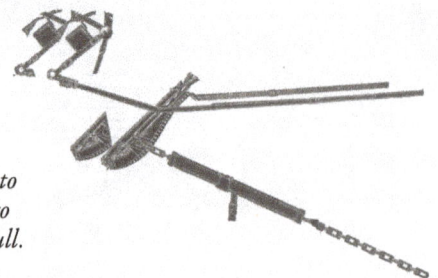

Stripped down to bare essentials to accomplish a pull.

If the belly band is too loose it won't hold the forward portion of the tug in line. As each and every horse's angle of shoulder is different, and as the head-set of a pulling horse may be more or less down or up, these factors will affect the angle of the shoulder at work. It is worth noting in this context that the comfort and efficiency of collar and hames can be affected by: 1. Proper fit (collar size, hame length, hame strap attachments, etc.) 2. Proper length, position and adjustment of belly band 3. The conformation of the animals.

From the standpoint of equine conformation as it relates to pulling, what we look for as optimal is a natural angle of the shoulder face (or collar seat) to match the angle of the pastern.

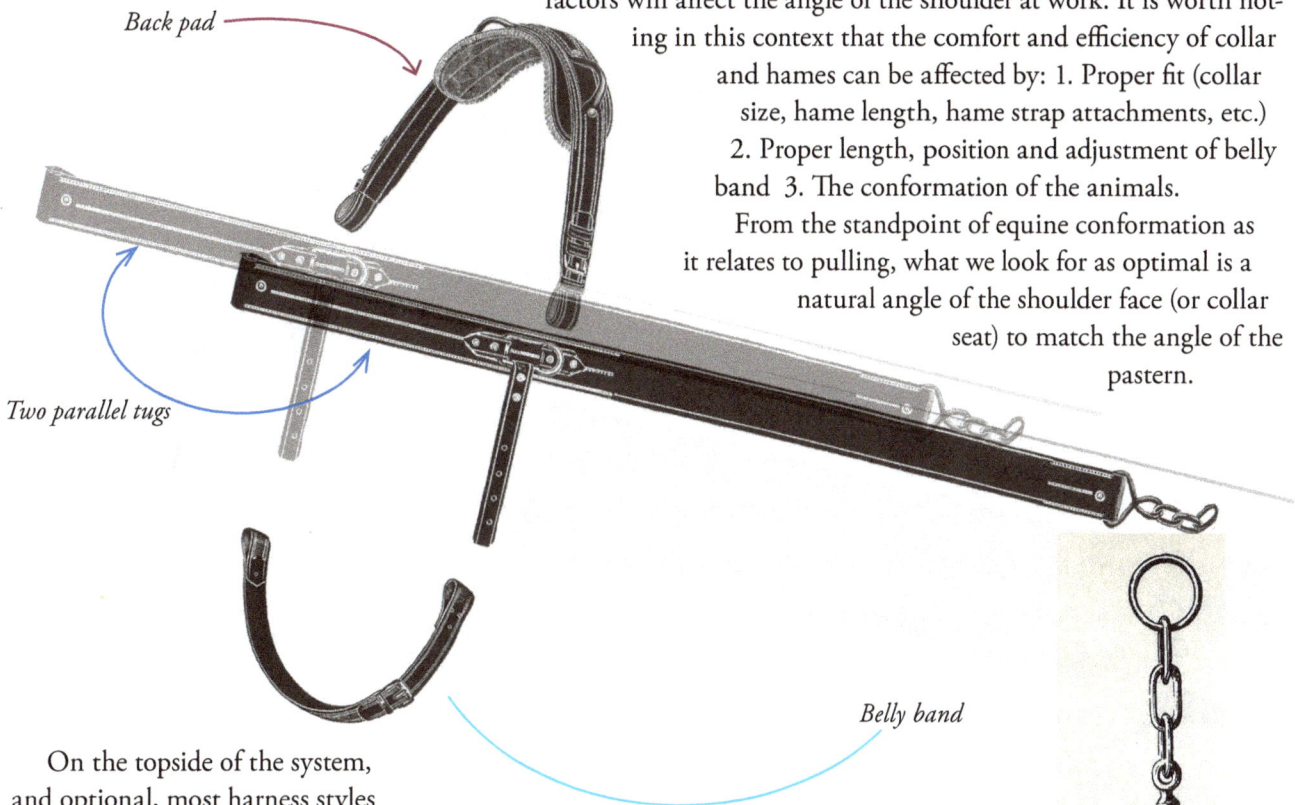

Back pad

Two parallel tugs

Belly band

Shaft loop strap

BUTT CHAINS.

On the topside of the system, and optional, most harness styles employ a back strap assembly that also locates the tugs in a similar fashion to the belly band; belly band from the bottom — back pad from the top. The tugs usually have a position loop for billet straps on the bottom and back pad attachments on top.

Shaft Hookup Variables

'Market Tug' style harness and some plow harnesses have no back pads (or some people say *saddles*). The back pad or saddle comes into play when horses are worked in shafts and need to carry the weight that the shafts balance from the cart or implement. Attached to the back pad sides are shaft loop straps which hang and carry the shafts.

Most tugs have tug or trace chains permanently attached at their ends. Some employ butt chains which are separate from the tug and feature end rings. These chains are typically seen threaded at the end of single trees. By hooking one ring or link on the butt chain hook at the end of the tug, and allowing the ring on the other end to come to hold on the end of the single tree, you are hooked full out. By pulling that second ring forward to hook alongside the other ring or link, you have halved the distance, or doubled back and with this shortening of length allowed for a more aggressive lifting action as the horses step forward. Butt chain harness is popular with woods or logging work.

This last link of butt chain slips over the hook at the end of the tug, after passing through hardware at the end of the single tree. The ring on this chain either stops the chain at full length or doubles back and hooks into the tug hook to halve the distance and lift the load.

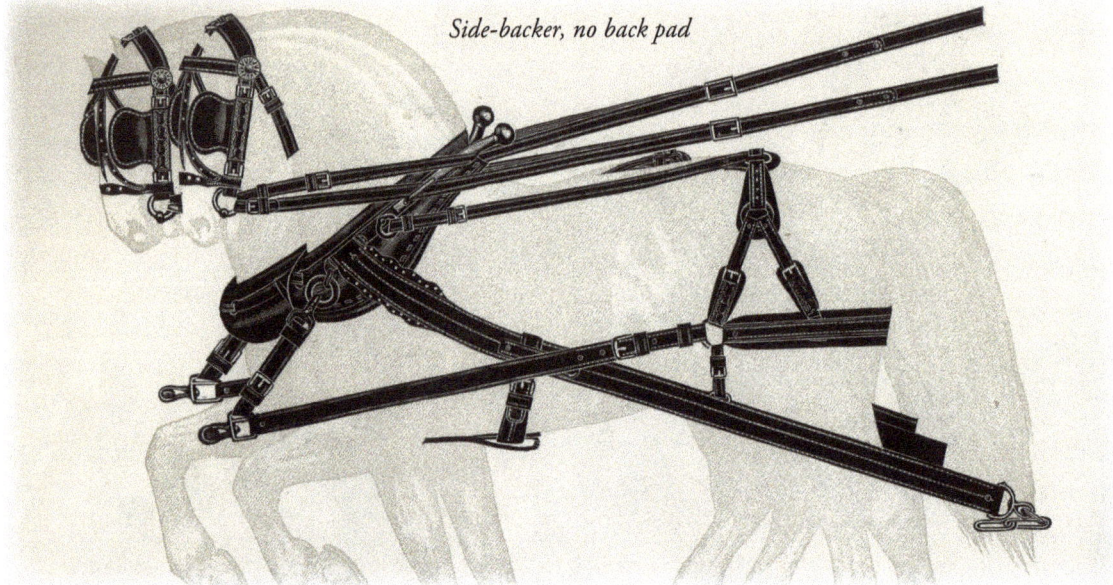

Side-backer, no back pad

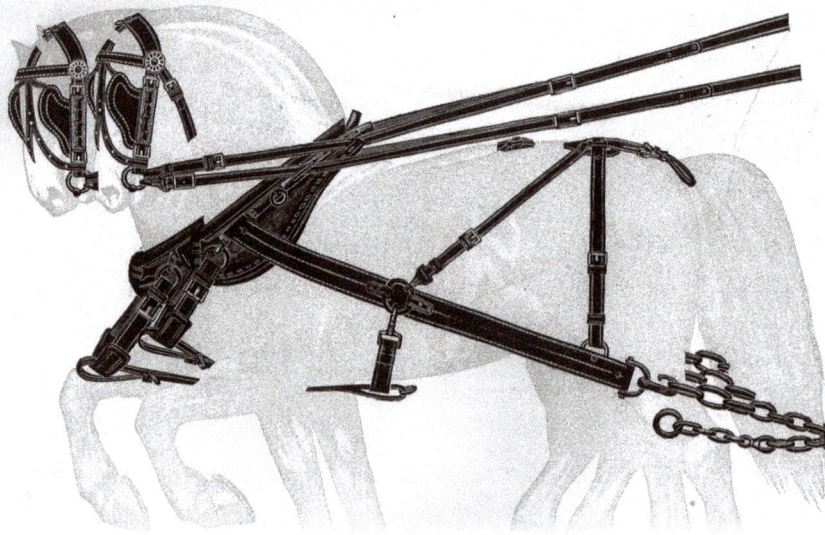

Instead of quarter straps (or some people call them hold-backs), the harness above uses side-backer straps.

Butt chain doubled back to shorten hitch and lift while pulling.

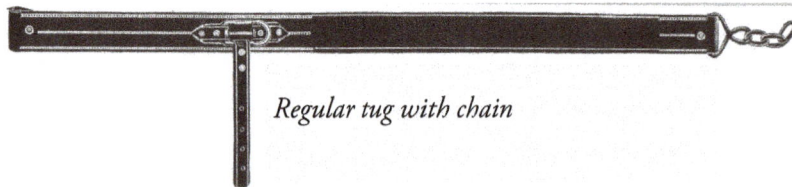

Regular tug with chain

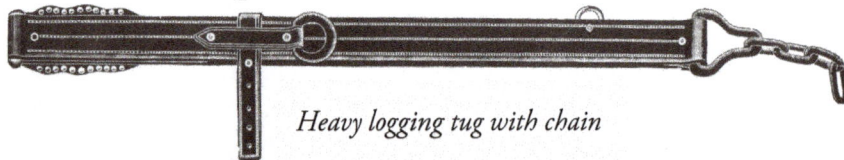

Single tree hooks here.

Heavy logging tug with chain

Butt chain tugs style top and bottom

Butt chains hook here.

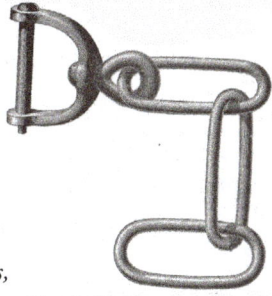

Trace chains,
usually four to five links.

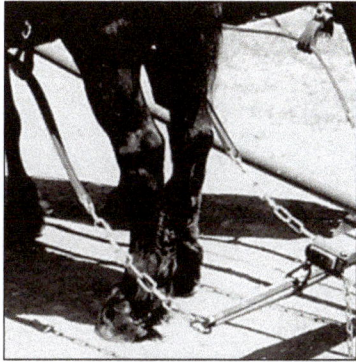

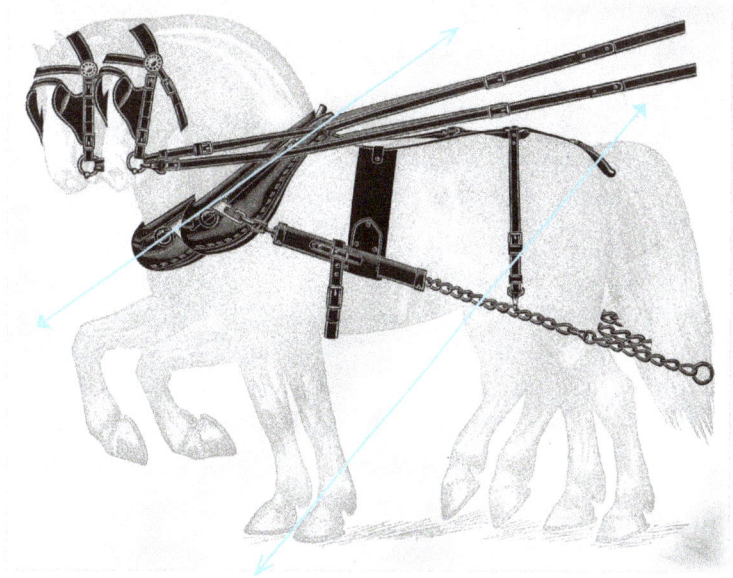

Every age and discipline has had its own aesthetic. Notice how in this old harness catalog image that the collar lays way back at an exaggerated angle, and that that angle is quite at odds with the pastern angle. Might look good to some, but it is the picture of discomfort and inefficiency. Most of the old harness catalog pictures are this way.

The more common tug chain harness, with chain fastened to the end of the tug, or chain tug harness (as pictured left, hitched to a White Horse Plow), is hitched by threading a chain link over the single tree end. You can hook at various lengths and have a similar shortening and lifting effect as the butt chain tugs. If you hook in the middle of the chain length, thread the link you will hook through the end link to avoid having extra length dangling.

No. 473 HOOKS AND DEES

No. 474 HOOK CLEVISES

TEAM TRACES

DOUBLE AND STITCHED TEAM TRACES, IMITATION HAND SEWED

		1¼ in.	1⅜ in.	1½ in.	1¾ in.
No. 1200	6 feet 2 inches long, per set..................................	$....	$....	$26.60	$30.50
No. 1200	6 feet 6 inches long, per set..................................			28.10	32.60

DOUBLE AND STITCHED IMITATION HAND SEWED

		1½ in.	1¾ in.
No. 1216	5 feet 8 inches long, per set..................................	$25.30	$29.00

SINGLE PLY TRACES

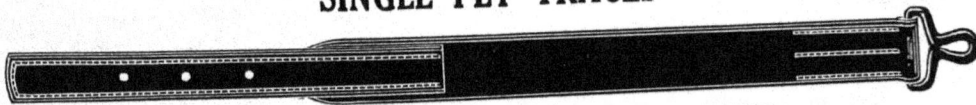

SINGLE PLY TEAM TRACES, CAMPBELL LOCK STITCHED

No. 2214	1½x2½-inch, single ply, double and stitched points, 2½-inch cockeyes, per set..................	$26.00
No. 2252	1¾x2¾-inch, single ply, double and stitched points, 2¾-inch cockeyes, per set..................	27.90
No. 2254	2 x3 -inch, single ply, double and stitched points, 3 -inch cockeyes, per set..................	32.10

WAGON TRACES

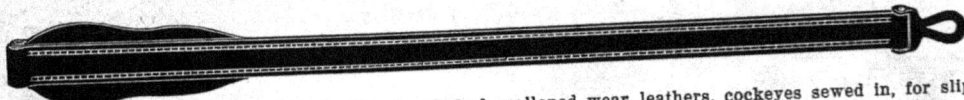

No. 933X 6 feet 10 inches long, Campbell lock stitched scalloped wear leathers, cockeyes sewed in, for slip tug harness.
1⅜-inch, per set ....$31.50   1½-inch, per set...............................$33.10

TEAM HAME TUGS

		1½ in.	1¾ in.
No. 1604X	X. C. or Japanned, solid		
	Without belly band billets, per set......	$10.80	$12.80
	With 1½ in. belly band billets, per set	12.50	14.60

FOR BOLT HAMES

No. 1216X 1½-inch metal hame tug plates, leather safe under-plates. X. C. or japanned, with 1½-inch belly band billets, per set $13.50

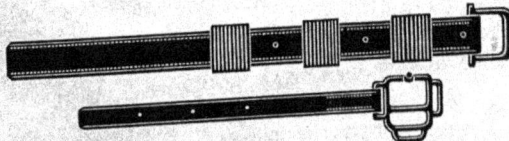

		1½ in.	1¾ in.
No. 1195X	Japanned, solid, for clip hames		
	Without belly band billets, per set......	$9.50	$11.20
	With 1½ in. belly band billets, per set..	11.40	13.00

CHAIN PIPES

CHAIN PIPES

	Length, inches	36
No. 2	Per set of four..........................	$8.30

Tugs versus Traces – *It is my conjecture that before WWI the word 'trace' in harness nomenclature referred to the complete pulling 'strap' such as is seen above top and on the next page. While the word 'tug,' commonly used in concert with 'hame' as in 'hame tug,' referred to the short portion of the two piece 'trace.' (Just above.) The longer rear segment was detachable. With so many regional peculiarities in function and form, many of the names of harness parts morphed and solidified. Tugs and Traces became, over time, synonymous with each other, while in some areas one name was preferred over the other.*

FARM TRUCK TRACES

6 FEET LONG, CAMPBELL LOCK STITCHED WITH NO. 65S CHAIN $27.90

No. 2700X	1½-inch, with 1¼-inch billets, straight wear leather. Per set	32.10
No. 2700X	1¾-inch, with 1¼-inch billets, straight wear leather. Per set	34.20
No. 2765X	1½-inch, with 1½-inch billets to loop. Per set	37.90
No. 2765X	1¾-inch, with 1½-inch billets to loop, 3 rows stitching. Per set	

6 FEET 2 INCHES LONG, CAMPBELL LOCK STITCHED, SOLID THREE PLY, 8 LINK SCREW DEE CHAIN
Gopher Grade

No. 2832X	1½-inch, billets 1½-inch wide with liner. Per set	$39.50
No. 2834X	1¾-inch, three rows of stitching, billets 1½-inch wide. Per set	43.90
No. 2942X	1½-inch unfinished, less billets and chain. Per set	33.60
No. 2943X	1¾-inch unfinished, less billets and chain. Per set	39.50

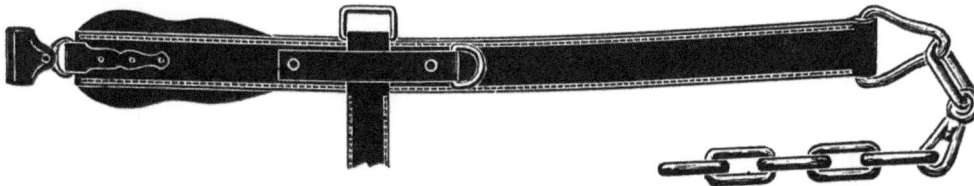

6 FEET LONG, CAMPBELL LOCK STITCHED SOLID THREE PLY, 8 LINK SCREW DEE CHAIN
Gopher Grade, 1½-inch Billets with Liner

No. 2808X	1½-inch, with 918 clips. Per set	$36.50
No. 2808X	1¾-inch, with 918 clips, three rows of stitching. Per set	40.50

6 FEET 3 INCHES LONG, CAMPBELL LOCK STITCHED, SOLID THREE PLY, 8 LINK SCREW DEE CHAIN
Custom Grade. No Loops or Billets

No. 2772X	1½-inch. Per set	$34.70
	1¾-inch, three rows of stitching. Per set	40.00

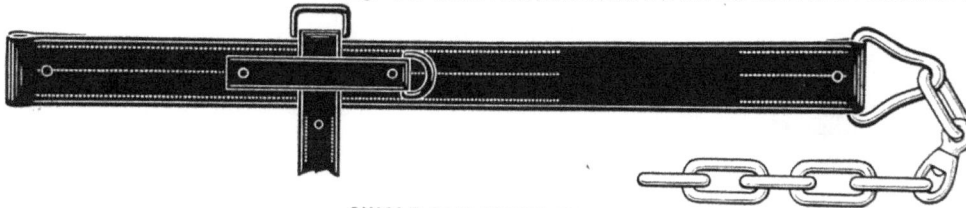

SINGLE PLY TRUCK TRACES

No. 2706	2½-inch, single ply, double and stitched at hame end, 6 link heel chain, 1½-inch belly band billets. Per set	
No. 2706	2¾-inch	$33.70
		36.60

HOOK AND DEE FARM TRACES

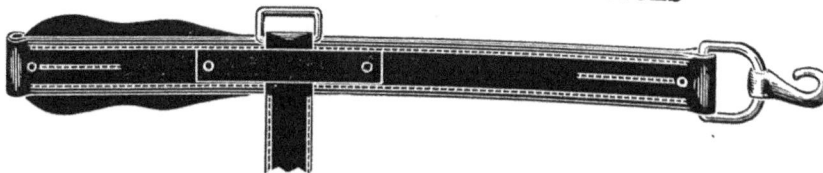

4 FEET 10 INCHES LONG, CAMPBELL LOCK STITCHED. SOLID THREE PLY
With No. 474 Screw Bolt Hook and Dees

	Per Set			Per Set
No. 2349X	1½-inch, with 1½ billets.....$30.90	No. 3140	2¼-inch with 1½-inch double and stitched billets with 2-inch liner, 2 rivets at hame end.......	
No. 2359X	1¾-inch, three rows of stitching with 1½ billets 38.60			
No. 2361X	2-inch, three rows of stitching, with 1½ billets 42.10	No. 3140	2½-inch, otherwise same as above.....	$43.50 / 48.00

Trace Chains, Cockeyes and Butt Chains.

Today, in work horse circles, there are two predominate forms of traces - those with chains affixed and those with hooks affixed to receive butt chains.

There was a third style of trace, seen on the previous page, which featured Cockeye hardware on the ends of which either chains or various other connections may have been attached. These are most uncommon today.

BACK BANDS

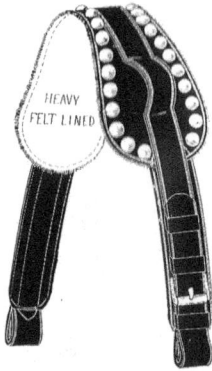

No. 42 5-inch pad, fe't lined, 2 inch body,   Per Doz. Pr.
1½-inch billet $10.00

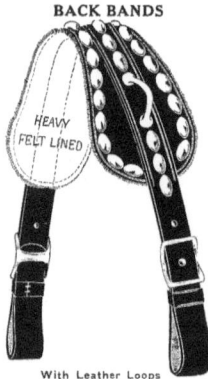

With Leather Loops
Price, per Pair
No. 25 4½-inch pad, felt lined, 1¼-inch bil'et..............$5.00
No. 26 5 -inch pad, felt lined, 1½-inch bi'let.............. 5.00

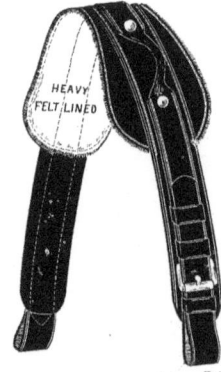

Price, per Pair
No. 38 5-inch pad, felt lined, 2-inch body, 1½-inch
billet, round side loops...........................$11.00

WEB BACK BANDS

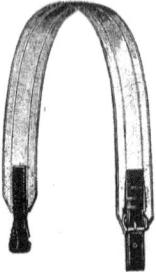

No. 1 4-inch No. 0 white back band web, 42 inches
long. 1¼-inch, stitched leather billets. With
No. 150 buckle ....Per Set of 2 $3.30
No. 2 4-inch No. 0 white back band web, 42 inches
long. 1¼-in. stitched leather billets. Set of 2   4.00
No. 3 Same as No. 2, with 1½-inch billets. Set of 2   4.50

The information on these catalog pages, especially pricing, is of course outdated. We've left this here for what it tells us about advertising, construction, and cultural preferences.

SNAP AND BUCKLE CRUPPER

Price. per Doz.
No. 214 ¾-inch, folded, buckle and snap..................$8.50
No. 215 ⅞-inch, folded, buckle and snap.................. 9.00

BUCKLE CRUPPERS
Price. per Doz.
No. 212 ¾-inch, folded, with center bar buck!es..........$7.00
No. 213 ⅞-inch, folded, with center bar buckles.......... 7.50

CRUPPER FORK ATTACHMENT, TEAM

Price. per Doz.
No. 35 1½-inch body; ⅞-inch forks; folded crupper,
to buckle $15.00

DUNCAN & SONS, Inc., SEATTLE

TEAM OR TRUCK TRACES—All Traces have 6 Link Toggles

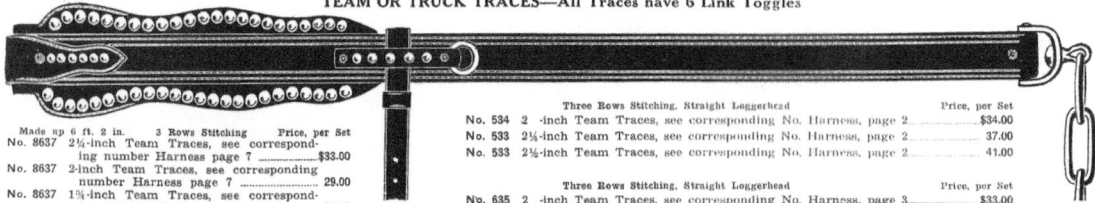

Made up 6 ft. 2 in.   3 Rows Stitching   Price, per Set
No. 8637 2½-inch Team Traces, see corresponding
number Harness page 7 ....$33.00
No. 8637 2-inch Team Traces, see corresponding
number Harness page 7.................... 29.00
No. 8637 1¾-inch Team Traces, see corresponding number Harness page 7.............. 25.50
No. 8637 1½-inch Team Traces, see corresponding number Harness page 7.............. 23.00

Three Rows Stitching, Straight Loggerhead   Price, per Set
No. 534 2 -inch Team Traces, see corresponding No. Harness, page 2.............$34.00
No. 533 2½-inch Team Traces, see corresponding No. Harness, page 2............. 37.00
No. 533 2½-inch Team Traces, see corresponding No. Harness, page 2............. 41.00

Three Rows Stitching, Straight Loggerhead   Price, per Set
No. 635 2 -inch Team Traces, see corresponding No. Harness, page 3.............$33.00
No. 635 2½-inch Team Traces, see corresponding No. Harness, page 3............. 36.00
No. 634 2½-inch Team Traces, see corresponding No. Harness, page 3............. 40.00

THREE ROWS STITCHING

No. 681 2¼-inch Logging Traces, see corresponding Harness, page 4......................$35.00

No. 681 2¼-inch Logging Traces, see corresponding number Harness, page 4 .... 38.00

No. 681 2⅜-inch Logging Traces, see corresponding number Harness, page 4 .... 41.00

LOGGING TRACES—5 Feet Long

Three Rows Stitching

No. 8635 2¼-inch Logging Traces, see corresponding number Harness, page 6 ...$27.00

No. 8635 2½-inch Logging Traces, see corresponding number Harness, page 6 ... 30.00

No.683xx 2½x66-inch, Extra Heavy and Strong, average thickness, ¾-inch;
Rivet Hooks and Dees, see corresponding number Harness, page 5... 44.00

HARVESTER TRACES

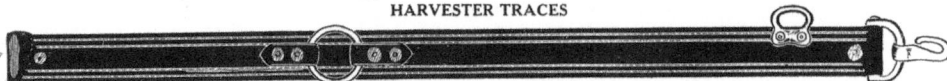

No. 8645 2 -inch Harvester Traces, see corresponding number Harness, page 11..$25.00
No. 8645 2¼-inch Harvester Traces, see corresponding number Harness, page 11 .. 27.00
No. 8645 2½-inch Harvester Traces, see corresponding number Harness, page 11.. 30.00

No. 8645 2 -inch Traces, shown on corresponding Harness, page 11, only with
2-hole hook 25.00
No. 8645 2¼-inch Traces, shown on corresponding Harness, page 11, only with
2-hole hook 27.00

No. 52 CHAIN PIPES with SAFES ON ENDS and STRAP and BUCKLE
Per Set of Four

	36-In.	42-In.
Length		
Price	$5.00	$5.00

No. 53 CHAIN PIPES with SAFES ON END; NO STRAP OR BUCKLE

	36-In.	42-In.
Length		
Price	$4.50	$5.50

SINGLE ORCHARD OR CULTIVATING TRACES No. 8650
Double and Stitched 3-Ply Wide Hame Piece as shown on Page 13
6 Feet 2 Inches, with 6 Link Toggles

No. 8650 1½-inch, price, per pair...............................$11.00
No. 8650 1¾-inch, price, per pair............................... 13.00
No. 8650 2 -inch, price, per pair............................... 15.00

ALL PRICES SUBJECT TO CHANGE WITHOUT NOTICE

TRACE DIMENSIONS

TEAM TRACES.
RAISED ROUND EDGE.

IMITATION HAND-STITCHED.

Per Set.
No. 304—1¼-inch, 6-foot, 6 inches.................... ......
No. 306—1⅜-inch, 6 foot, 6 inches.................... ......
No. 308—1½-inch, 6 foot, 6 inches.................... ......

HAND-STITCHED.

No. 324—1¼-inch, 6 foot, 6 inches....................
No. 325—1⅜-inch, 6 foot, 6 inches....................
No. 326—1½-inch, 6 foot, 6 inches....................

TEAM TRACES.
CREASED EDGE.

IMITATION HAND-STITCHED.

Per Set.
No. 410—1½-inch, 6 foot.................... ......
No. 1414—1½-inch, 6 foot, 3 inches....................
No. 413—1½-inch, 6 foot, 6 inches....................
No. 461—1¾-inch, 6 foot....................
No. 1461—1¾-inch, 6 foot, 6 inches....................

HAND-STITCHED.

No. 514—1½-inch, 6 foot....................
No. 1514—1½-inch, 6 foot, 3 inches....................
No. 2514—1½-inch, 6 foot, 6 inches....................
No. 520—1½-inch, 6 foot, 2-ply....................
No. 1520—1½-inch, 6 foot, 3 inches, 2-ply....................
No. 522—1½-inch, 6 foot, 6 inches, 2-ply....................
No. 561—1¾-inch, 6 foot....................
No. 1561—1¾-inch, 6 foot, 6 inches....................

TEAM TRACES—CONTINUED.

Per Set.
No. 600—2¼-inch, single strap, 1½-inch double and stitched points, clip cockeyes ....
No. 603—2¼-inch, single strap, 1½-inch double and stitched points, triangular cockeyes ....
No. 605—2½-inch, single strap, 1½-inch double and stitched points, triangular cockeyes ....
No. 609—2¾-inch, single strap, 1¾-inch double and stitched points, triangular cockeyes ....

PINERY HARNESS TRACES.
WITH DEE AND CHAIN.

Imitation hand-stitched, 6 foot, 3 inches, with heel chains and billets, for clip hames.

Per Set.
No. 1480—1½-inch, 2-row stitched.................... ......
No. 1481—1¾-inch, 3-row stitched....................
No. 1484—1¾-inch, 3-row stitched....................
No. 1488—2 -inch, 3-row stitched....................

Imitation hand-stitched, 6 foot, 3 inches, with heel chains and billets, for bolt hames.
No. 480—1½-inch, 2-row stitched....................
No. 481—1¾-inch, 3-row stitched....................
No. 484—1¾-inch, 3-row stitched....................
No. 488—2 -inch, 3-row stitched....................
No. 489—2¼-inch, 3-row stitched....................

Hand-stitched, 6 foot, 3 inches, with heel chains and billets, for clip hames.
No. 1580—1½-inch, 2-row stitched....................
No. 1581—1¾-inch, 3-row stitched....................
No. 1587—1¾-inch, 3-row stitched....................
No. 1588—2 -inch, 3-row stitched....................

Hand-stitched, 6 foot, 3 inches, with heel chains and billets, for bolt hames.
No. 580—1½-inch, 2-row stitched....................
No. 581—1¾-inch, 3-row stitched....................
No. 584—1¾-inch, 3-row stitched....................
No. 588—2 -inch, 3-row stitched....................
No. 589—2¼-inch, 3-row stitched....................

BACK BANDS OR PADS

No. 2.   No. 408.

LONG TUG TEAM PADS AND SKIRTS.

No.   1—Harness leather housings, with 1¼-inch layer and billets, fancy spotted ....

No.   2—Harness leather housings, with 1¼-inch layer and billets, fancy spotted, felt lined....................

No. 408—Harness leather housings, felt lined, with 1¼-inch doubled and stitched billets, for either single or crotch back straps. ......

No. 1408—Swell shape, single strap housings, felt lined, fancy spotted, with 1¼-inch Market strap skirts, for single or crotch back straps............ ......

xc.

No. 1408.

Lise Hubbe's well harnessed ladies.

From the complex to the simple, tugs or traces provide the linkage to get the load pulled.

TEAM PADS.

UNIVERSAL TEAM PADS.

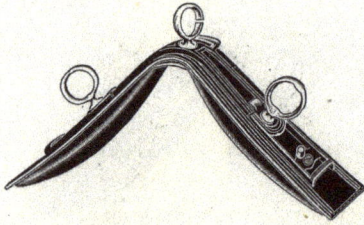

MOLINE TEAM PADS.
STITCHED BOTTOM.

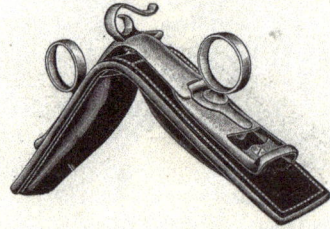

MOLINE TEAM PADS.

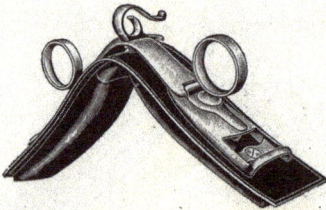

MOLINE TEAM PADS.
SPOTTED HOUSINGS.

BADGER TEAM PADS.

IRON CITY TEAM PADS.

No. 050 Badger—XC trimmed, harness leather housings, harness
leather top layer, with dees........................ ......

No. 500 Badger—XC trimmed, cherry leather housings, harness leather
top layer, with dees................................. ......

No. 14 Iron City—XC trimmed, plain leather housing, harness leather
bottom, with dees for market straps.................. ..

VICTORIA TEAM PADS.

No. X3 Victoria—New fancy swell beaded housing, XC yoke, hook and
terrets, japanned loop end plates, patent leather housing,
harness leather bottom pad........................ ......

No. 14½ Iron City—XC trimmed, plain leather housing, harness
leather bottom ....

SKIRTED TEAM PADS.

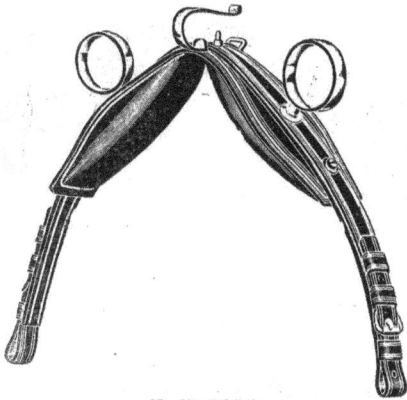

No. 420 and 520.

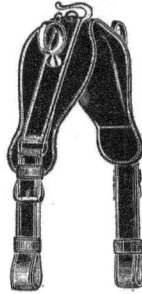

No. 785 and 789.

No. 788.

No. 300 and 302.

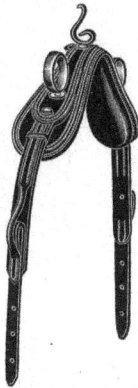

No. 304.

No. 335.

No. 409 and 413.

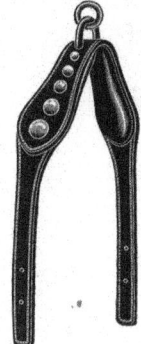

No. 608 and 610.

DUNLAP'S BUTTON CHECK HOOKS.

No. 2 SAFETY CHECK HOOKS.

Even with horses and mules in harness, we haven't been able to escape the worthless parading of class distinctions. Besides the fancy breeding, and often contrived postures of 'better' horses, the harness and its 'over-the-top' ornamentation was a perfect place for 'better' people to distinguish themselves from the fray. These 'appointed' team pads, with their terrets and hooks and buttons, serve the same purpose as the back bands and saddles you see on pages 114, 115 with the one structural exception; they gave a reinforced framework for the buttons and hooks sometimes employed in fastening overchecks.

LOOP BOLT PAD HOOKS.

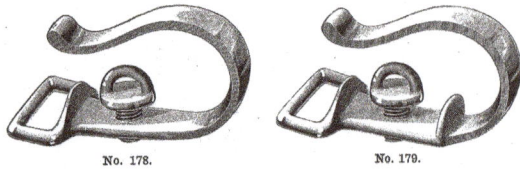

No. 178.   No. 179.

LOOP POST AND CHAMPION PAD HOOKS.

No. 182.   No. 262.

PEERLESS CHECK HOOKS.

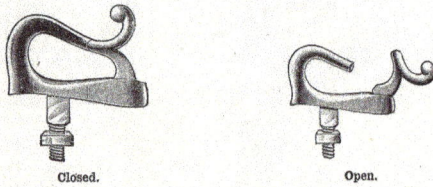

Closed.   Open.   No. 659.   No. 30 Pedestal.

CLINTON CHECK HOOKS.

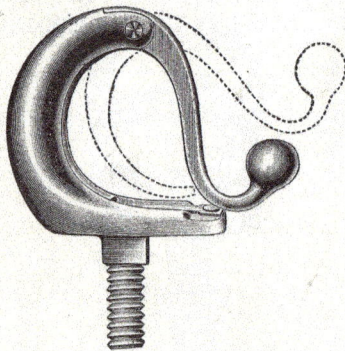

McKINNEY CHECK HOOKS.

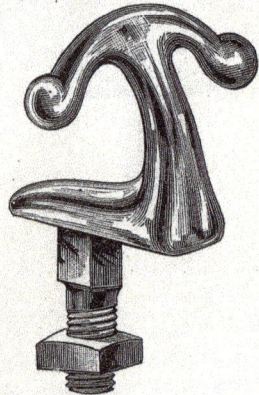

DECORAH CHECK HOOKS.

DROP HOOKS AND TERRETS.

No. 700.   No. 701.

No. 8 MOLINE TREES.

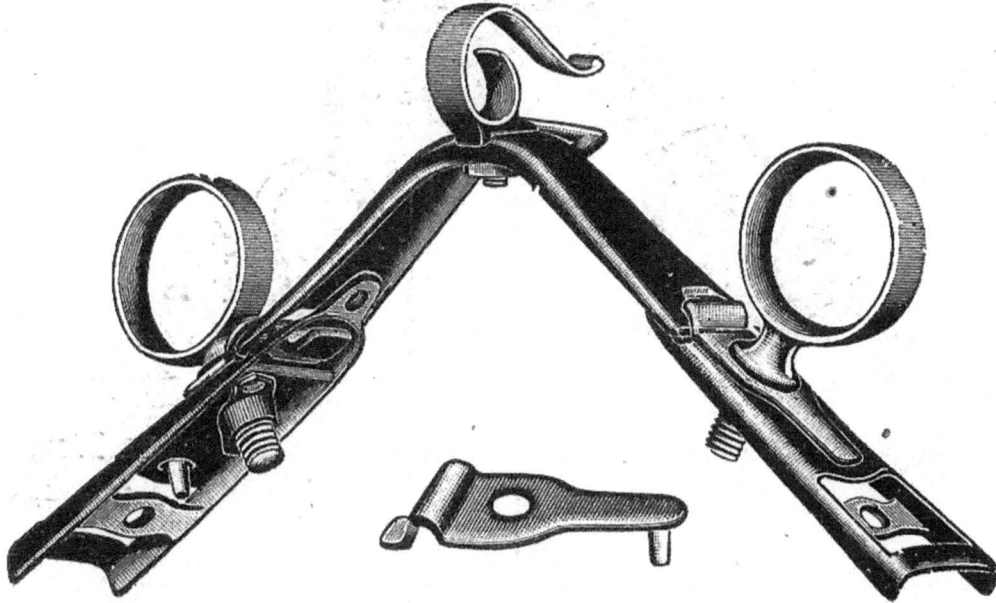

No. 8 MOLINE TREE PARTS.

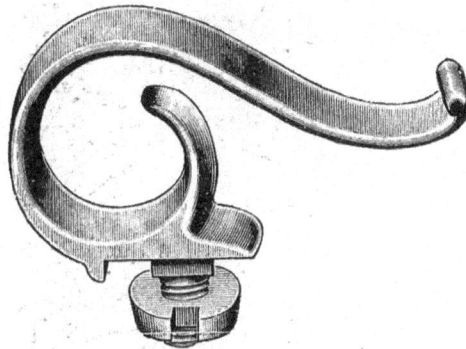

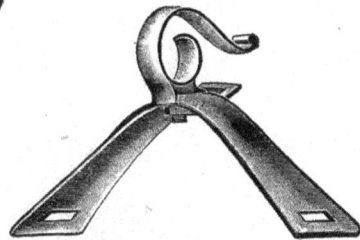

Tree Terret.   Tree Hook.   Tree Yoke.

Tree Jockey.   Tree Hinge.

Author with three of Mike McIntosh's Belgians plowing.

X

Hitch point single tree to triple tree

Illustrating draft path from a single horse's tugs.

UNDER THE HEADING OF A BIT TOO COMPLICATED: In the complex rope and pulley system for a twelve horse hitch, the dotted line represents the path of the tug, the end of which is attached to a rope running through a pulley and then forward (beneath) to the lead horse tug, off picture to right. Multiply by 16 or 24.

Tug path with swingle (brichen) tree

Notice that as the cable lifts, the swingle tree wants to raise up. Here is a classic example of how the belly band (A) breaks the lifting line of the tug (B), maintaining the proper angle from hame to belly band and eliminating any upward pull of collar on throat.

Marvin Brisk with haul back horse.

Twelve tugs on six-abreast Percherons with tugs hooked long to drag evener for harrow.

BRIGHT TRACE.

BRIGHT BUTT OR HALF TRACES.

TRACE CHAINS

Wrought iron, bright, welded links and ring, with malleable iron swivel.

Western Standard, straight link, with swivel and ring.
Western Standard chains contain one link less per foot than indicated.

Length, feet	$6\frac{1}{2}$	$6\frac{1}{2}$	7	7	7	7	7	7	7	7	7	7	7	$7\frac{1}{2}$
Links, per foot	6	8	8	10	12	8	10	12	14	16	8	10	12	10
Wire Gauge Nos	2	2	2	2	2	1	1	1	1	1	0	0	0	0
Diameter of wire, inch	$\frac{1}{4}$	$\frac{1}{4}$	$\frac{1}{4}$	$\frac{1}{4}$	$\frac{1}{4}$	$\frac{9}{32}$	$\frac{9}{32}$	$\frac{9}{32}$	$\frac{9}{32}$	$\frac{9}{32}$	$\frac{5}{16}$	$\frac{5}{16}$	$\frac{5}{16}$	$\frac{5}{16}$

As on page 108, the stripped-down, southern style chain-tug plow or logging harness (right) features straight-shot check reins around hames, and no breeching. The bridle has 8" tall pigeon wing blinders, no nose band, no gag swivels. The back pad and belly band hang off the chain piping. The back end is simple, with hip straps and a crupper. There is no pole strap (or martingale) and no breast straps. This would have been an economy harness best suited for flatland work.

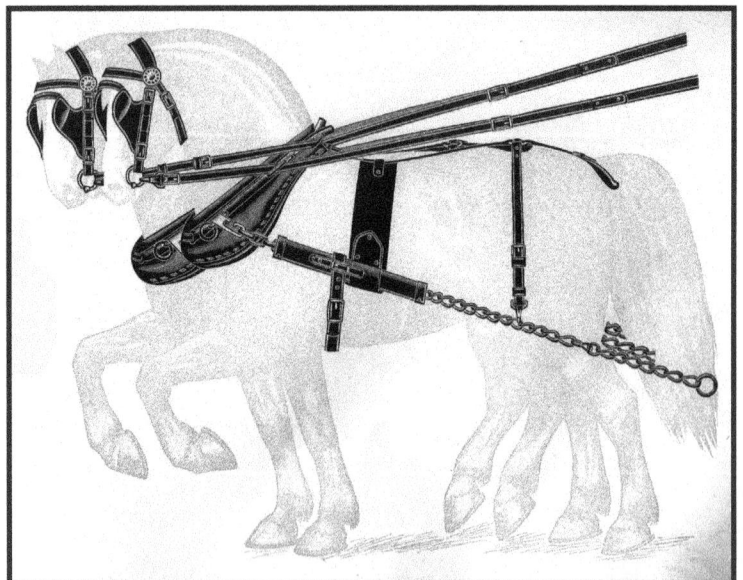

WAKEFIELD CLIP AND BOLT ATTACHMENT.

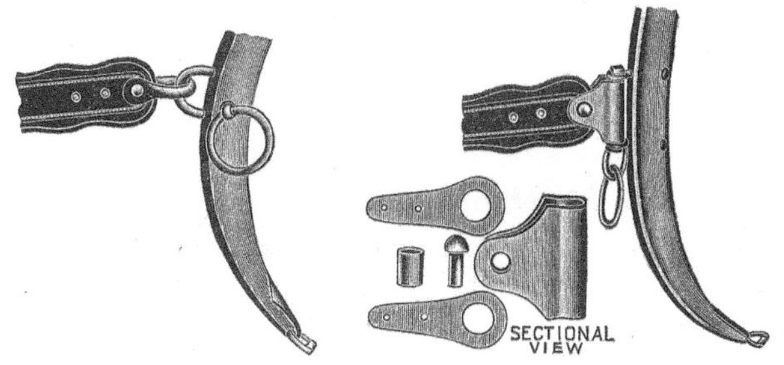
SECTIONAL VIEW

EUREKA PATENT HAME CLIPS.

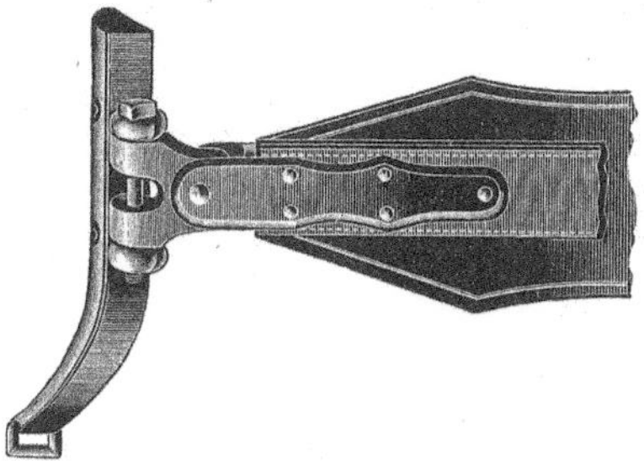

BUNKER'S PATENT HAME AND TRACE ATTACHMENT.

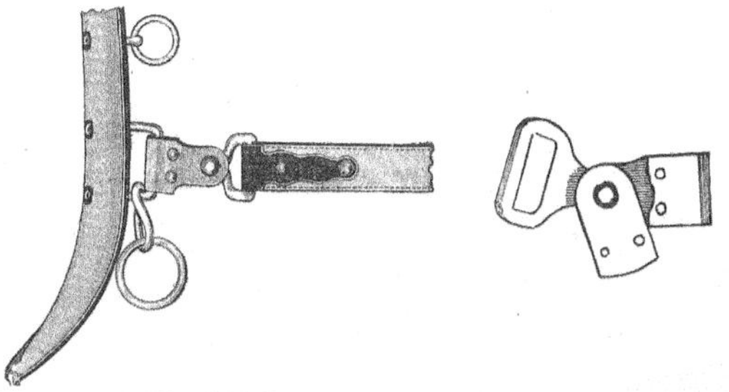

GRAY'S DRAFT LINKS AND CLIPS.

O. L. HAME CLIPS.

	Per Doz.
No. O. L.—Japanned, one leg iron hame clips........................	$ 88
No. O. L.—Nickel, one leg iron hame clips..........................	1 32

"H. B." REPAIR CLIPS AND LOOPS.

	Per Gross.
Japanned, H. B. repair clips and loops.............................	$ 6 66
Japanned, H. B. repair clips, without loops.........................	6 00

NO. 37 PATENT COMBINATION CLIPS.

	Per Doz. 1¼	1½-Inch.
No. 37—Japanned	$ 77	$ 88

PHILLPOTT'S PATENT WEAR CLIPS.

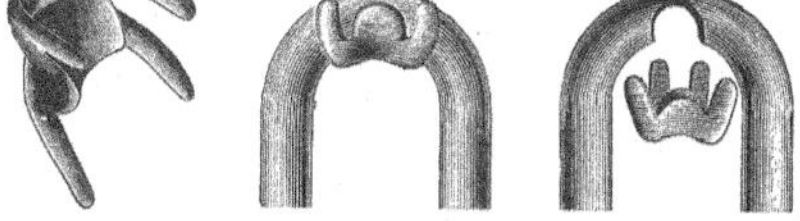

	Per Gross.
Polished.	$ 2 76

HAME STAPLES.

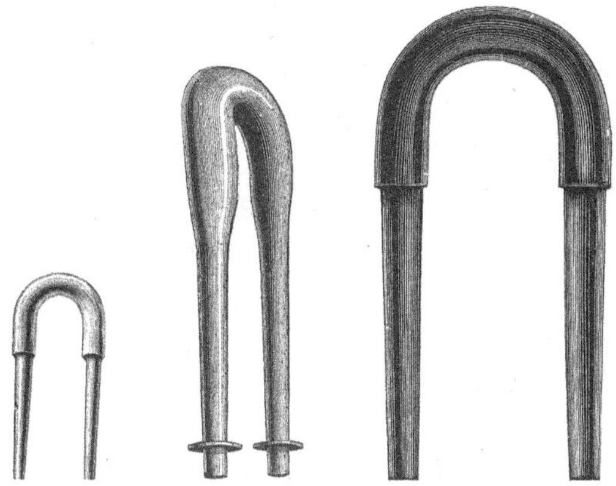

No. 950. No. 960. Nos. 957, 958 and 959.

	Per Doz.
No. 950—Polished, 5/16 iron, for oval iron wood coach hames.........	$ 33
No. 950—Japanned, 5/16 iron, for oval iron wood coach hames.........	60
No. 960—Polished team, ⅜ iron, with washers.......................	28
No. 962—Polished team, long and wide, ⅜ iron, with washers........	28
No. 958—Polished, extra long and extra wide, ⅜ iron, with washers...	33
No. 959—Polished, extra long head, ⅜ iron, with washers, for Hayden hold back plates	42
No. 957—Polished, 7/16 iron, for heavy Concord hames..............	52

SQUARE HAME STAPLES.

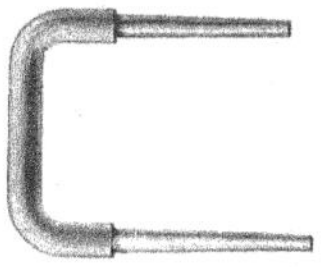

	Per Doz.
No. 450—Polished, 1¾-inch	$ 44
No. 460—XC plate, 1¾-inch	52
No. 5—Polished Concord, 1¾-inch	44
No. 8—XC plate, oval iron wood coach, 1½-inch..................	70

COOPER'S CLIPS

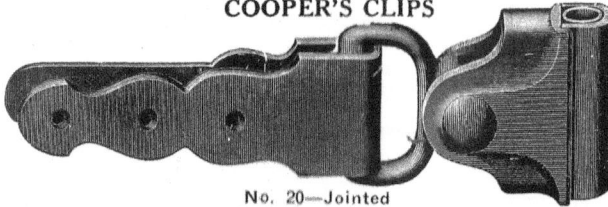

No. 20—Jointed
Price, per Dozen

	1½	1¾	2	2¼	2½
No. 21 Jointed	$3.50	$4.00	$4.40	$5.00	$5.70

COOPER'S CLIPS

No. 40—Jointed

	2x2	2¼x2½	2½x2½
No. 40 Jointed	$10.40	$11.40	$12.20

COOPER'S CLIPS

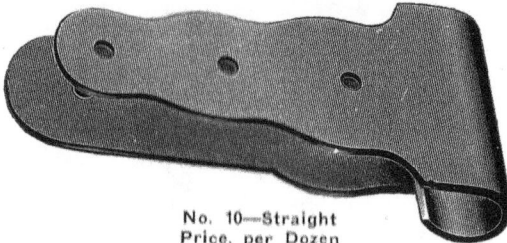

No. 10—Straight
Price, per Dozen

	1½	1¾	2	2¼	2½
No. 10 Straight	$1.40	$1.60	$1.70	$1.90	$2.10

BACK STRAP RING

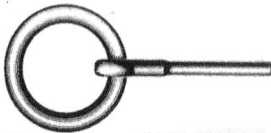

No. 137 REPAIR RINGS

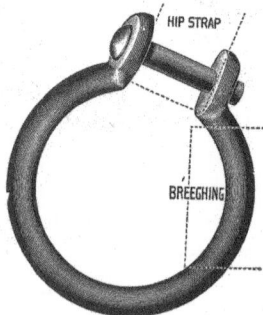

HIP STRAP

BREECHING

HAME TRIMMINGS

BOLT ROLLERS

Price, per Doz.
Flange Rollers $1.80
Plain Rollers 1.50

WROUGHT IRON HAME CLIPS—FOR WOOD HAMES

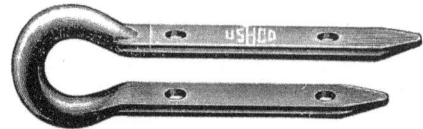

Price, per Doz.
No. 982   ⅜-inch $1.30

CONCORD HAME LINE RING AND STUD

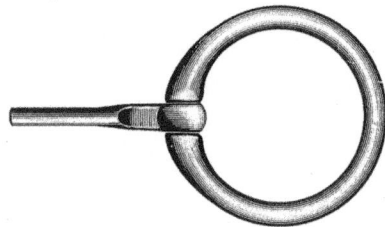

Price, per Doz.
Polished $1.20
Nickel 3.50
Brass 3.50

HAME BOLTS

COCKEYES

No. 465 SCREW COCKEYES

No. 455 TRIANGULAR COCKEYES

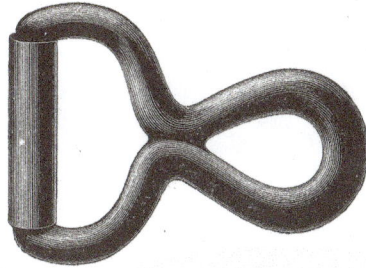

No. 466 REPAIR COCKEYES FOR CLIPS

HARDY'S DETACHABLE COCKEYES

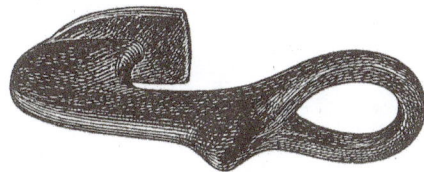

No. 980 UTILITY COCKEYES

No. 440 CLIP COCKEYES

No. 21 JOINTED CONCORD CLIPS

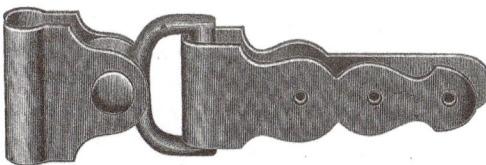

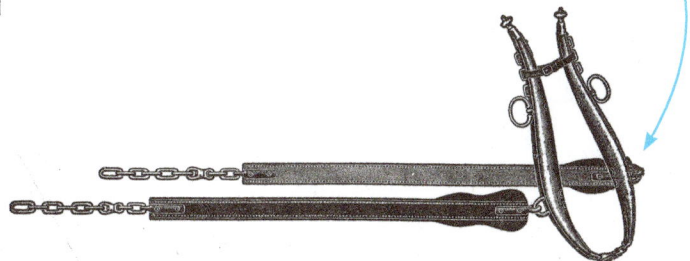

No. 20 HEAVY ROLLER TRACE SNAP

No. 448 HEAVY SWIVEL TRACE SNAP

No. 46 SLIDE AND SNAP

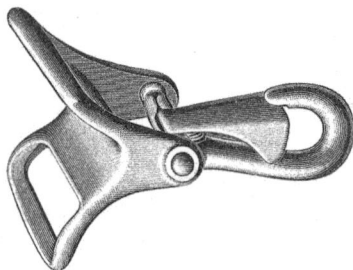

No. 617 SLIDE AND SNAP

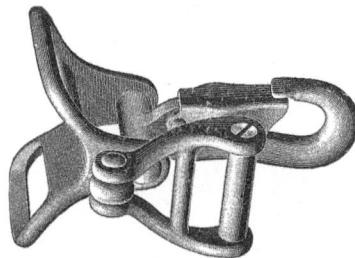

Lazy straps hang from the brichen and hold up the traces. Without them the trace (or tug) might, in certain working postures, cause the horse to step over and perhaps get tangled up. Though they are not critical to the operation of the harness they do gather things up and out of the way.

TEAM LAZY STRAPS
Smooth Finish.

Lazy straps should be adjusted so that they do not break the angle of the trace when pulling. If they are too short, and the horse is doing a dead drag load, they could put a lot of unwanted pressure on the hips.

No. 10.   No. 25.   No. 30.   No. 32.   No. 35.

Tugs in play, soil responding. Photo by Kristi Gilman-Miller.

Chapter Five

Brichens, Breast Straps, Pole Straps, Quarter Straps, Trace Carriers, etc.

A drawing by the author from his volume, "Training Workhorses / Training Teamsters."

The backing and braking structure for the western basket-brichen team harness.

Though it is secondary in the overall function and purpose of the harness, it is potentially the more complex and variable portion of the system. It is useful to keep in mind that these combined straps are specific and vital to its named operation along with the related essentials of keeping a rolling vehicle or implement from running up on the heels of the animal or team and holding the front end of the tongue or pole up off the ground.

Breast strap attached to the bottom hame rings.

Back strap attached to mid hame ring.

Trace carrier

Hip drops

Brichen

Neckyoke

Parts of the backing and braking system.

Pole strap

Quarter strap

Lazy (or mud) strap

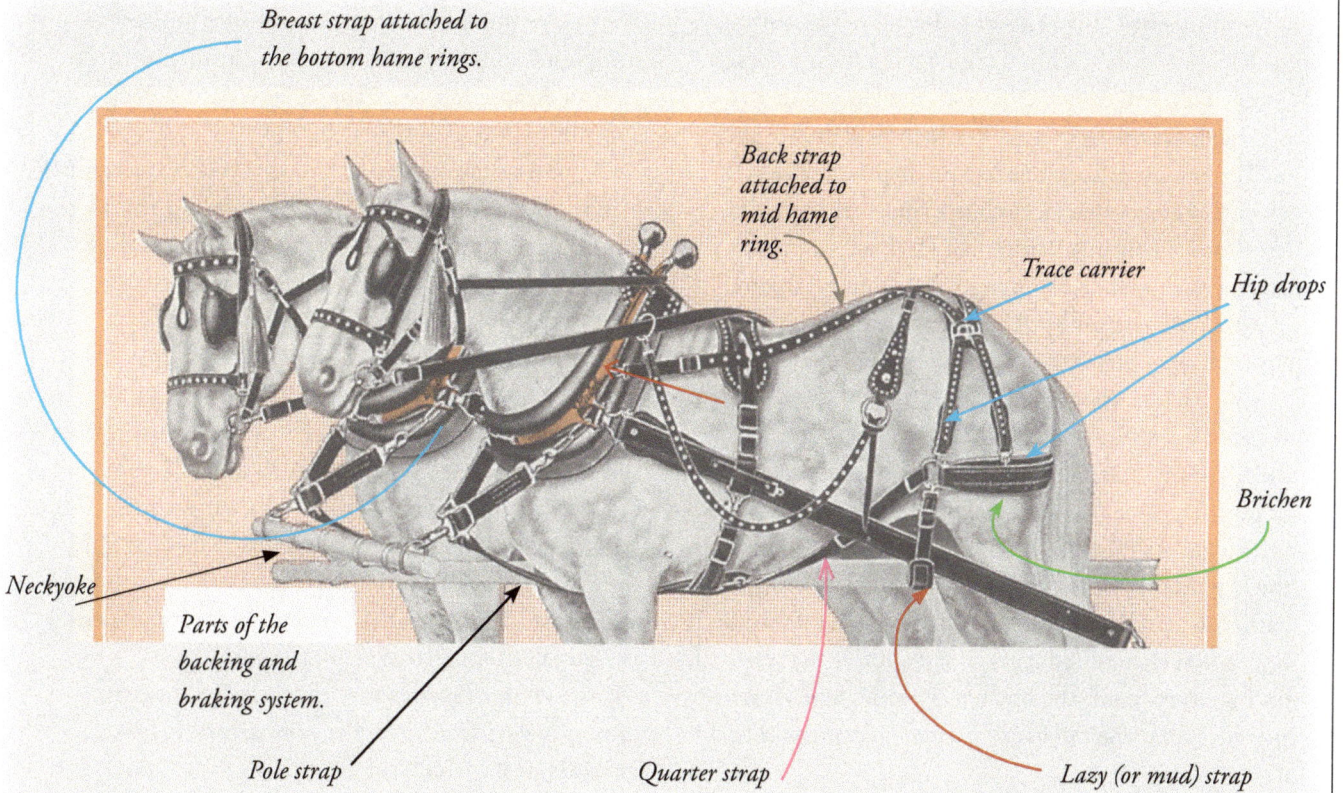

Special Note: *The quarter strap/pole strap adjustment is most important. There is a 'sweet spot' that has to be 'found,' otherwise when the animals step ahead the neckyoke might come free of the pole. You need just the right 'tension.' Also, it is important that the pole strap go over the loose fitting belly band.*

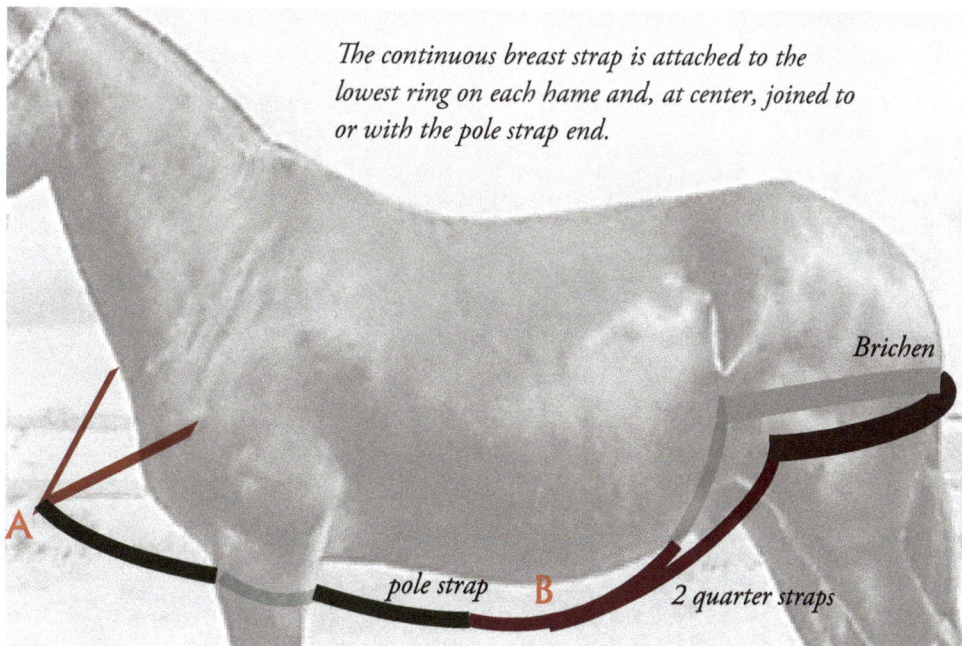

The continuous breast strap is attached to the lowest ring on each hame and, at center, joined to or with the pole strap end.

Brichen

A

pole strap   B   2 quarter straps

When the harnessed horse steps back he tightens the breast strap and pole strap *A* and also the quarter straps and brichen *B*. In this way he either backs up the load or holds it from rolling forward.

As illustrated, the brichen or breeching assembly of the western style team work horse harness typically fastens forward, over the back, to the hames by way of two **back straps** forming a V attachment and support.

Though often seen in carriage rigging, for work harness there are few occasions a single strap may be employed – running straight up the back to the top of the collar – but this, in many working situations, can be most uncomfortable for the animal as it puts contradictory pressure on the top of the collar.

The back straps come together over the hips to form a spider assembly which rests atop the hips and feeds the **hip drop straps** down to hold the brichen strap in position. (See illustration page 129.) The 'fit' of the brichen affects the comfort of the animals to both hold the load and to back it. If the brichen is too low it will gather the back legs in a way that is awkward for the horses and therein reduces the strength of the move. If, on the other hand, the brichen is too high, it wants to ride up under the tail causing unwanted surprise and discomfort.

The **brichen strap** is usually a half circle, hung by that aforementioned spider assembly with one, two or three hip drop straps. The best brichen strap construction features a doubled and rolled heavy strap which encircles the rump. A single flat thick piece of suitable leather can be made to work as a brichen but its sharp edges can remove hair, irritate and eventually might cause sores on the animal, therefore the rolled (or folded) edge is much preferred.

When the animal backs it puts pressure around the rump and on the brichen which in turn tightens the quarter straps, and then, the pole strap and neckyoke assembly. (Illustration page 129.)

A most notable exception to this basket (or rump) brichen is the **Yankee (or hip) Brichen**. In this case the brichen strap rides just above the tail and works over the hip and down to form the quarter straps. The Yankee (or hip) Brichen is held in place by a crupper that encircles the tail. I have used this style with good success, with one important if obscure exception. The Yankee Brichen does not work best for horses hitched to a buckrake as the pole strap attaches forward and down low to the rake axle. The buckrake can be a heavy implement to back up and may cause the horses to lower their back ends to gain advantage. This can cause some discomfort as the hip brichen does not give enough support.

The **spider assembly**, with the basket brichen, almost always features a piece of hardware, **trace carriers**, from which tugs may hang when not in use. (See illustrations page 129.) The ends of the brichen strap also carry **lazy (or mud) straps** which hold the tugs or traces from dragging 'in the mud.'

The **quarter straps** fasten to the ends of the brichen strap and move forward to the **pole strap**. These quarter straps should be adjustable in length so that they may be customized for the individual animal and keep the straps loose but up relatively close to the belly of the animal to avoid it getting a leg over. As quarter straps may on occasion have to hold

The author drives a four-abreast. The team on the left are harnessed with a Yankee (or hip) Brichen, the team on the right have the conventional basket brichen. Photo by Kristi Gilman-Miller.

a great deal of weight they should be made of the best leather. (Quarter strap snaps should be fastened face away from the horse's belly skin to avoid irritation.)

Breast straps or chains come in several types and configurations. The most common are the heavy leather straps. These may thread through hame ring and back to self, with or without a steel slide to receive the neckyoke ring or a snap for attaching to the neckyoke. Often a combination snap, at the juncture of pole strap and breast strap, is employed to fasten to the tongue. In some climates breast chains are preferred.

Pole straps and their attaching assemblies come with some variety. Notice in drawing, on page 129, that the pole strap is the front leg of a *Y* attaching the brichen all the way forward to the **neckyoke** or yoke chains. On the back end of the pole strap there is a strong ring with a cover flap of leather to protect the belly of the animal. The quarter strap snaps attach to this ring. The pole strap runs over the loose fitting belly band and forward between the front legs. It is then attached in some fashion to the breast strap which, as illustrated, holds the pole strap, and therefore the pole itself, up.

These two photos by Fuller Sheldon feature Tom Odegard's Belgians. Top photo illustrates perfect position for brichens as this four-abreast harvest potatoes. Just above, cultivating four rows of corn. Note that a long specialty neckyoke is em-ployed by the two straight-ahead, alert horses. In this case the 'team harness' is on the two center horses. The outside pair have no part in the backing and braking system. They are along to pull the implement and covet the corn.

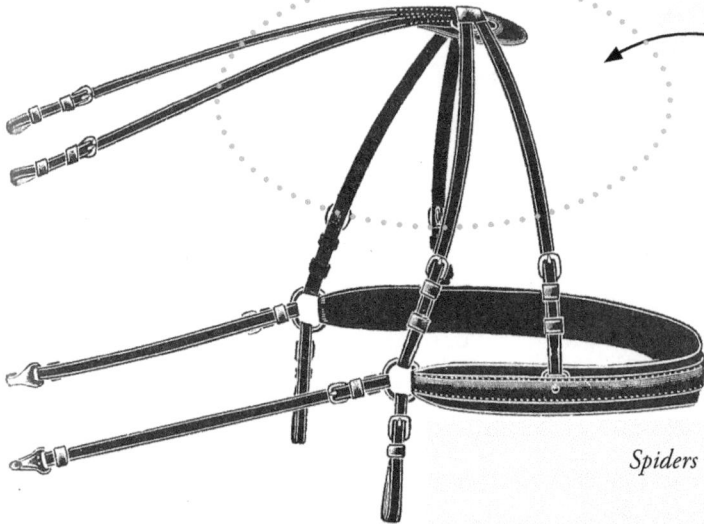

Left: two strap brichen with double backstraps and snapped quarter straps.

Below left: two strap brichen with single back strap, and conwayed quarter straps.

Below: two strap brichen with double backstraps and quarter straps.

Spiders

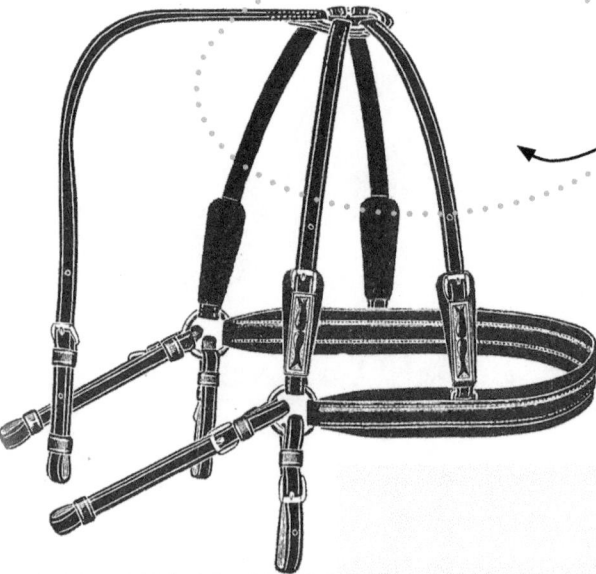

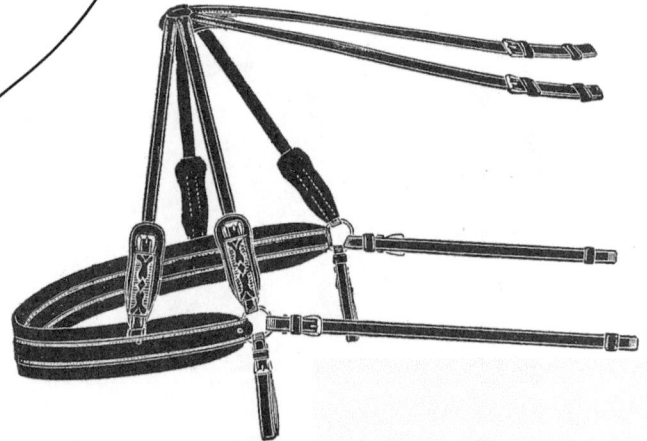

Below: harness catalog illustration showing pole straps hanging by a collar strap from the collars. Breast strap has one snap and features a rolling neckyoke snap.

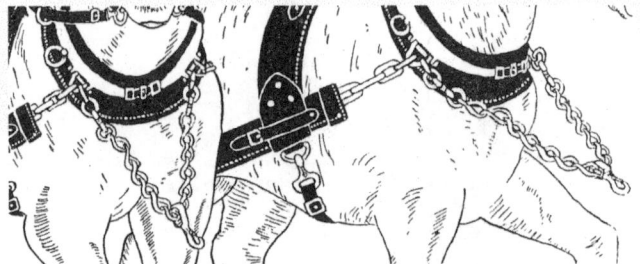

Above: breast chains on Southern chain tug harness.

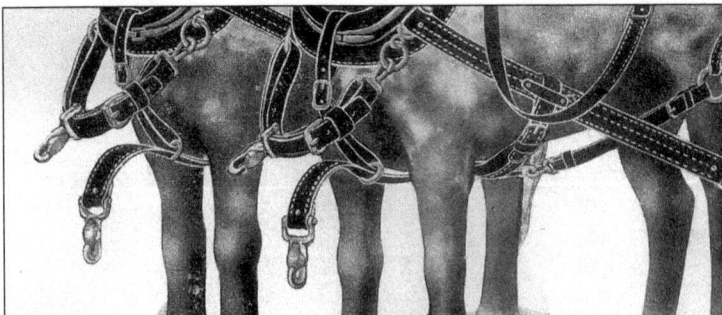

Combination pole-strap breast strap snap.

Combination snaps hold neckyoke and pole up. (Neckyoke in this case is bolted to pole end.) Mower cannot roll up on horse's heels. Bud Dimick mowing at Singing Horse Ranch in 1988.

The author driving Renee Russell's Percherons, note bolt-on neckyoke.

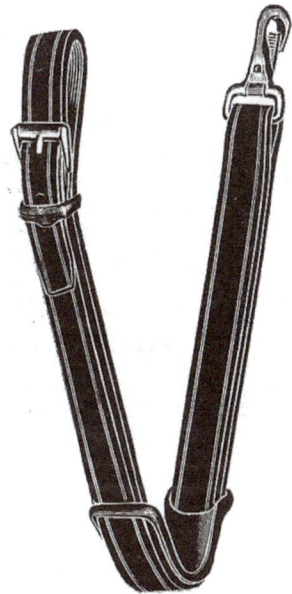

NECKYOKE SNAPS
Patented.

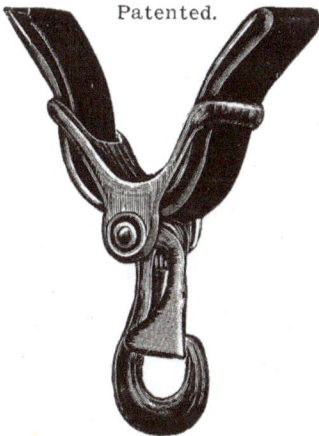

No. 46

Left, breast strap with steel slide.

Center, breast strap with two heavy top snaps.

Right, all leather breast strap.

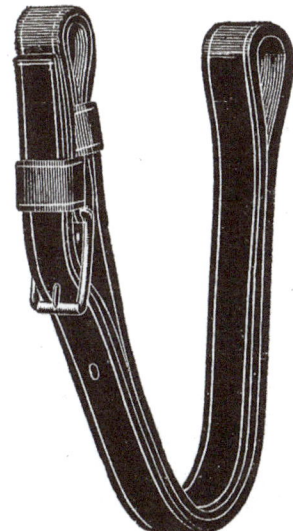

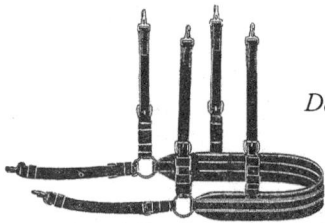

Detachable brichen

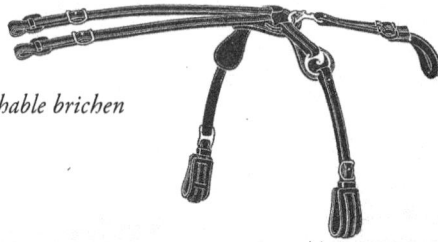

BREAST STRAP SLIDE

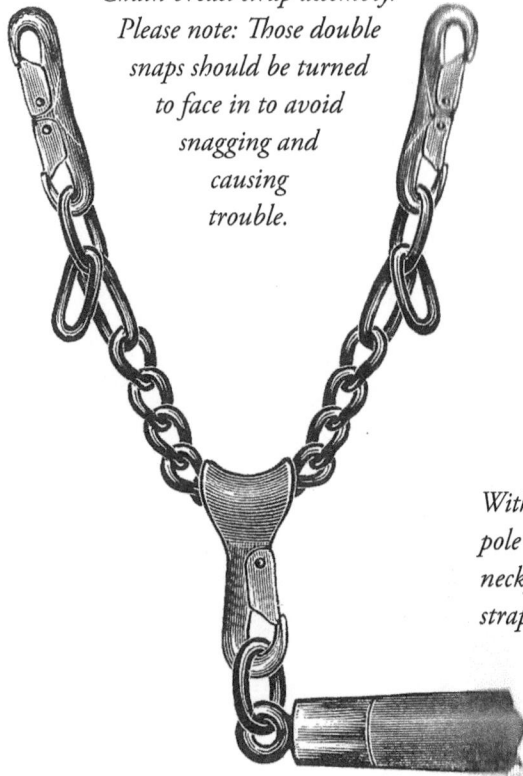

*Chain breast strap assembly.
Please note: Those double
snaps should be turned
to face in to avoid
snagging and
causing
trouble.*

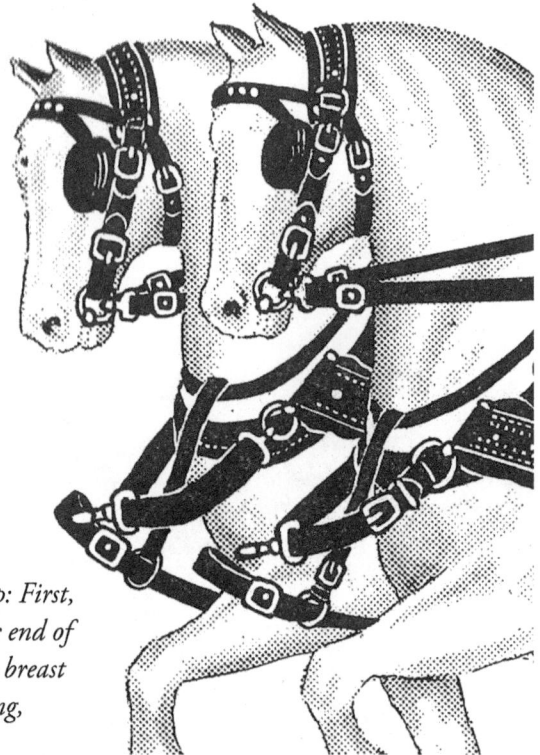

*With this style set up: First,
pole strap slides over end of
neckyoke, then snap breast
strap to neckyoke ring,*

*No snap option: pole strap
over end of neckyoke,*

*then the breast strap passes
through neckyoke end ring.*

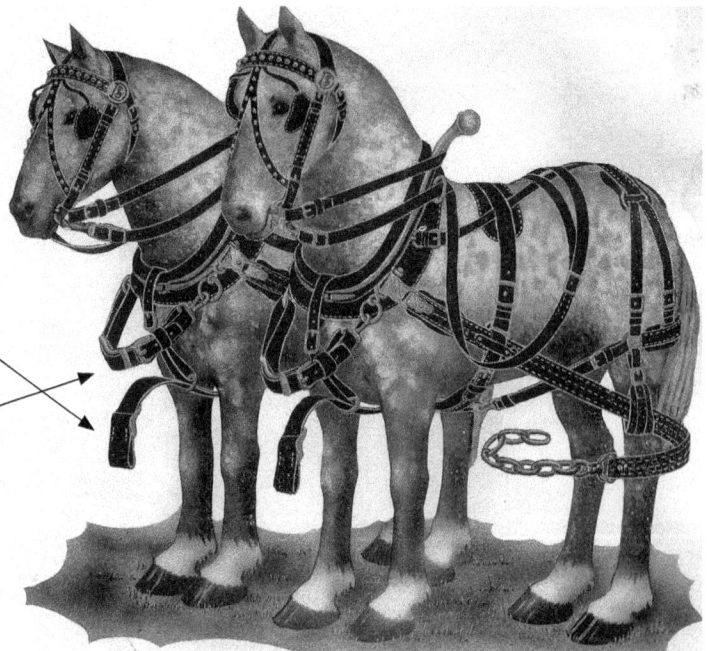

BREAST SNAP BUCKLE AND SNAP

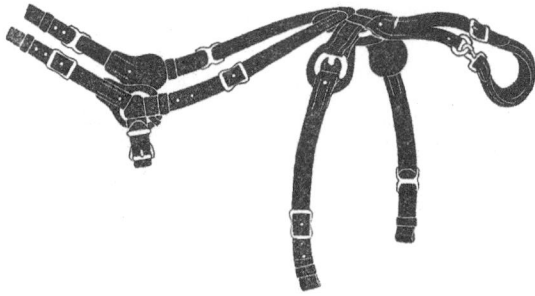

HARVESTER HIP STRAPS

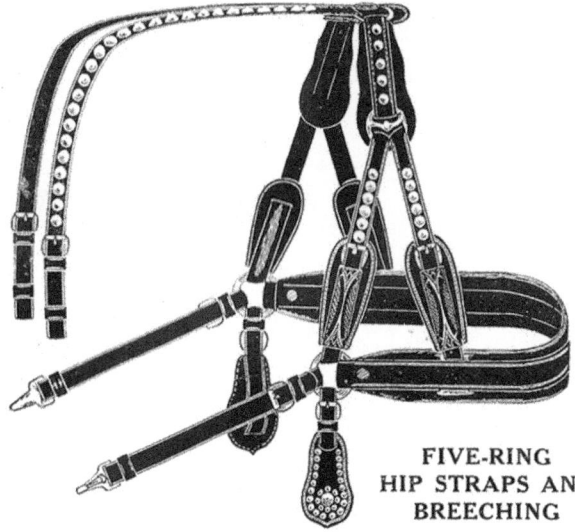

FIVE-RING
HIP STRAPS AND
BREECHING

BREECHING BODIES

HIP SRAPS

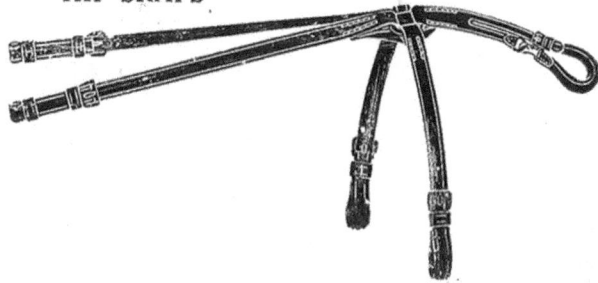

Detachable Breeching (Brichen)

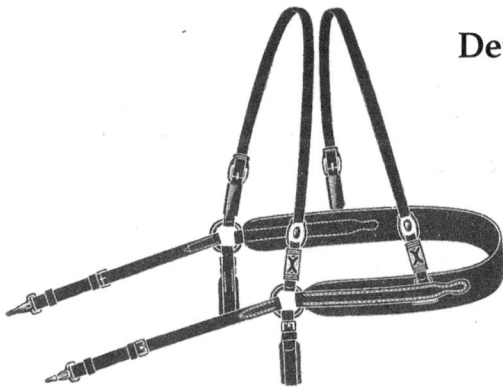

No. 25.

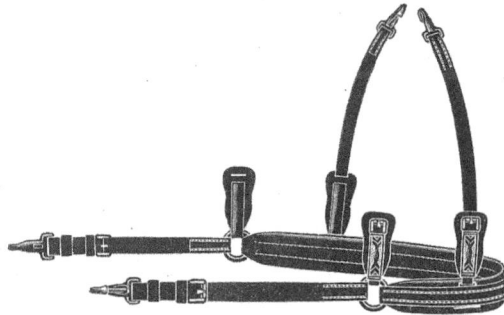

Nos. 50 and 51.

CRUPPERS.

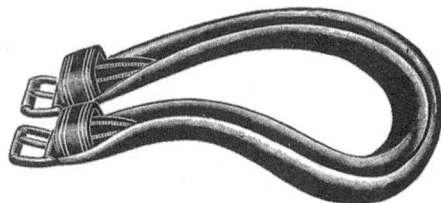

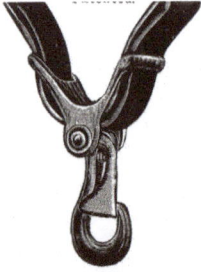

When using snaps on breast straps, combined with pole straps or not, the snaps should face inwards on the team. This reduces the chances of bits getting hooked into the snaps when a horse rubs.

Side note: For comfort and work efficiency, the neckyoke width should be the same as the doubletree / evener length.

HIP BREECHING

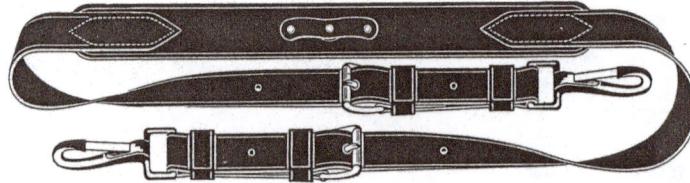

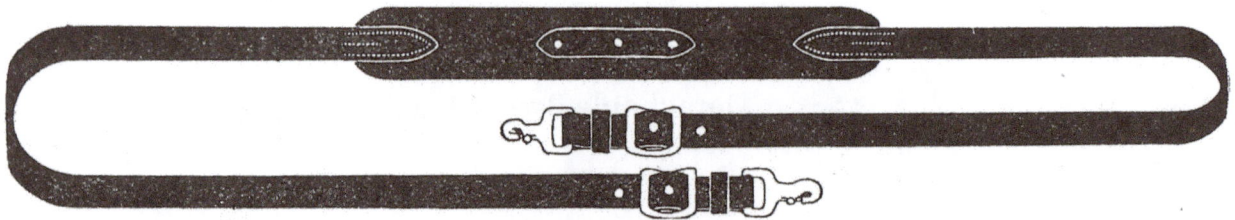

YANKEE BREECHING

Cali and Lana are wearing Yankee Brichens, the straps of which curve round under the belly and hook to pole straps. The retraint and backing come from the top of the rump rather than around the rump like the basket brichen.

NO. 5. GIANT TRACE SNAP

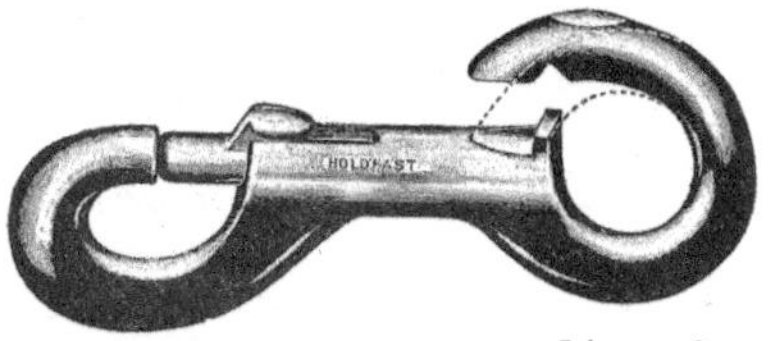

		Price, per Doz.
No. 5	⅝x5½-inch	$7.00

YANKEE CHAIN SNAP

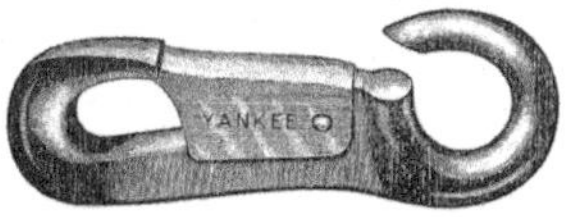

		Price, per Gross
No. 825	⅝x3 -inch	$21.40
No. 826	½x3¾-inch	30.00
No. 827	⅝x4⅝-inch ... Per Doz.	6.50

OPEN EYE BOLT SNAP

		Price. per Gross
No. 520	XC plate; snaps, ⅝x3⅛-inch	$16.30
No. 521	XC plate; chain snaps. ⅝x3½-inch	20.70
No. 522	XC plate; trace snaps, ½x4-inch	20.20
No. 48	XC plate; heavy trace snaps, ½x4-inch	41.00

DOUBLE BOLT SNAP

		Price. per Gross
No. 510	Light weight XC plate, 3-inch	$30.00
No. 511	Medium weight XC plate, 3¾-inch	36.00
No. 512	Heavy weight, 4⅝-inch	44.80

SWIVEL YANKEE ROPE SNAP

No. 830

Price, per Gross

Inches	⅝	¾	⅞	1
Per Gross	$27.60	$30.30	$41.10	$57.70

POLO SNAP

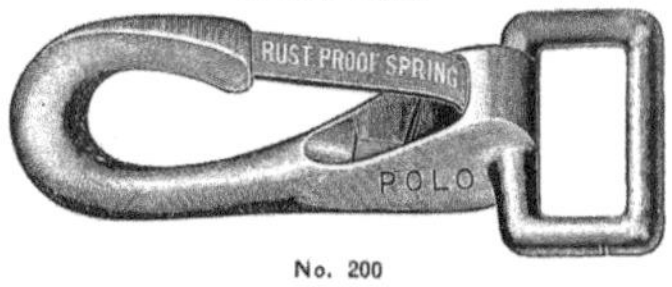

No. 200

Price, per Gross

Inches	⅝	¾	⅞	1	1⅛
XC Plate	$7.00	$7.00	$7.00	$7.00	$8.80
Inches	1¼	1½	1¾	2	
XC Plate	$10.50	$12.00	$14.00	$16.00	

ROUND EYE SWIVEL

No. 50

Inches	⅜	½	⅝	¾	⅞	1
No. 50 XC	$23.60	$24.00	$25.00	$27.30	$37.10	$52.00

ROUND EYE SNAP

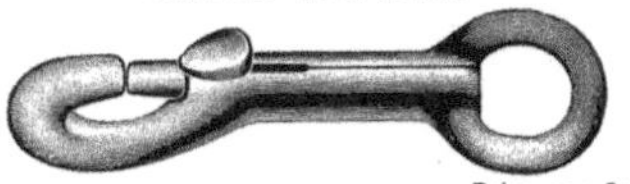

Price, per Gross

Inches	½	⅝	¾
No. 19	$16.00	$16.90	$17.70

TROJAN NO. 25. BOLT SNAP

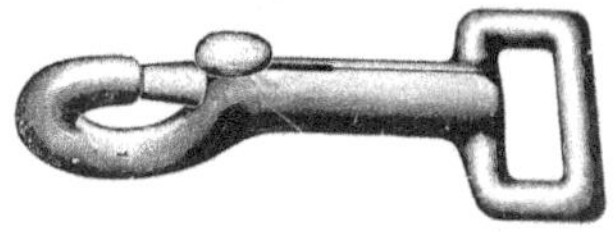

No. 25. Trojan Bolt. Per Gross

Inches	⅝	¾	⅞	1	1⅛	1¼
No. 25	$9.50	$9.50	$9.50	$9.50	$12.00	$14.40
Inches				1½	1¾	2
No. 25				$16.40	$18.00	$19.80

BIT SNAP

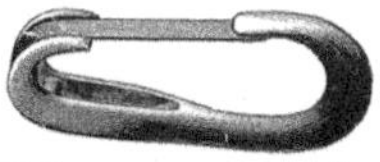

No. 421 Japanned ... Per Gross $14.10

BIT SNAP

No. 452 Japanned ... Per Gross $16.80

SWIVEL BABY SNAP

		Price, per Gross
No. 337	Round, ¼-inch. eye, nickel	$15.00
No. 337	Round, ⅜-inch, eye, nickel	16.30
No. 337	Round, ½-inch, eye, nickel	17.50

NO. 340. BAG SNAP

Price, per Gross

Inches	½	⅝	¾	⅞
XC	$6.90	$7.70	$8.50	$9.20
Nickel	9.20	10.20	11.40	13.00

DUNCAN & SONS, Inc., SEATTLE

BREAST STRAP ROLLER SNAP

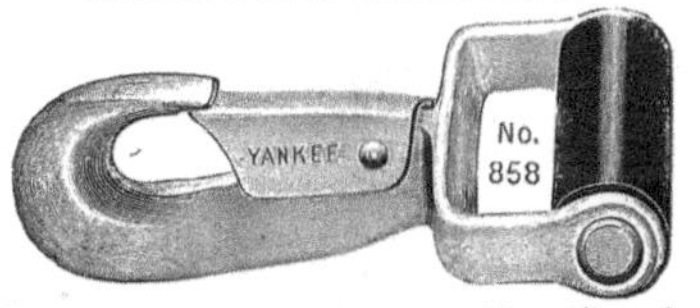

Inch		1½	1¾	2
No. 858	Per Doz.	$3.00	$3.50	$5.00

One Dozen in Box

SWIVEL TRACE SNAP

Inch	1½	1¾	2	2¼	2½
Per Dozen	$9.10	$9.80	$10.60	$11.90	$12.90

CENTER BREAST CHAIN SNAP

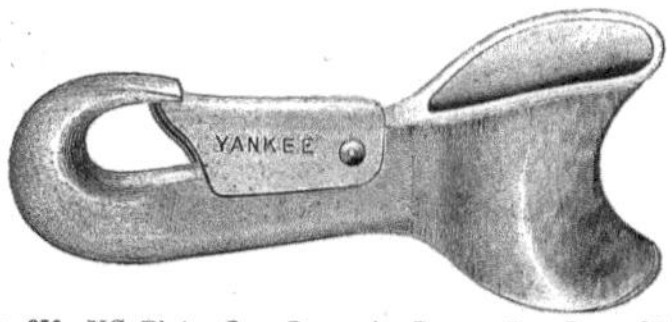

No. 850 XC Plate, One Dozen in Box ... Per Doz. $5.80

DOUBLE CHAIN SNAP

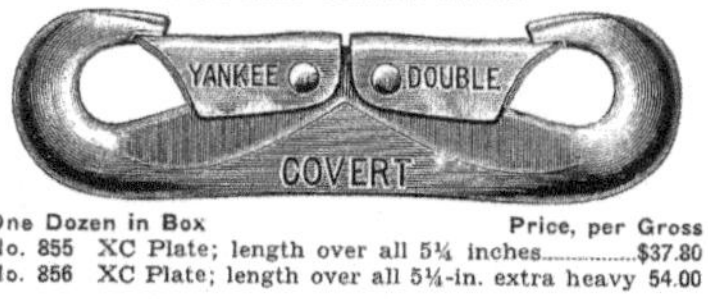

One Dozen in Box		Price, per Gross
No. 855	XC Plate; length over all 5¼ inches	$37.80
No. 856	XC Plate; length over all 5¼-in. extra heavy	54.00

HUBBARD HARNESS SNAP

No. 54 Hubbard

Sizes ⅝, ¾, ⅞ and 1-inch ... Price, per Gross. $23.30

BREECHING SNAP

No. 515 Price, per Doz.

Inch	1½	1¾	2	2¼	2½
Breeching Snap	$9.10	$9.80	$10.60	$11.90	$12.90

In all my research, through the many harness catalogs from the past, I have found lots of variation in the naming of harness parts and hardware. Most on this page are named for function.

BREAST SNAP BUCKLE AND SNAP

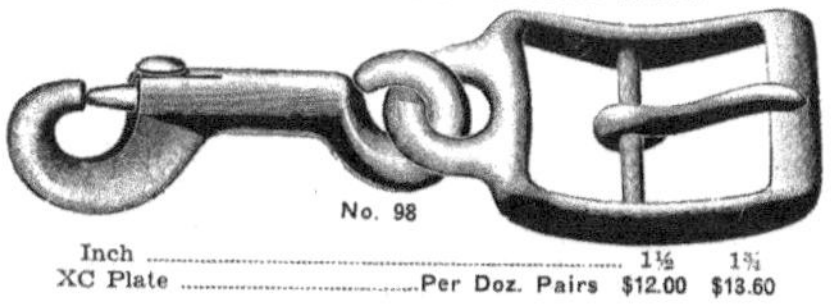

No. 98

Inch		1½	1¾
XC Plate	Per Doz. Pairs	$12.00	$13.60

NO. 62. THIMBLES

For Rope—Inch ... Per Dozen	1½	⅝	¾
XC Plate	$2.50	$2.80	$3.50

NO. 281. ROUND EYE SNAPS AND THIMBLES

Price, per Gross

Size Thimble—Inch	½	½	⅝
Size Snap—Inch	⅝	¾	¾
XC Plate	$46.00	$50.50	$60.00

NO. 3141. CATTLE SNAPS

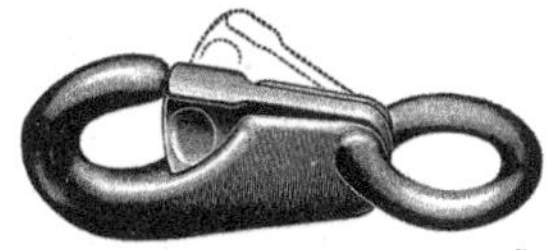

Per Dozen

Size of Eye, ¾-in.; length overall, 3½-in.
Cadmium Plated ... $3.00

NO. 333. SWIVEL EYE BULL SNAPS

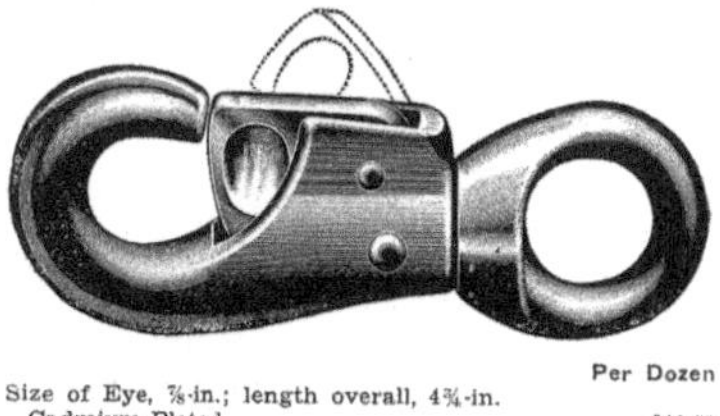

Per Dozen

Size of Eye, ⅞-in.; length overall, 4¾-in.
Cadmium Plated ... $10.00

ALL PRICES SUBJECT TO CHANGE WITHOUT NOTICE

No. 428 JONES'

Miscellaneous Snaps

No. 830 YANKEE ROUND EYE

Nos. 755 AND 855 DOUBLE CHAIN SNAP

No. 20 COVERT'S

No. 19 COVERT'S BOLT SNAPS, ROUND EYE

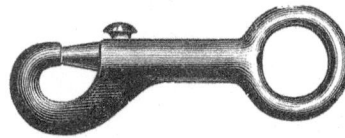

Nos. 47 AND 147 SECURITY

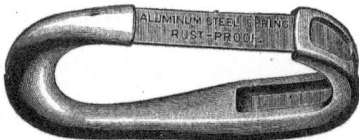

No. 421 BUFFALO

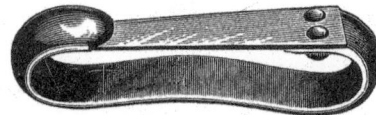

No. 50 COVERT'S SWIVEL SNAPS

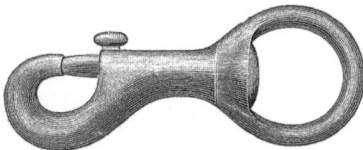

Nos. 510 and 512 DOUBLE SNAPS

OPEN EYE SNAPS

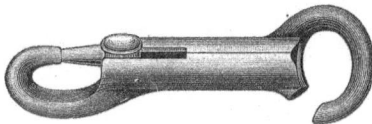

Nos. 520 to 522.

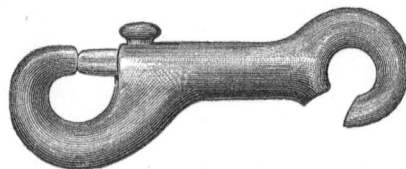

No. 5.

EXPRESS HOLDBACKS

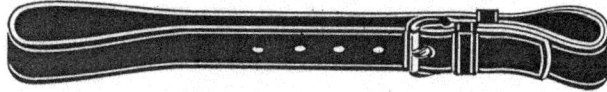

Best Quality, Sewed Leather Loop, Long Lap.

PRICES PER DOZEN

No. 85, 1¼ inch by 4½ feet ....$14.00   No. 94, 1½ inch by 4½ feet $16.50

TEAM CHOKE STRAPS

Best Quality, Single Strap, With Collar Strap Slide.

PRICES PER DOZEN

No. 711, 1½ inch ....$16.50   No. 712, 1¾ inch $18.90

Fair Quality, Single Strap, With Collar Strap Slide.

No. 811, 1½ inch ....$13.80   No. 812, 1¾ inch $16.20

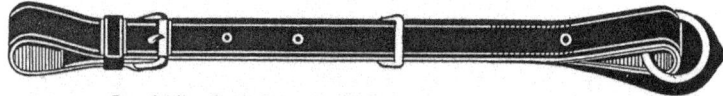

Best Quality, Single Strap, With Fender, With Collar Strap Slide.

PRICES PER DOZEN

No. 911, 1½ inch ....$19.10   No. 912, 1¾ inch $21.50

Best Quality, Single Strap, Buckle and Billet, With Collar Strap Slide.

PRICES PER DOZEN

No. 505, 1½ inch ....$19.00   No. 506, 1¾ inch $22.00

Best Quality, Heavy Double and Stitched, With No. 447 Snap, With Collar Strap Dee.

PRICES PER DOZEN

No. 1051, 1½ inch ....$24.30   No. 1052, 1¾ inch $27.90

Best Quality, Heavy Double and Stitched, With No. 454 Swivel Snap, With Collar Strap Dee.

PRICES PER DOZEN

No. 411, 1½ inch ....$25.60   No. 412, 1¾ inch $29.20

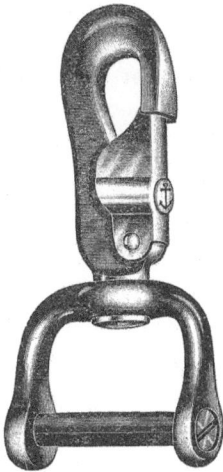

On the forward end of the pole strap (a.k.a. choke strap or martingale) you may find either: 1. A large loop of the doubled leather to slip over the end of the neckyoke. Or 2. A heavy snap to attach to the end ring of the neckyoke. Or 3. A combination snap which attaches breast strap to pole strap with one snap to neckyoke ring.

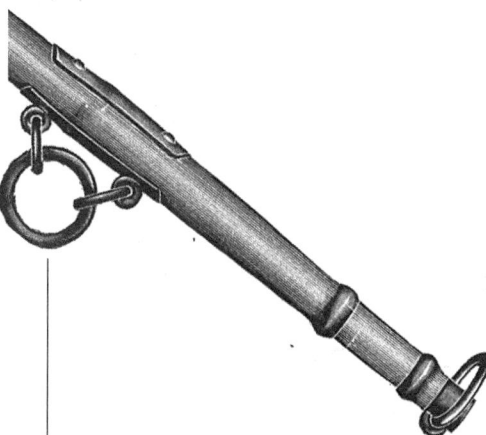

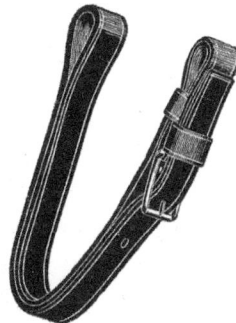

DEES

No. 325 HARNESS DEES

No. 153 BREECHING DEES

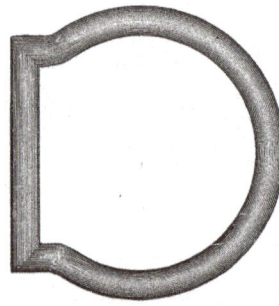

No. 490 SCREW DEES

No. 373 HULL'S POLE STRAP DEES

Side-backer notes: The side-backer style harness has tug straps running forward along the outside of the horses, and these fasten to a front yoke assembly that looks like a small double tree. Some farmers prefer this style harness saying that it allows less 'swing' of the pole in backing. It weighs more because of the additional straps. And it takes a little more to rig.

D-Ring or Boston Truck harness is a further development in the side-backer principle. It has tugs and backer tugs split and attached at a large D ring allowing that, after hitching to yoke at front, the front portion of the backer tug may be tightened to have a taut, paired, triangulation (dotted red lines beloiw)which takes all weight off the back and adds to the precision possible with turns and backing. Les Barden and Ken Demers, many years ago, explained to me that this system, when properly hitched, created an individual comfortable 'basket' of harness for each horse.

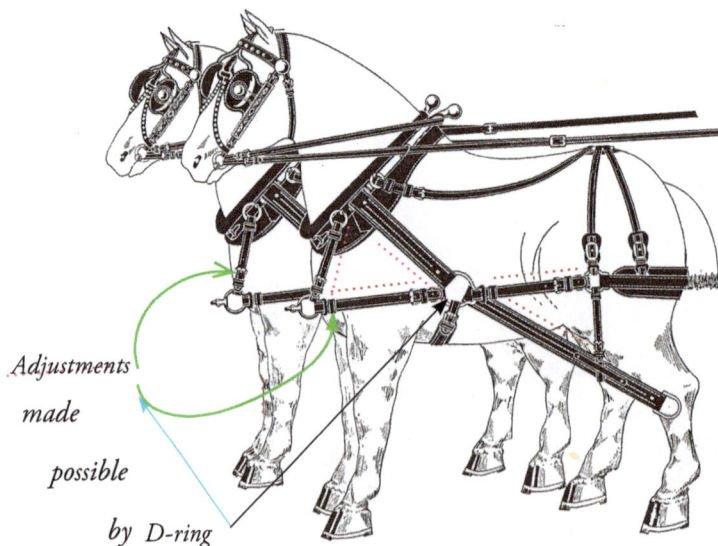

Adjustments made possible by D-ring

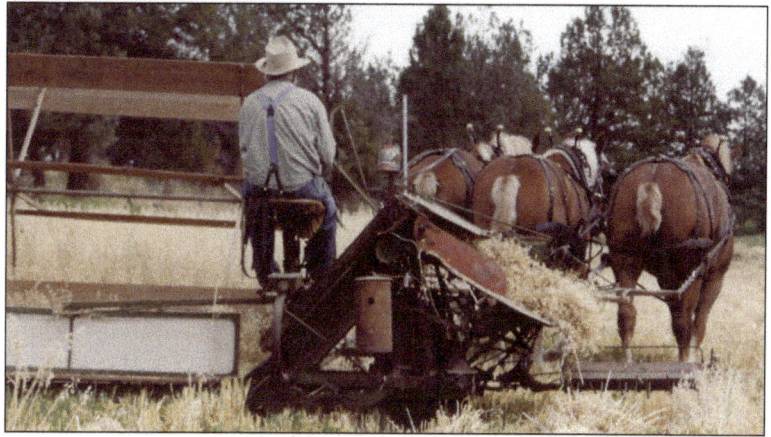

The author driving three McIntosh Belgians on a grain binder. Note the brichen positions.

The brichen on the grey horse is an irritant in this position, because it's too high.

In this old advertising image below, for JD manure spreaders, the team has two breast straps snapped to a bolt-on neckyoke. There are no quarter straps nor pole straps. The backing pressure is on the top of the collar and on back to the cruppers. Far less comfortable.

SNAPS

No. 447. CHOKE STRAP SNAPS

		Per Dozen	
	Size, inch	1½	1¾
No. 447	X. C.	$3.70	4.00
	Rubber finish	3.70	4.00
	2 Dozen in box.		

No. 454. SWIVEL CHOKE STRAP SNAP

		Per Dozen	
	Size, inch	1½	1¾
No. 454	X. C.	$6.70	7.40
	Rubber finish	6.70	7.40
	2 Dozen in box.		

No. 85 BREAST STRAP ROLLER SNAPS

		Per Dozen		
	Size, inch	1½	1¾	2
	Polished	$4.80	5.30	5.90
No. 85	X. C.	5.30	5.90	6.40
	Rubber finish	5.30	5.90	6.40
	1 Dozen in box.			

No. 97. ROLLER SNAP WITH LOOP

		Per Dozen		
	Size, inch	1½	1¾	2
No. 97	X. C.	$8.00	8.80	9.60
	Rubber finish	8.00	8.80	9.60
	½ Dozen in box.			

No. 80. YANKEE SWIVEL TRACE SNAPS

		Per Dozen		
	Size, inch	1½	1¾	2
No. 80	X. C.	$8.00	8.60	9.40
	Rubber finish	8.00	8.60	9.40
	1 Dozen in box.			

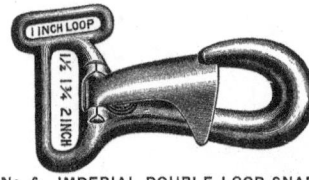

No. 6. IMPERIAL DOUBLE LOOP SNAPS

		Per Dozen
	Size, inch	1¾
No. 6	Rubber finish	$4.50
	3 Dozen in box.	

No. 514. TEAM TRACE SNAPS

		Per Dozen		
	Size, inch	1½	1¾	2
No. 514	X. C.	$6.60	7.30	8.00
	Rubber finish	6.60	7.30	8.00
	1 Dozen in box.			

No. 515. TEAM TRACE SNAPS

		Per Dozen		
	Size, inch	1½	1¾	2
No. 515	X. C.	$8.00	8.60	9.40
	1 Dozen in box.			

PRICES SUBJECT TO CHANGE WITHOUT NOTICE

TEAM BREECHING POLE STRAPS

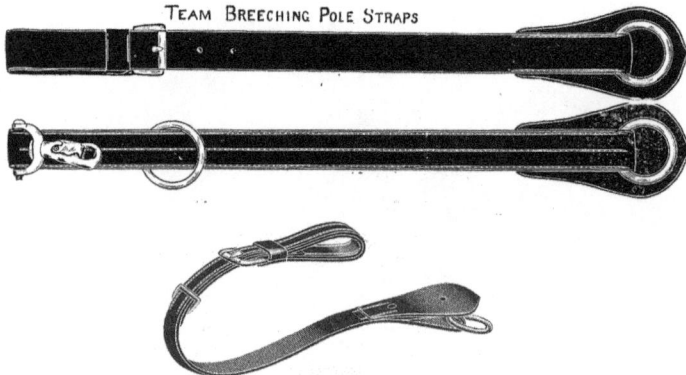

BREAST STRAP SLIDE

The information on these catalog pages, especially pricing, is of course outdated. We've left this here for what it tells us about advertising, construction, and cultural preferences.

No. 187. YANKEE DOUBLE SNAPS

5¼ inches long, heavy, for neck yoke rigs.

Per Dozen

No. 187 X. C. $3.20
No. 187 Polished 2.80
2 Dozen in box.

No. 98. BREAST STRAP BUCKLE SNAPS

Per Dozen

Size, inch 1½
X. C. $4.50
1 Dozen in box.

No. 850. YANKEE CENTER BREAST CHAIN SNAPS

Per Dozen

X. C. $5.60
Polished 5.00
1 Dozen in box.

No. 1616. ROLLER SNAP AND SLIDE

X. C. or Rubber Finish

Per Dozen

Size, inch 1½   1¾
No. 1616 ....9.60   10.40
1 Dozen in box.

No. 18. IMPERIAL DOUBLE SNAP

Per Dozen

No. 18 Rubber Finish, extra heavy for short neck
yokes $4.80
2 Dozen in box.
Pat. July 8, 1890.

No. 46

REICHERT'S COMBINED NECKYOKE SNAPS

X. C. or Rubber Finish

Per Dozen

Size, inch 1½   1¾   2
No. 46 ....$6.40   6.90   7.50
1 Dozen in box.

No. 346

ROLLER SNAP AND SLIDE

X. C. or Rubber Finish

Per Dozen

Size, inch 1½   1¾
No. 346 ....$8.00   8.80

No. 246

REICHERT'S COMBINED NECKYOKE SNAPS WITH SCREW MARTINGALE LOOP

X. C. or Rubber Finish

Per Dozen

Size, inch 1½   1¾
No. 246 ....$9.60   10.40
1 Dozen in box.

Back in the day there were hundreds of varieties of harness snaps and buckles. Over the last fifty plus years, any time I find a rotten harness my first job is to remove the hardware and place it in a bucket of old diesel to soak until I am ready to return and appraise its condition and value. I have five gallon buckets full of assorted serviceable hardware gathered over this half century. Don't need it all, but I'll be darned if I'm going to let it go to a landfill.

No. 442 ROLLER SNAPS

No. 25 LOOP ROLLER BREECHING SNAPS

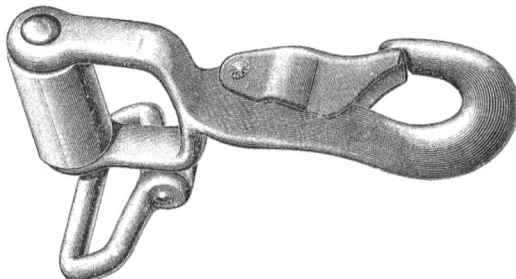

TRACE CARRIERS

No. 379
KIERNAN'S TRACE CARRIERS

**COOPERS
TRACE CARRIERS
No. 386**

No. 1
STOUTS TRACE CARRIERS
To Rivet

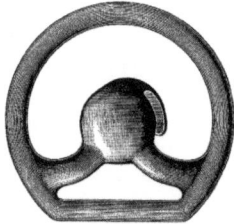

No. 398
TRACE CARRIER
For Hip Strap

No. 394
TRACE CARRIERS
Can be used with or without
Crupper Attachment. Both snaps
should be upside down to prevent
the lines from catching.
NO LINE CATCHER. NEAT AND
LOW

No. 670
TRACE CARRIERS FOR HIP DROP

No. 667. TRACE CARRIER

No. 229
IMPERIAL SCREW TRACE CARRIER

No. 510
DROP TRACE CARRIER

COOPER'S IMPROVED TRACE CARRIERS

MOLINE TRACE CARRIERS

TRACE CARRIERS.

No. 386 COOPER'S TRACE CARRIERS.

To Sew on

	Per Doz.
Japanned, to sew on	$ 55
XC plate, to sew on	55
Nickel on composition, to sew on	4 88
Solid brass	4 88

No. 379 BEST OUT TRACE CARRIERS.

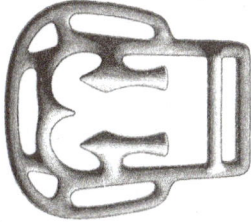

	Per Doz.
Japanned	$1 18
XC plate	1 18

No. 8 KANSAS TRACE CARRIERS.

	Per Doz.
Japanned	$ 67
XC plate	67

No. 385 COOPER'S TRACE CARRIERS.

To Rivet on.

	Per Doz.
Japanned, to rivet on	$ 55
XC plate, to rivet on	55

No. 391 TRACE CARRIERS.

	Per Doz.
Japanned	$ 67
XC plate	67

No. 50 KIERNAN'S TRACE CARRIERS.

	Per Doz.
Japanned, 3-inch	$ 67
XC plate, 3-inch	75

STOUT'S TRACE CARRIERS.

No. 1 to Rivet.

	Per Doz.
No. 1—Japanned, to rivet	$ 67
No. 1—XC plate, to rivet	67
No. 1—Nickel plate, to rivet	3 00
No. 1—Solid brass, to rivet	4 88
No. 2—Japanned, to sew	67
No. 2—XC plate, to sew	67

No. 389 REYNOLDS' TRACE CARRIERS.

	Per Doz.
Japanned, 1¼-inch	$ 67
Japanned, 1½-inch	1 27
XC plate, 1¼-inch	67
XC plate, 1½-inch	1 27

No. 19 L. C. TRACE CARRIERS.

PAT. OCT. 16. 94.

		Per Doz. Pair.
	⅞ 1	1¼ 1¼-In.
XC plate, L. C. patent trace carriers	$ 50 $ 50	$ 62 $ 62

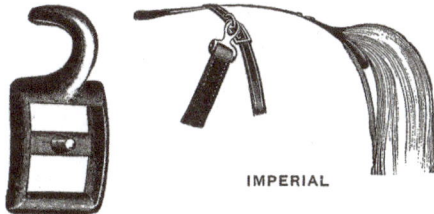

IMPERIAL

J

This Flowers Brabant stallion and mare have their tugs hanging on the trace carriers.

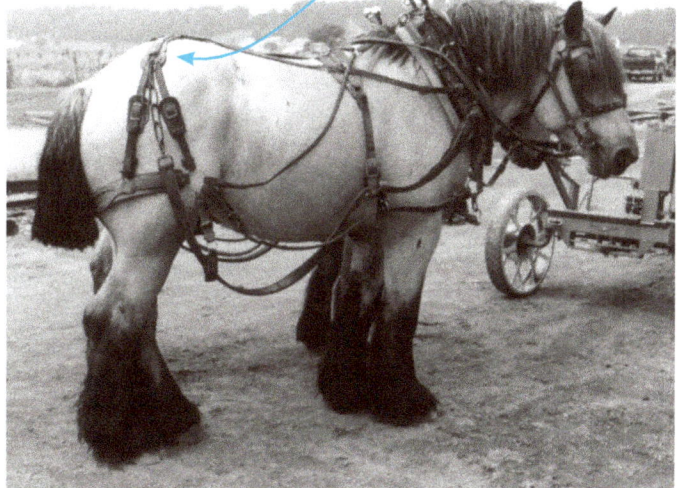

R. S. TRACE CARRIERS

Size, inches	1¼	1½
Nos.	**352**	**353**
Japanned ....Per Dozen,	$1.60	1.92
XC plate ....Per Dozen,	1.80	2.12
Nickel on composition..Per Dozen,	5.30	5.60
Solid brass ....Per Dozen,	4.70	5.00

No. 512 NUBRITAIN TRACE CARRIERS

Size, inches	1¼	1½
Nos.	**512**	**512**
Japanned ....Per Dozen,	$1.60	1.92
XC plate ....Per Dozen,	1.80	2.10
Nickel on composition..Per Dozen,	5.40	5.70
Solid brass ....Per Dozen,	4.70	5.00

No. 378 TRACE CARRIERS

Per Dozen

No. 378—Japanned, 3 inch................$2.20
No. 378—Japanned, 3½ inch.............2.88
No. 378—XC plate, 3 inch................2.50

No. 394 DIPPERT'S TRACE CARRIERS

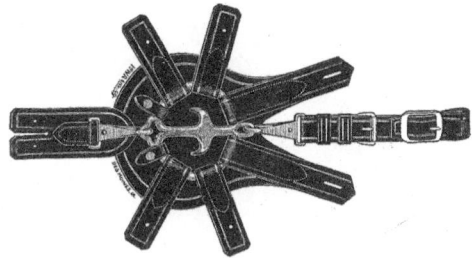

Per Dozen

No. 394—Japanned, 3¼ inch ring........$3.40
No. 394—XC plate, 3¼ inch ring.........3.40

No. 395 DIPPERT'S TRACE CARRIERS

No. 390 WILSON'S TRACE CARRIERS

Per Dozen

No. 395—Japanned 1 loop..............$1.96
No. 395—XC plate, 1 loop................1.96
No. 390—Japanned 2.04
No. 390—XC plate 2.04

No. 19 IMPERIAL TRACE CARRIERS

Inches	⅞	1	1⅛
Japanned ....Gross Pairs,	$13.80	13.80	17.40

The information on these catalog pages, especially pricing, is of course outdated. We've left this here for what it tells us about advertising, construction, and cultural preferences.

No. 193 CONDEN'S PERFECTION LOOPS

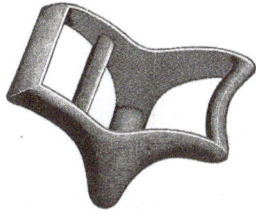

No. 210 CONWAY LOOPS

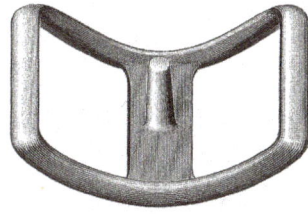

LOOPS

No. 247 TWIN HARNESS LOOPS

No. 209 TWO-FOLD HARNESS LOOPS

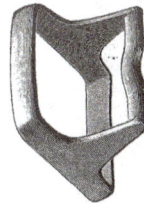

No. 207 FLEXIBLE HARNESS LOOPS

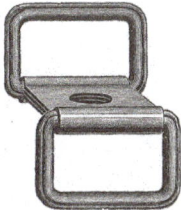

No. 229 HITCH STRAP LOOPS

Clear view of breast straps to neckyoke. Horse Progress Days 2022 scene captured by Jerry Hunter.

Chapter Six

Lines, Spreaders & Center Rings

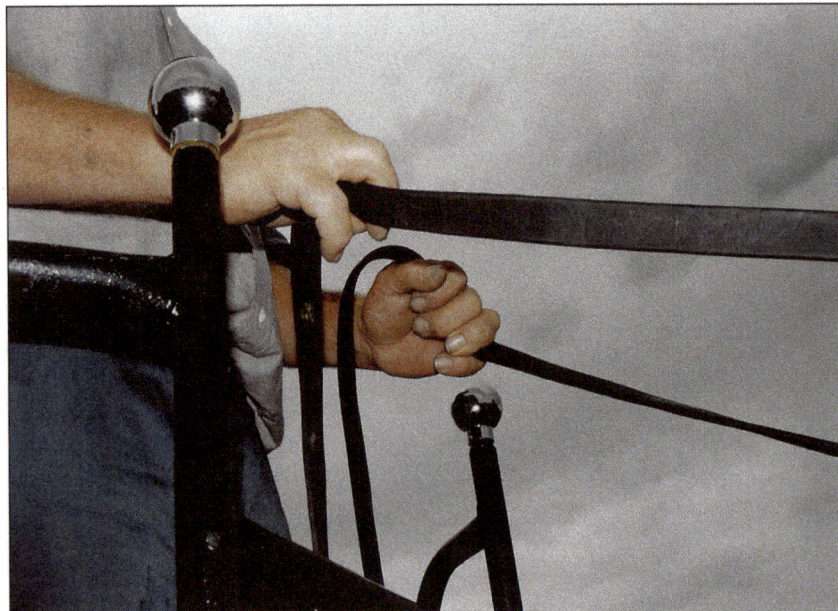

Two horses or mules working side by side are generally referred to as a team. The customary procedure for 'driving' them is to employ team lines which, fastened to the outside of each of two bits, provides the teamster the ability to apply varying pressure. With experience, training and maturity the teamster might learn to softly send messages to each equine, through light pressure at the corners of their mouths as to preferred direction, speed, and halt.

Team lines are one of the more obvious tactile proofs that a fourth dimension exists if only for those who trust and depend on their usefulness. Bear with me on this: with full-sized team harness you might have two twenty-foot long lines, each transected by a short adjustable check line. Lay these two mirror lines on the ground and see how it is that the cross checks, always on the inside, can be adjusted to shorter than or longer than the main line. (Most commonly longer by inches.)

Look over your shoulder at the two animals you will be asking to work together. The mare may be shorter than the gelding, otherwise they are a perfect match, unless of course you take into account that the mare walks slower than the taller gelding. Imagine them standing side by side, in harness, with the main controlling features of their positions being their hitch setup AND those two lines.

See in this drawing shown from overhead (next page) that the long, continuous, main-line portion runs along the outside of each animal, through the top ring on the outside hame and on to the outside of the bit. The short, adjustable, check line passes through the top inside hame ring on each animal before crossing over to the inside of the bit. OR, the check line runs through a 'spreader,' which allows the horses to stand slightly further apart, before moving on to the opposing animal's inside bit. OR, the check line runs through the spreader

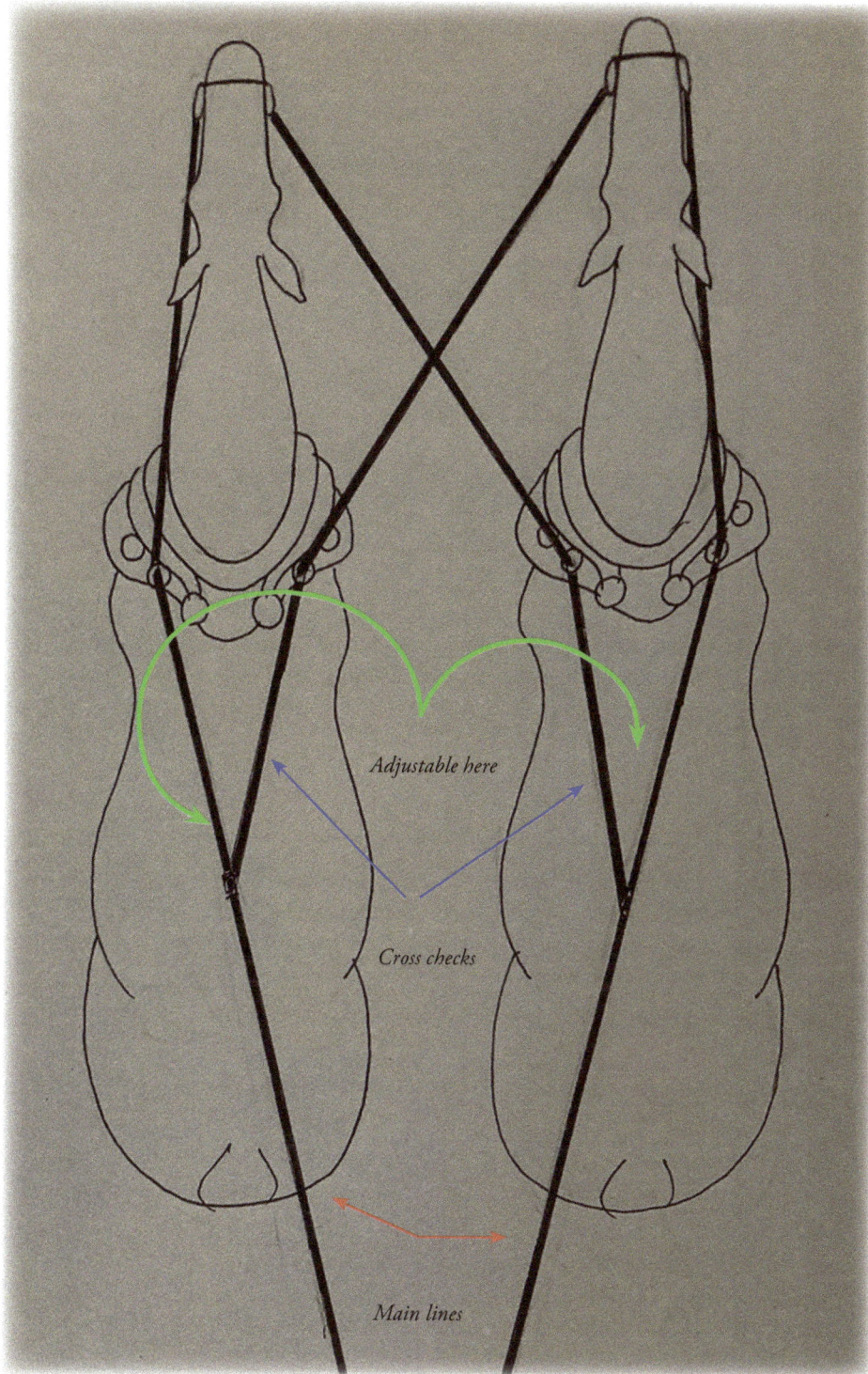

Adjustable here

Cross checks

Main lines

Team lines threaded through the top hame rings. This arrangement holds the team near one another.

Experienced and observant teamsters know that minute adjustments in cross-check to main-line buckling can have dramatic effect on the alignment of heads and necks where precision work is most valued (for example on a walking plow.) Balance and comfort are not always attained with symmetry. The furrow horse, down in, may cause the cross-check to pull the land horse's head over unnaturally. Sensitive hands and fine tuning of the lines, that's the ticket.

Three different ways to fasten adjustable cross checks to main lines. Top: this is sometimes referred to as an 1890 buckle wherein the cross check is passed through a side D and riveted to itself. The adjustable portion employs a conway that is quite safe. Middle: this is a standard conway buckle. The cross check has a hole punched in the end and this then is held in the conway by a raised pin and the pressure of the main line. This is the most common arrangement and far and away the least satisfactory as it is too easy for the end of the cross check leather to split and release itself from the lines. Bottom: the very best arrangement. A buckle is riveted to the end of the cross check which has the mainline passed through it.

AND a *center ring* which 'gathers' the two lines at their crossing to keep the geometry 'collected.' More later.

This author likens the world of a teamster to that of a puppeteer with marionette strings in hand. But, of course and with certainty, the animals are not stiff puppets but rather sentient beings responding best to the subtlest of touches. Pull back hard on the lines and destroy possibility even as you might get the animals to stop or back up. Whisper to them and then touch the lines with a feather and the best animals will react with the soft and sure movement you require.

Structurally, team lines are commonly used when rigging larger field and wagon hitches, such as four-up, six-up and eight-up employing multiples of team spans. (See diagrams page 154.)

The same can be said of *abreast* hitches as one common approach is to add stub cross-checks to a basic team line setup.

Also, I know of teamsters who drive single horses and have customized quick ways to remove the cross-checks altogether resulting in a set of single lines.

Once the geometry and dynamics are understood

there are interesting variations available to the use of team lines, for example the header (see page 164) and the buckrake.

(Next page.) Two piece team lines, where the front portion of the lines have a D ring on the end and are snapped into the hand parts, are not commonly used unless the teamster is putting together various hitches and wants to save time with changes. For example, a farmer who would wish to have a longer line to rig a three or four-abreast with team lines may remove the standard length hand part and exchange with a set that is sufficiently long for the hitch and work configuration.

I have spoken of what is possible in a gentle relationship with working equines, where the lines are *lines of communication* best kept *open* without abusive pressure. The lines are akin to strings on a musical instrument, they require a perfect tension, one which constantly reassures and reminds the animals of the position the teamster requires. Constant reassurement can be music.

The center ring or drop, while not absolutely essential to the employ of a team harness, does provide a useful service. A team standing may cause the cross-checks to go slack. Should this occur while a horse turns its head, such as the right hand horse in this illustration, the loose check could slip in under the end of the tongue and cause confusion. When a center ring is in place it gathers that slack where the lines cross and reduces the chances of a misshap.

With two-piece lines, the snap should always face upwards. Even so, it could hang up on something and prevent the lines from working. Taping this joint with electric tape is a simple precaution.

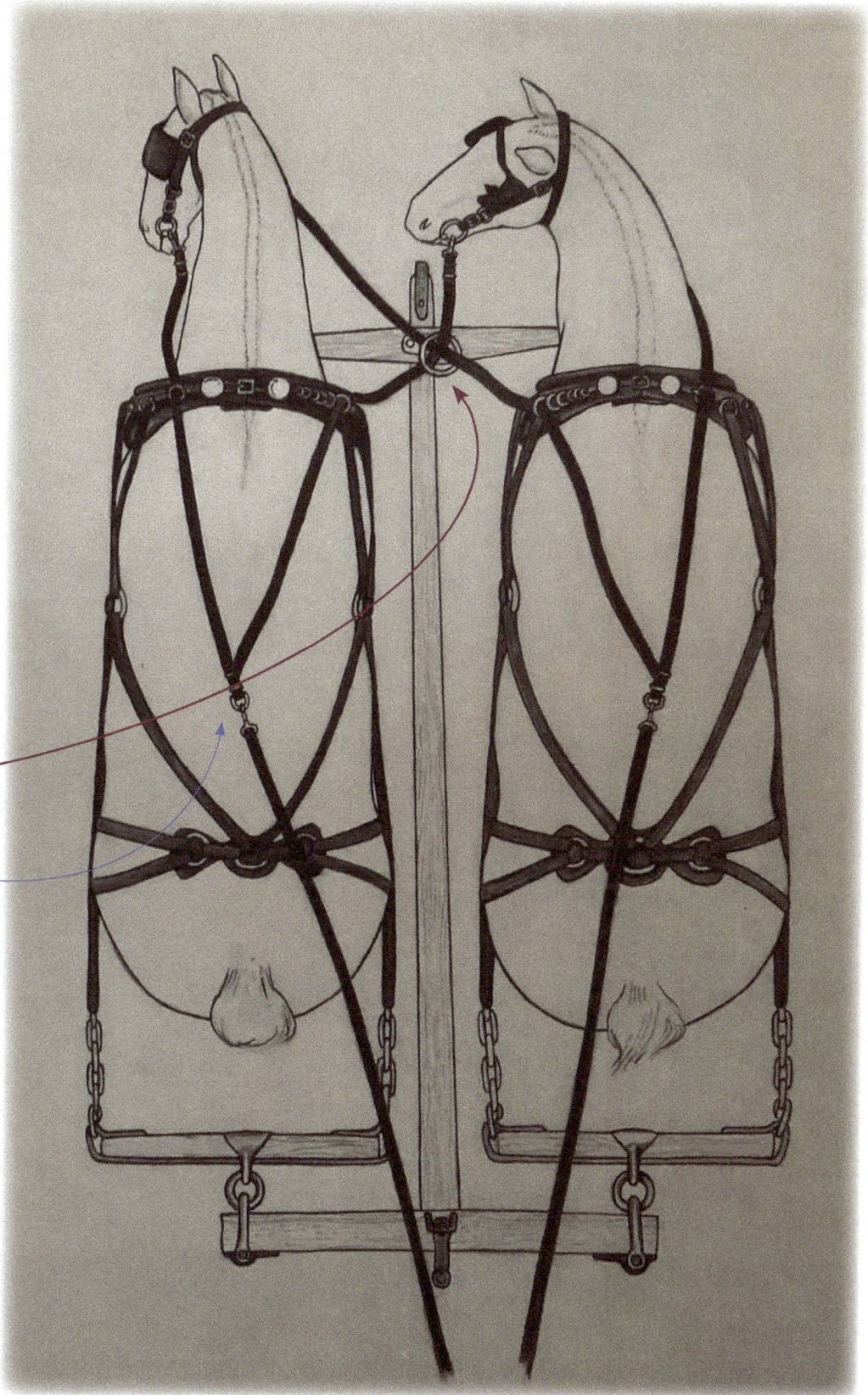

All original drawings in this book have appeared in other Work Horse Library titles by the author and artist L.R. Miller.

No. 193 CONDEN'S PERFECTION LOOPS

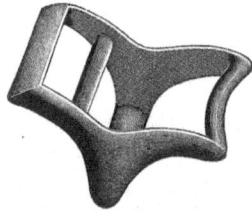

No. 210 CONWAY LOOPS

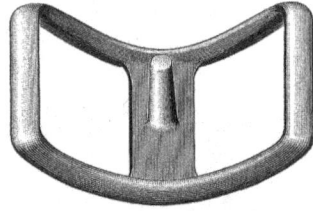

BARREL ROLLER HARNESS BUCKLES

No. 50

NO. 12 WIRE BENT HEEL HARNESS BUCKLES

No. 12

The Conway buckle, above and below, depends entirely on the strength of the perforated end piece of line leather. Too often it can break loose.

CONWAY LOOP BUCKLE
No. 210

No. 188 REIN BUCKLES

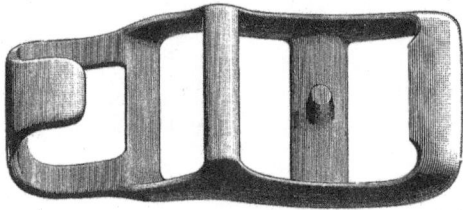

NO. 209. TWO-FOLD LOOP

No. 170 CENTER BAR REPAIR BUCKLES

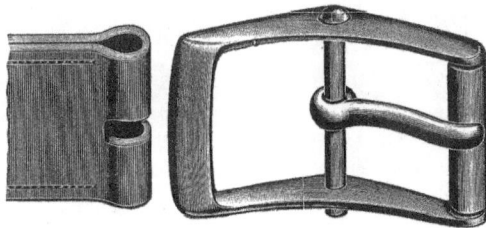

No. 93 UNION LOOP BUCKLES

BUCKLES

No. 140 LOOP HARNESS BUCKLES

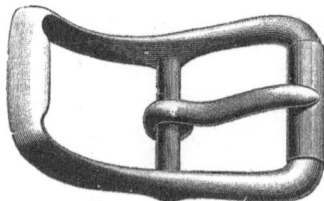

No. 150 LOOP ROLLER BUCKLES

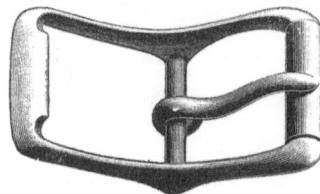

Leather or *Beta/Biothane* lines are fastened to bits either by hardware snaps or by buckle assemblies as shown in these two illustrations.

If hardware snaps are used be aware that an animal rubbing against another animal or rigging may snag its line snap where it does not belong causing confusion that, with less than well-trained animals, might result in panic. Also please note that snaps can fail or break.

Buckled leather does not come apart until you want it to, and it does not break, unless of course the leather is rotten.

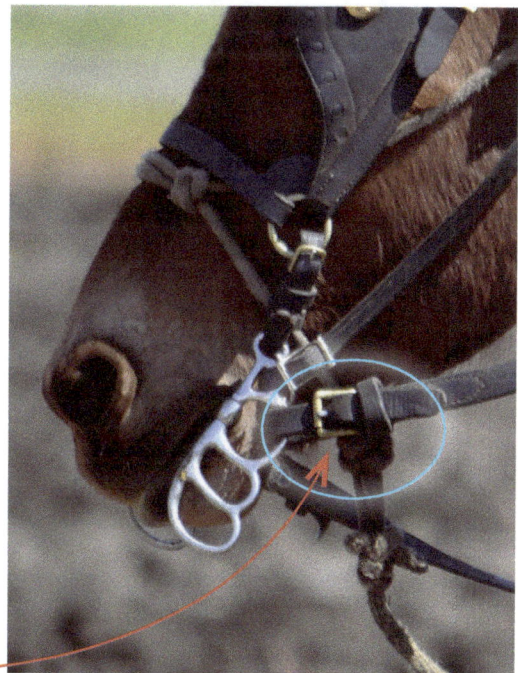

To the right: here is a
six-up hitch with 3 sets
of team lines employed,
each set a different
length back to the plow
or wagon (or what
have you) and to end
up with three lines in
each teamster hand.

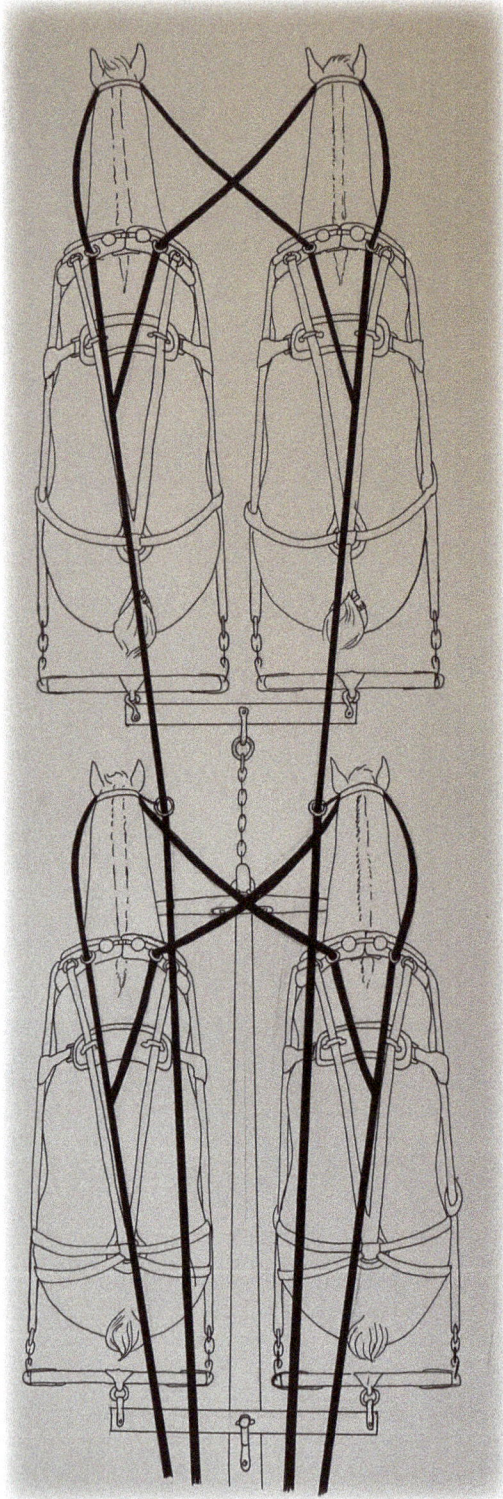

Notice above that with this four-up hitch two sets of
team lines are employed, the difference between them
being their lengths. The lengths are determined by the
size of the animals and the distance the teamster is from
the hitch. The teamster in this case will be driving with
two lines in each hand.

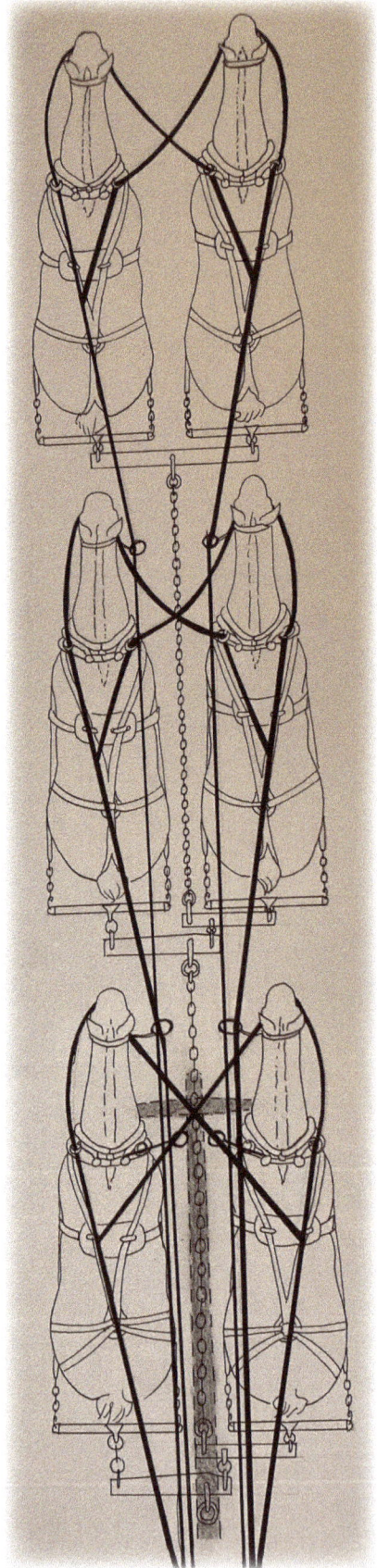

Renee Russell's Percheron team, hitched to a mower, is using a center ring with a heart drop.

Also, please notice that the breast strap snaps, at the neckyoke rings, are facing in to avoid a bit or strap hanging up should a horse lower its head and rub.

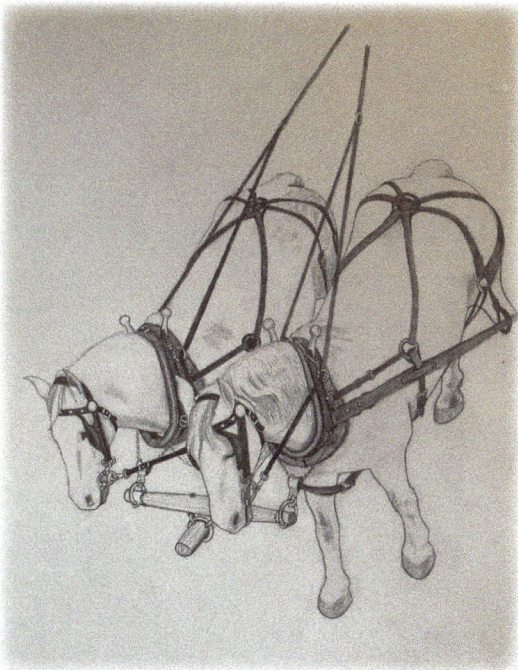

A Fjord team at Horse Progress Days. Notice lines buckled into levered bits, rope halters beneath bridles, wide set blinders, comfortable head set, no check reins, adjustable collar, collar pads, and what look to be Beta (synthetic) lines. The breast straps/pole straps employ combo snaps.

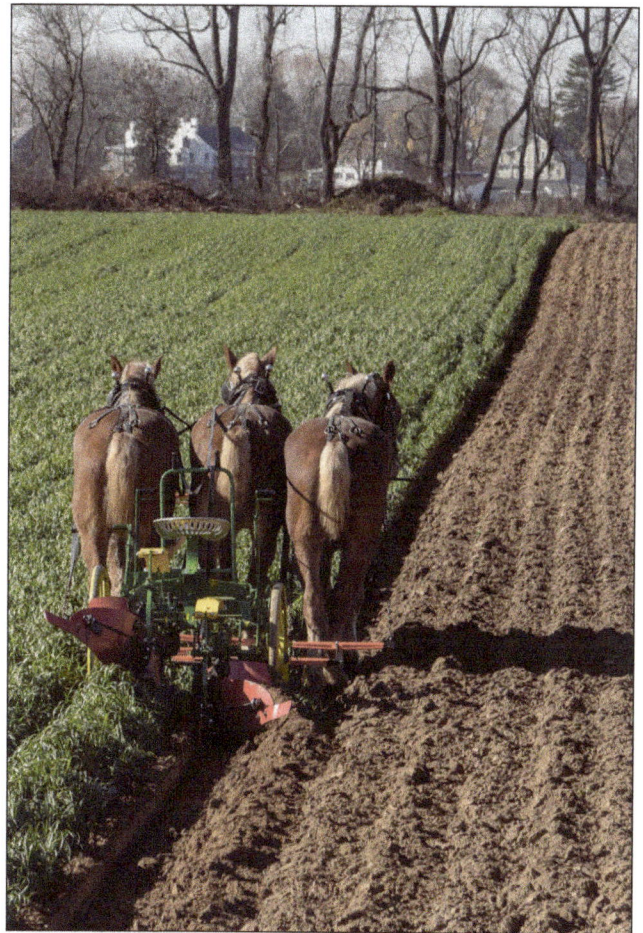

The illustration above shows the angle of the lines that might be involved if a team is being driven from up above, either from a high seat on forecart or wagon, or in the case of unusual work such as Khoke Livingston's root cellar digging, where the team might be down in a hole and the teamster up on the edge. (see page 157)

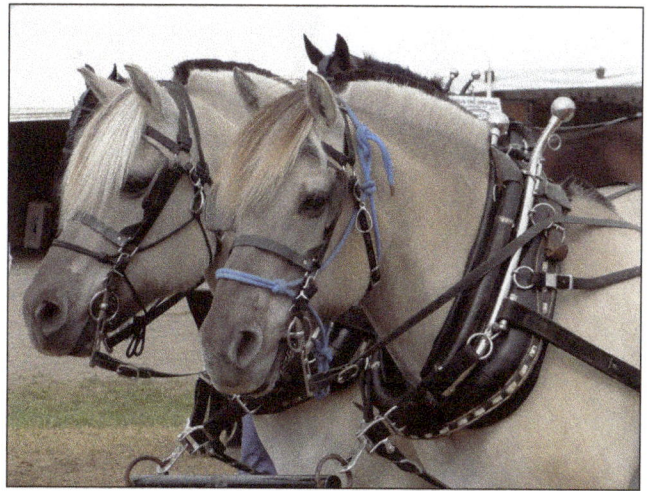

Three Pennsylvania Amish Belgians hitched to a new White Horse two-way riding plow. Team lines on left side pair with furrow horse tied in with jockey stick.

Line weights & dimensions:
Commonly, leather driving lines have been made from hot-stuffed harness leather anywhere from 3.5mm to 5mm thick and 1¼" to ¾" wide. In modern times, lines are also made of synthetic polyurethane materials, the most serviceable of those being BETA. The VERY BEST lines I have had in my hands were a gift. They were made of 'bridle' leather, similar to 'harness' leather in strength but with a nicer finish. They are light weight.

Three and four-abreast hitches (and those even wider, such as the six-abreast shown on the baler on page 20) may be set up with team lines and separate individual check (or stub) lines or, in the case of some Amish communities, jockey sticks.

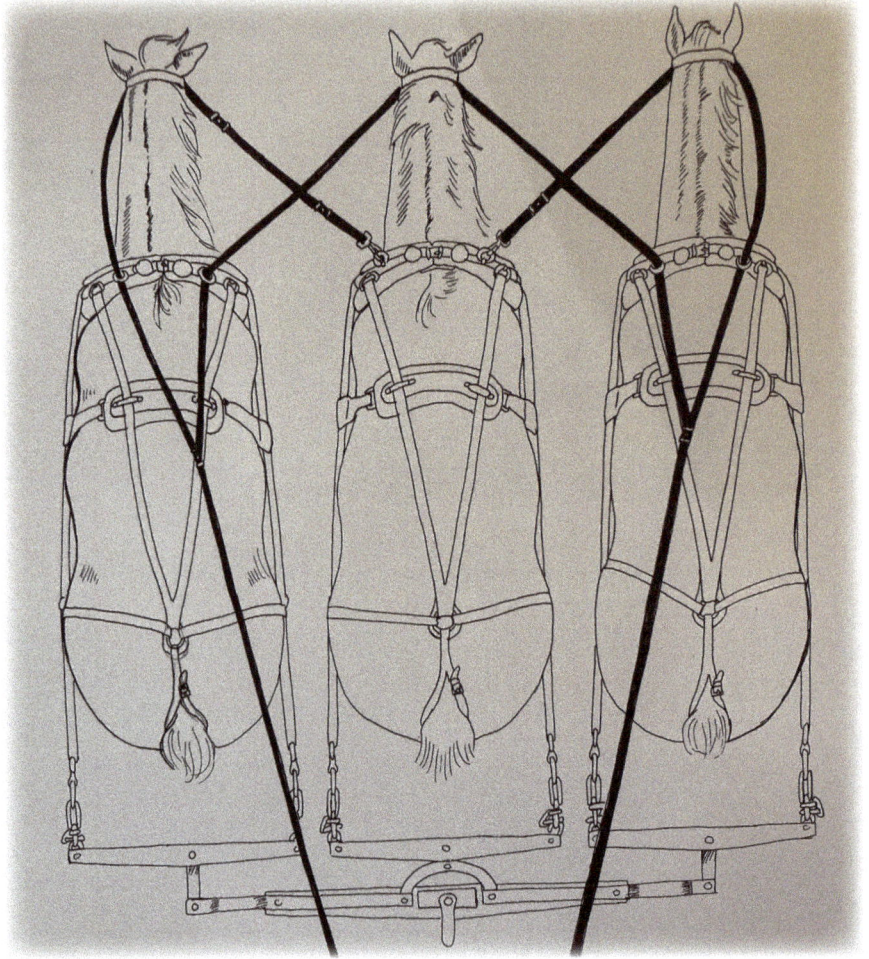

Team lines usually come in lengths of 18', 20' or 24'. For full-sized horses on a walking plow, or hitched to a seed drill and driven from behind, 20' may suffice, 18' is too short.

Two-piece line setups are handy as it is possible to have hand sets of various lengths allowing the teamster to get that perfect length for the job.

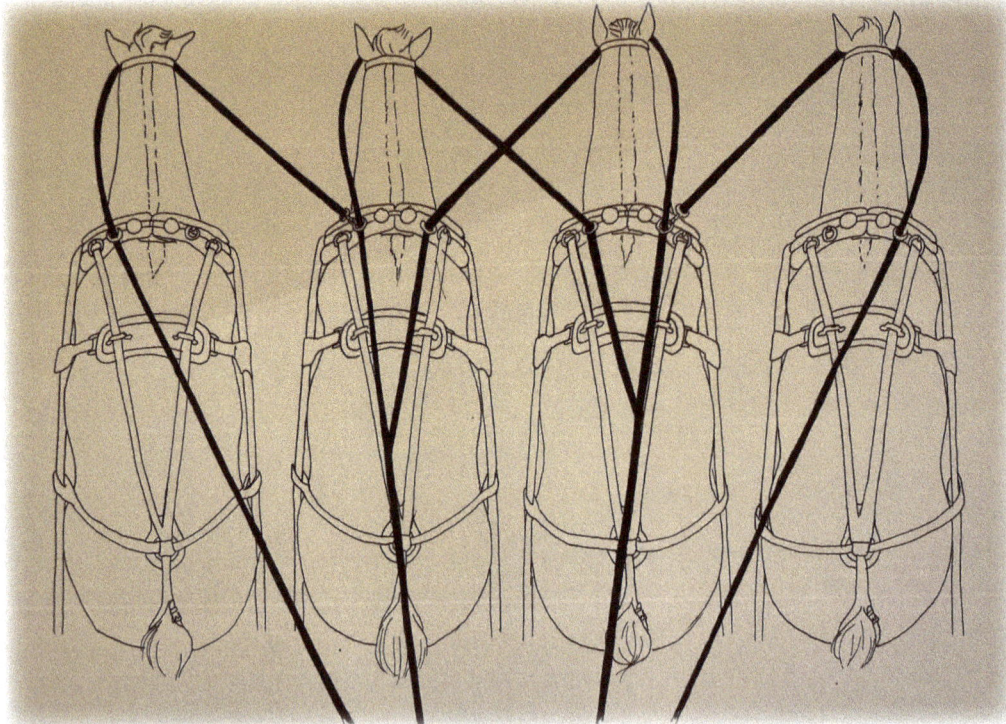

Khoke Livingston is using his team to excavate their root cellar. He has a full length pair of driving lines and works from up on the edge of the dig. Photo by Ida Livingston.

Hitch long
for clearance
on turns

Danger
Point

The approximate
difference in
line length
when turning

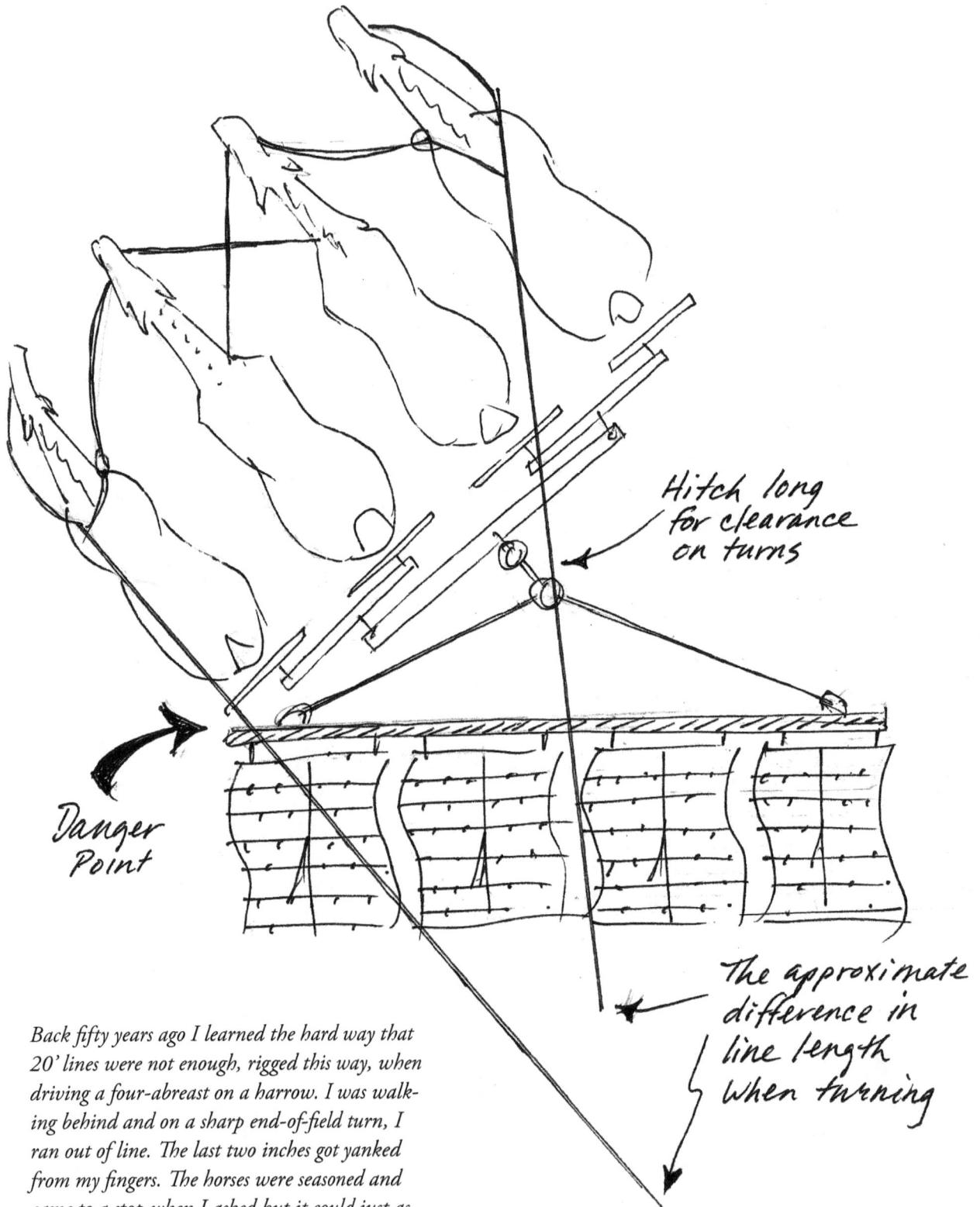

Back fifty years ago I learned the hard way that 20' lines were not enough, rigged this way, when driving a four-abreast on a harrow. I was walking behind and on a sharp end-of-field turn, I ran out of line. The last two inches got yanked from my fingers. The horses were seasoned and came to a stop when I asked but it could just as easily have been a wreck.

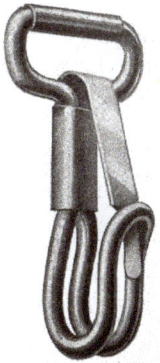

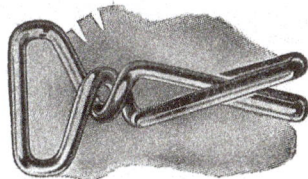

KLING SNAPS

*Spring and slide bolt style snaps.
The spring is preferred for lines.*

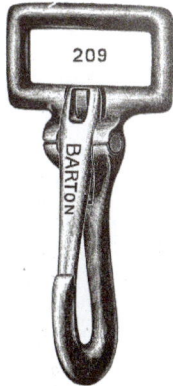

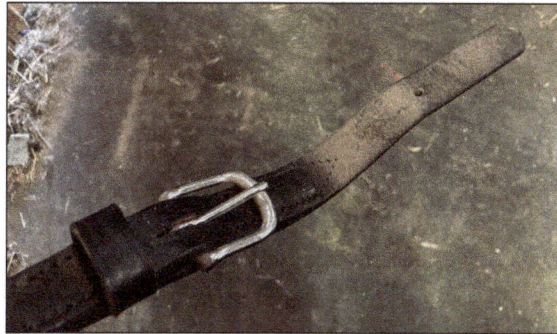

209

BARTON

*This is one of my lines and my preferred
setup to fasten to bit rings.*

Double Team Lines

No. 1—⅞, 1, 1⅛, 1¼ inches, 18, 20, 26 and 32 feet, Laps Sewed and Riveted.

No. 1—With Riley Loops and Cleveland Snaps, ⅞, 1, 1⅛, 1¼ inches, 18, 20, 26 and 32 feet, Laps Sewed and Riveted.

No. 3—With Noble Billet Buckles and German Snaps, ⅞, 1, 1⅛, 1¼ inches, 18, 20 and 26 feet, Laps Sewed and Riveted.

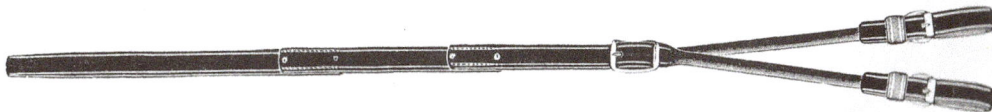

No. 60—Round, 1-inch Hand Parts and Billets, 18 feet, Laps Sewed and Riveted.

ALL LAPS STITCHED, 1880 BUCKLE AND 290 SNAPS
Double and Stitched Coupling

ALL LAPS STITCHED, CONWAY LOOP BUCKLES AND SNAPS
Double and Stitched Coupling

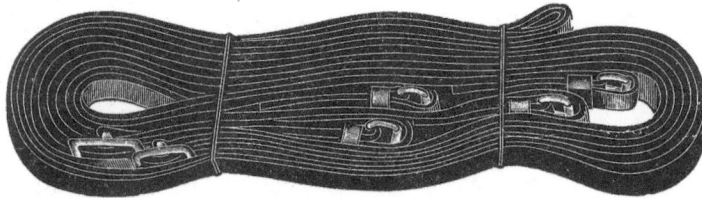

CAMPBELL LOCK STITCHED, SIX TO THE INCH, THREE INCH LAPS
Double and Stitched Coupling. Gopher Grade.

HAND PARTS

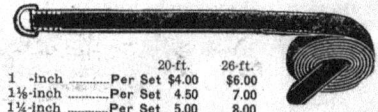

		20-ft.	26-ft.
1 -inch	Per Set	$4.00	$6.00
1⅛-inch	Per Set	4.50	7.00
1¼-inch	Per Set	5.00	8.00

TWO PIECE LINES:

The check portions usually have a snap at the end into which the hand parts snap. As above, hand parts were sold in 20' and 26' lengths but the clever farmer can always construct different lengths of hand parts to suit the work.

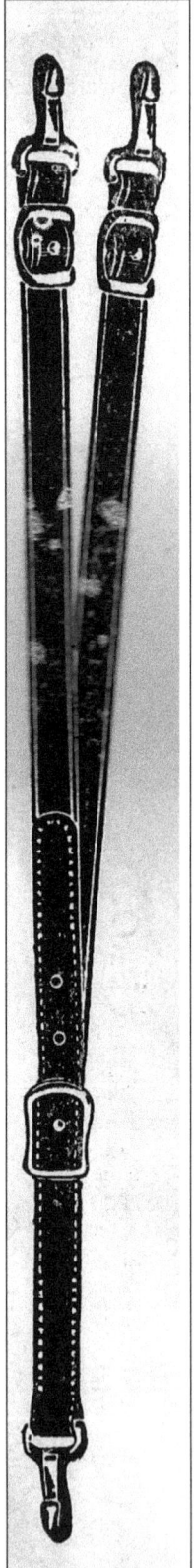

No. 209 TWO-FOLD HARNESS LOOPS

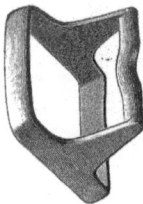

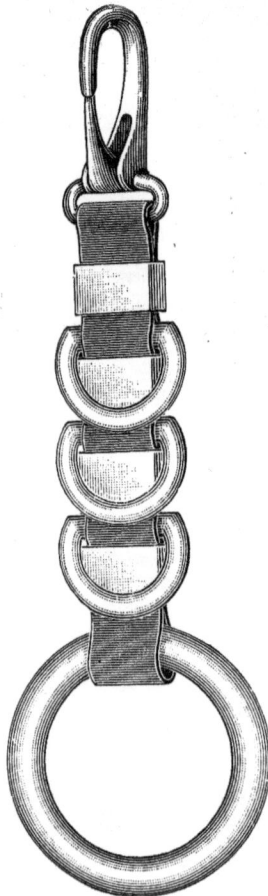

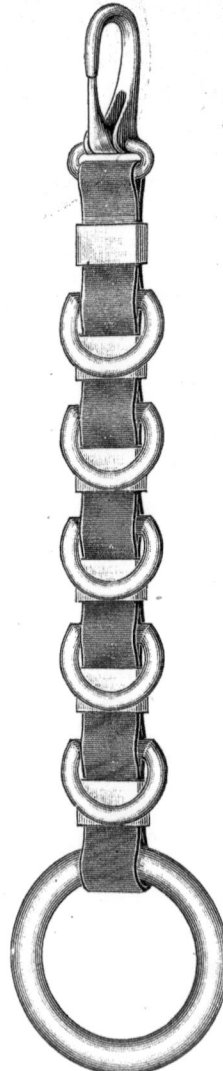

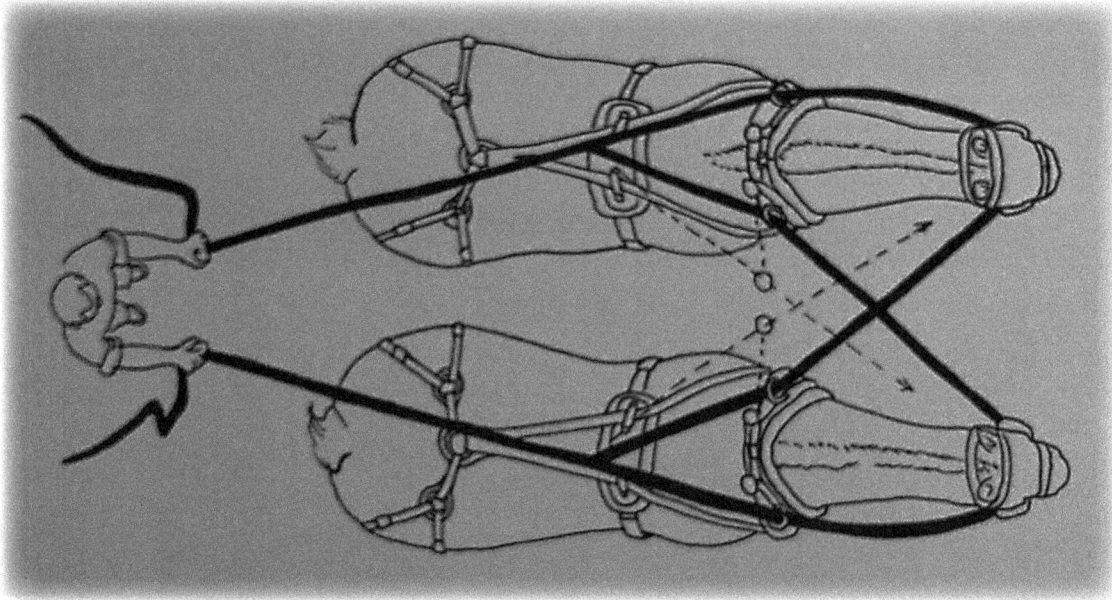

The longer the spreader the further apart you will be able to spread your team.

My mentor, Ray, had a short mare and a tall gelding as a team. He would use two different lengths of spreaders to even out the line when plowing.

No. 37.   No. 38.   No. 39.

LINE ROLLERS

No. 320.

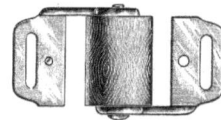

Adams.

Inches	1	1⅛	1¼
No. 320—XC plate, iron roller..........Per Gross,	$16.00	16.00	....
Adams—XC plate, wood roller..........Per Gross,	20.00	20.00	20.00

No. 41—Imperial.

No. 41—Imperial, Japanned or XC plate, wood roller .... Per Gross $16.00
Line rollers packed one-quarter gross in a box.

IMPERIAL SPREAD ROLLERS
Per Gross
No. 41 X. C. $16.00
¼ Gross in box. Roller is 1⅜ inches wide. Will take
1 and 1⅛ inch line.

	Per Dozen	
	Nickel	Brass
No. 23 1¾ balls......................	$9.50	$8.50
No. 24 2 balls	10.50	9.50
1 Dozen in box.		
No. 302 Bridges only	3.00	3.00
1 Dozen in box.		

No. 304 ADAM'S SPREAD ROLLERS
Per Gross
Japan, 1   inch $18.70
Japan, 1⅛ inch 19.30
Japan, 1¼ inch 20.00
¼ Gross in box.

LINE SPREADER TERRETS
Per Gross
No. 320 X. C., 1 inch $18.70
No. 320 X. C., 1⅛ inch 19.30
No. 320 X. C., 1¼ inch 20.00
¼ Gross in box.

SPREADERS

Spreader straps came and come in all sorts of designs and appointments. They could and can be as plain as a leather strap with ring affixed. Or fancied with colored rings, brass, nickel or chrome, or your   imagination. They should not be confused with center hearts shown below.

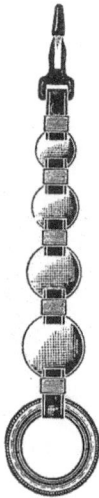   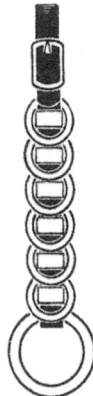   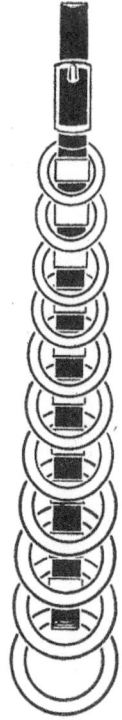

No. 85   No. 524   No. 500   No. 506   No. 514   No. 614   No. 502

SPREAD STRAPS.

Per Doz.

No. 495—⅝-inch, with 4 assorted color Zylonite rings and loops....... ......

No. 496—⅝-inch, with 6 assorted color Zylonite rings and loops....... ......

No. 700—Fancy nickel or brass spotted with rings.................... ......

CENTER HEARTS.

No. 62—Double Japanned, one side Fancy, with Nickel Ornaments, Duranoid Rings.

No. 10—Red or Black Patent Leather, Nickel or Brass Ornaments, Zylonite Rings.

No. 16.

No. 17.

SPREADERS OR HIP DROPS.

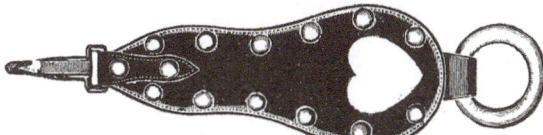

No. 13—Nickel Ornaments, Zylonite Rings.

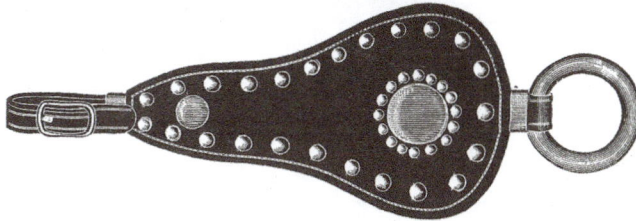

No. 14—Nickel or Brass Ornaments, Zylonite Rings.
No. 15—Nickel or Brass Ornaments, Zylonite Rings.
Red or Black Patent Leather

SPREAD STRAPS

No. 76.   No. 78.

Per Dozen

No. 76—13 inches long ⅝ inch black leather strap, 6 assorted sizes, white Zylonite rings, white celluloid slide loop, No. 5 white Zylonite ring at end, Security snap.................$......

No. 78—16 inches long, ¾ inch black leather strap, 10 assorted sizes white Zylonite rings, white celluloid slide loops, No. 5 white Zylonite ring at end, buckle and billet........ —

No. 86.   No. 180.

Per Dozen

No. 82—12 inches long, ⅝ inch black leather strap, 6 No. 00 white Zylonite rings, white celluloid slide loops, No. 5 white Zylonite ring at end, Security snap...................$......

No. 84—16 inches long, ¾ inch black leather strap, 10 No. 00 white Zylonite rings, white celluloid slide loops, No. 5 white Zylonite ring at end, Security snap.....................

No. 86—20 inches long, ⅝ inch black leather strap, 14 No. 00 white Zylonite rings white celluloid slide loops, No. 5 white Zylonite ring at end, Security snap.....................

No. 180—14 inches long, ¾ inch black leather strap, round harness leather circles, metal spotted, with large oval metal center, No. 4 white Zylonite ring at end, Security snap, nickel or brass trimmed................................

SPREAD STRAPS

No. 291

No. 292

No. 293

No. 295

Per

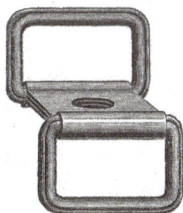

Mike McIntosh demonstrating how a corner is made with four-abreast on a header. Two teams are used, each with its own set of lines. Mike asks his left team to swing 'haw,' while holding his right hand team still.

Then he steps the left team forward while backing his right team. The header 'comes around' on the crazy wheel beneath him.

The author with Lana and Cali on the buckrake. Each side of the team set of lines have been flipped over so that the solid line is inside and the cross check to the outside. This way when the teamster pulls back on the lines he is effectively stopping each horse, individually. To steer the team, the author slows or stops the one horse and allows the other to step ahead. The action is a similar dynamic to the header.

Lines Telling Stories

This gentleman is driving a five-abreast with a single jerk line.

When ground driving and skidding
logs (or anything to be drug along the
surface), it is critical you handle
lines in a safe and smart way.
Imagine what might happen
if the teamster were to step
into the dangling loop
of the line, or if the
line should wrap
around a side-
ways roll of
the log?

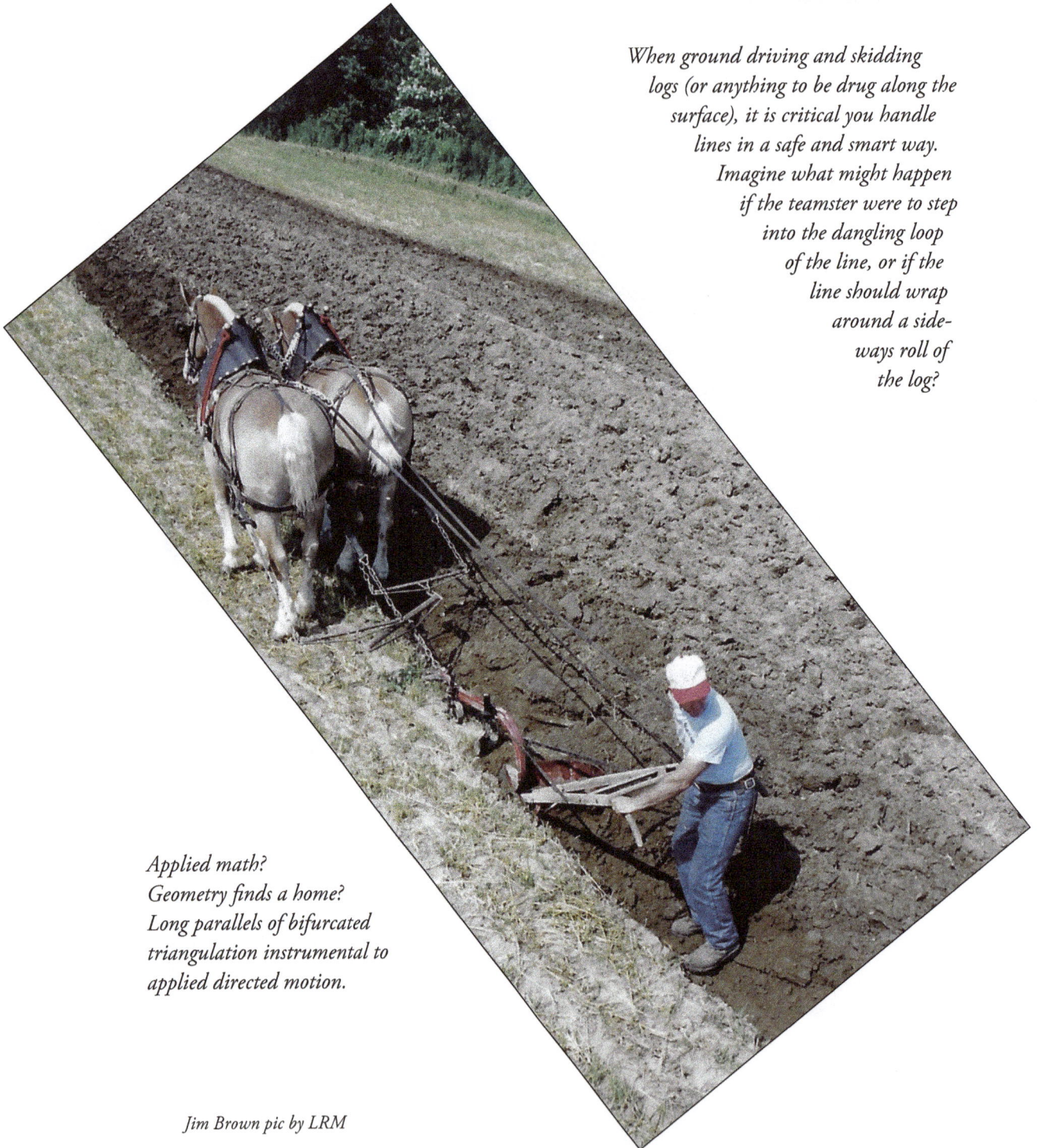

Applied math?
Geometry finds a home?
Long parallels of bifurcated
triangulation instrumental to
applied directed motion.

Jim Brown pic by LRM

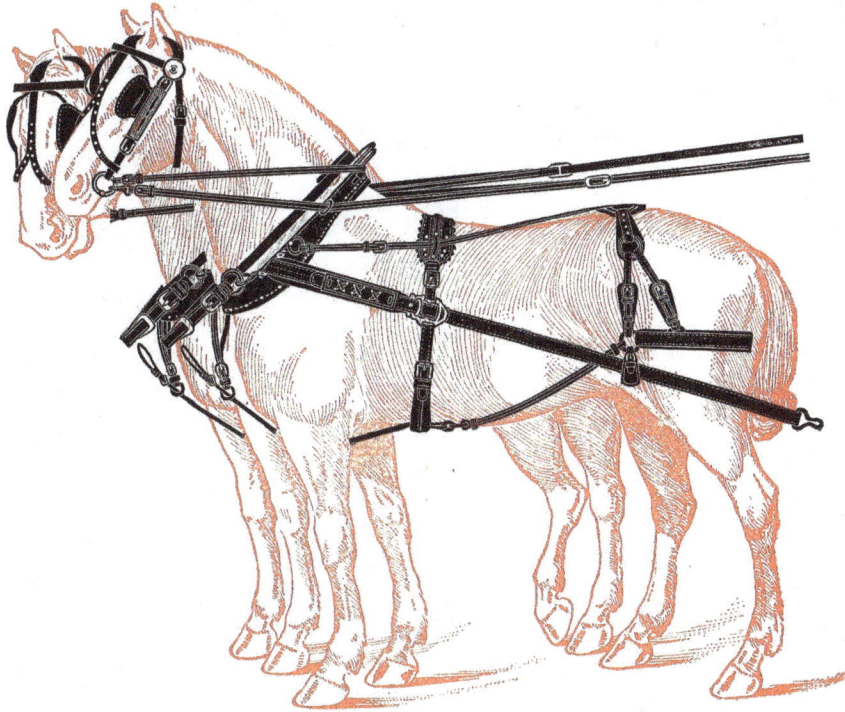

Chapter Seven

Harness Designs:
Variables & Variations

Doing the research for this volume, deep into the selection process for these many, many harness illustrations of variations, I was struck how the nature and quality of the illustrations tend to shape our judgement of the harness itself, its suitability, functionality, etc. Above, from the Wallace & Smith catalog, the handsome image tells me that this harness is just what it needs to be and nothing more. While, on page 161 the illustration suggests that the harness is overbuilt, heavy, sharp-edged, stiff and perhaps clunky. Either of those assessments may be off the mark. Most of us prefer to make important decisions about what harness to acquire based on firsthand evaluation. And that is right and proper. But in this day and age it may

not be possible. The harness maker you choose may be a thousand miles away. Therein lies one of the reasons I chose to include so many diagrams; hopefully to help harness makers and harness buyers make decisions. I like to imagine that Joe in Maine has a copy of this book when he writes to Ezra in Missouri, who also has a copy of this book. Joe says 'I want a harness like the one on page ?' and Ezra writes back, 'what about combining that with what you see on page ?' Wishful thinking? Perhaps.

One thing that stands out when we have this opportunity to see illustrations from many different catalogs is how the artistry varied so widely. Some of these images are excellent, some are not. Some of the drawings give

Above, a Pioneer Equipment ad photo with four head in two-strap brichen harness.
Below, White Horse Equipment ad photo of three head in brichen-less plow harness.

a rock solid sense of the harness design while others may be a bit misleading. But of course, these are drawings and because of that we see them as limited. I would like to make a case against the photographs and for the best drawings. Here are two modern harness catalog photos which I ask you to compare to these two early twentieth century drawings on this page. I for one can see the parts and how they go together much easier with the drawings.

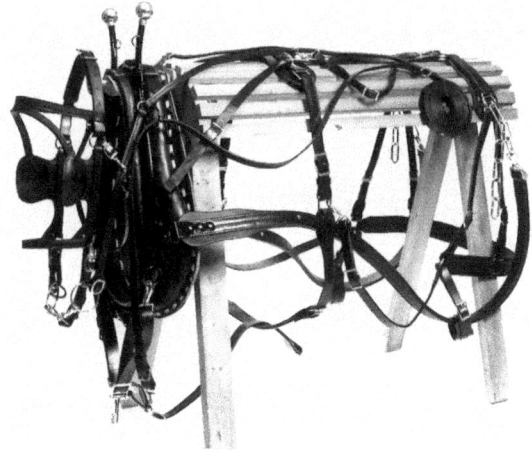

On the next page are two old catalog drawings that are unfortunate and perhaps confusing. If you are someone considering putting together a harness catalog for your shop I encourage you to consider the best, accurate line art for the parts, pieces and harness themselves. It just might save some disagreements and confusion with mail order customers.

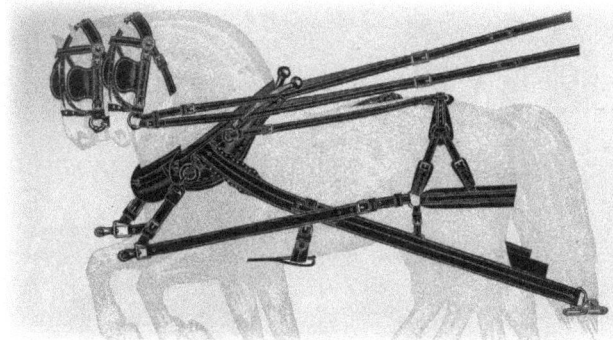

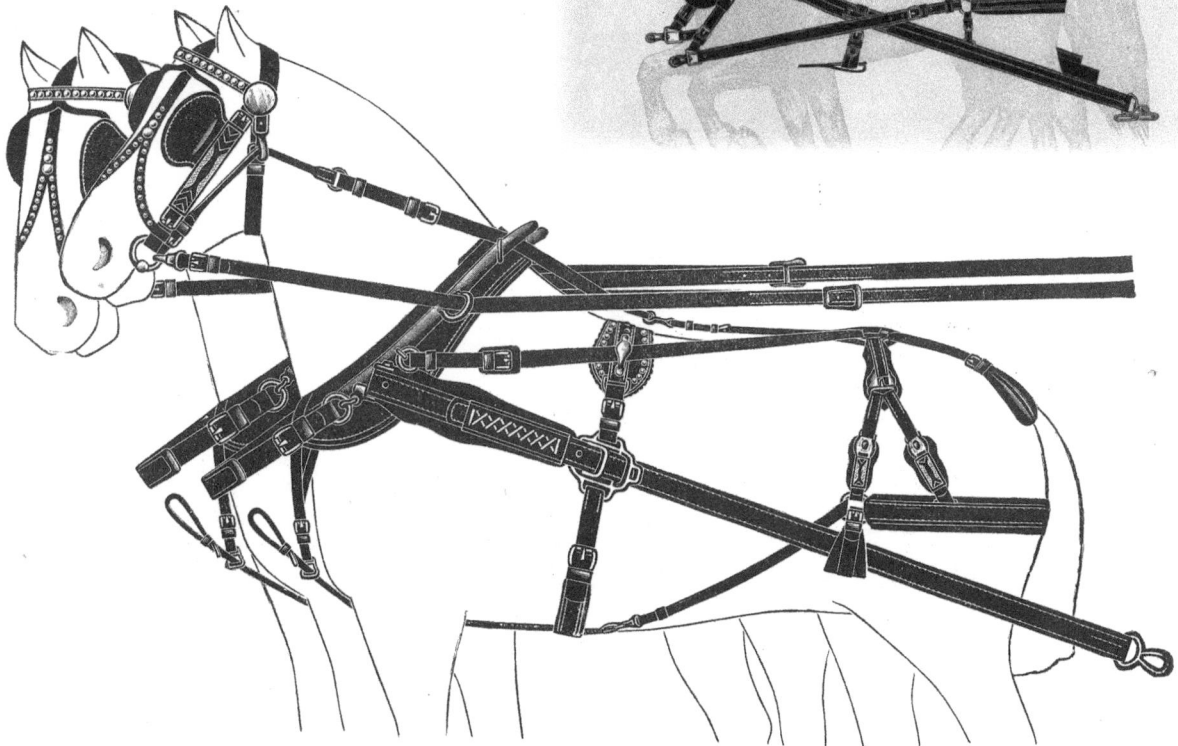

CUSTOM MADE FANCY SPOTTED WESTERN TEAM HARNESS

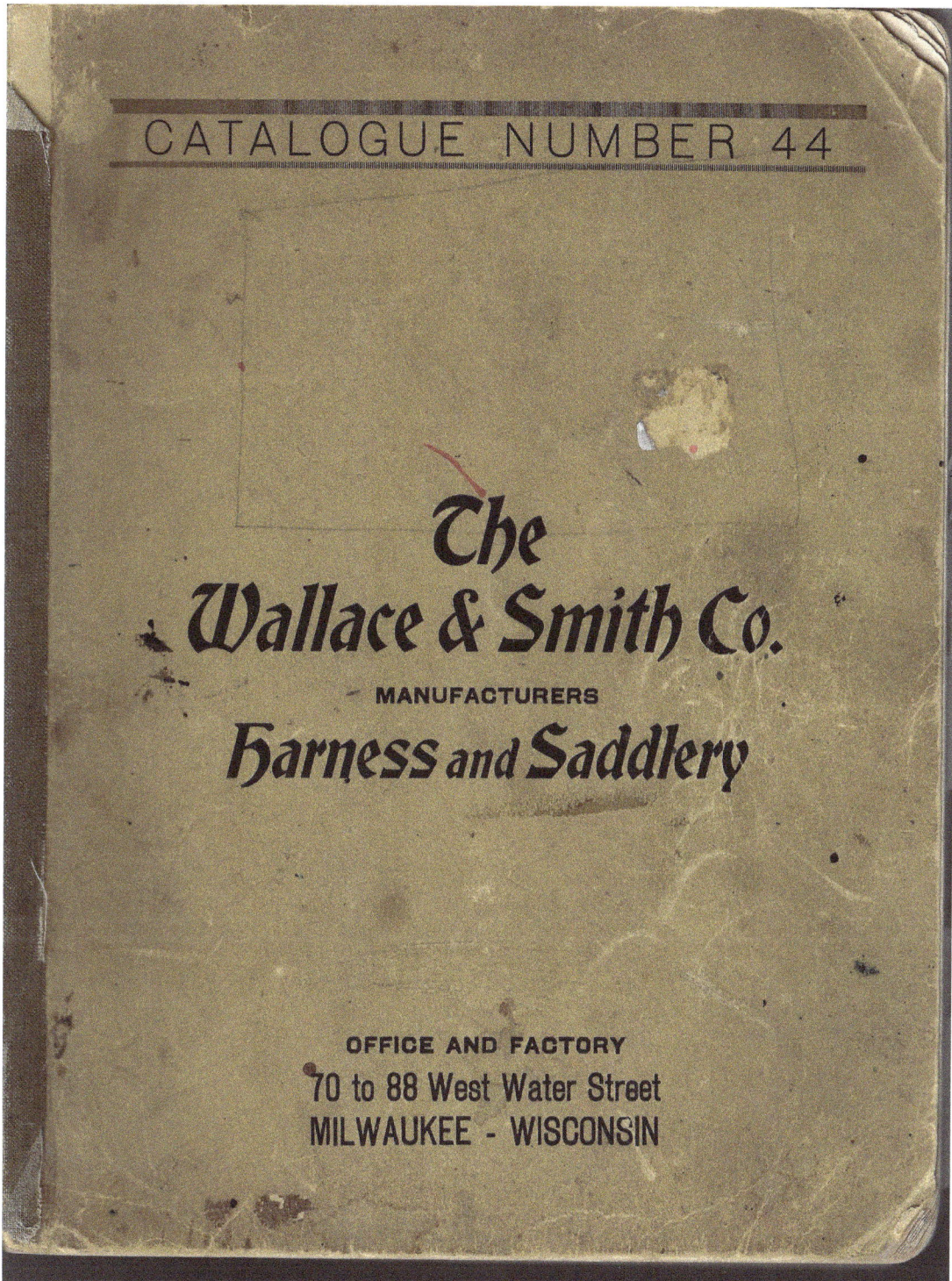

CATALOGUE NUMBER 44

The
Wallace & Smith Co.
MANUFACTURERS
Harness and Saddlery

OFFICE AND FACTORY
70 to 88 West Water Street
MILWAUKEE - WISCONSIN

It is difficult to determine how many harness supply companies were located in the U.S. from 1906 to 1914, but it was unimaginable that a city of any size didn't have at least one large outfit and likely more. Since 1970 when I began collecting horse-power literature, I have come across several dozen companies which had their own catalogs. The one company that made the biggest impression was Wallace & Smith. Within this book, there are illustrations from two different W&S catalogs. The 1906 volume was far and away the most beautiful, however, as information goes, this 1914 edition is nearly equal.

THE WALLACE & SMITH CO

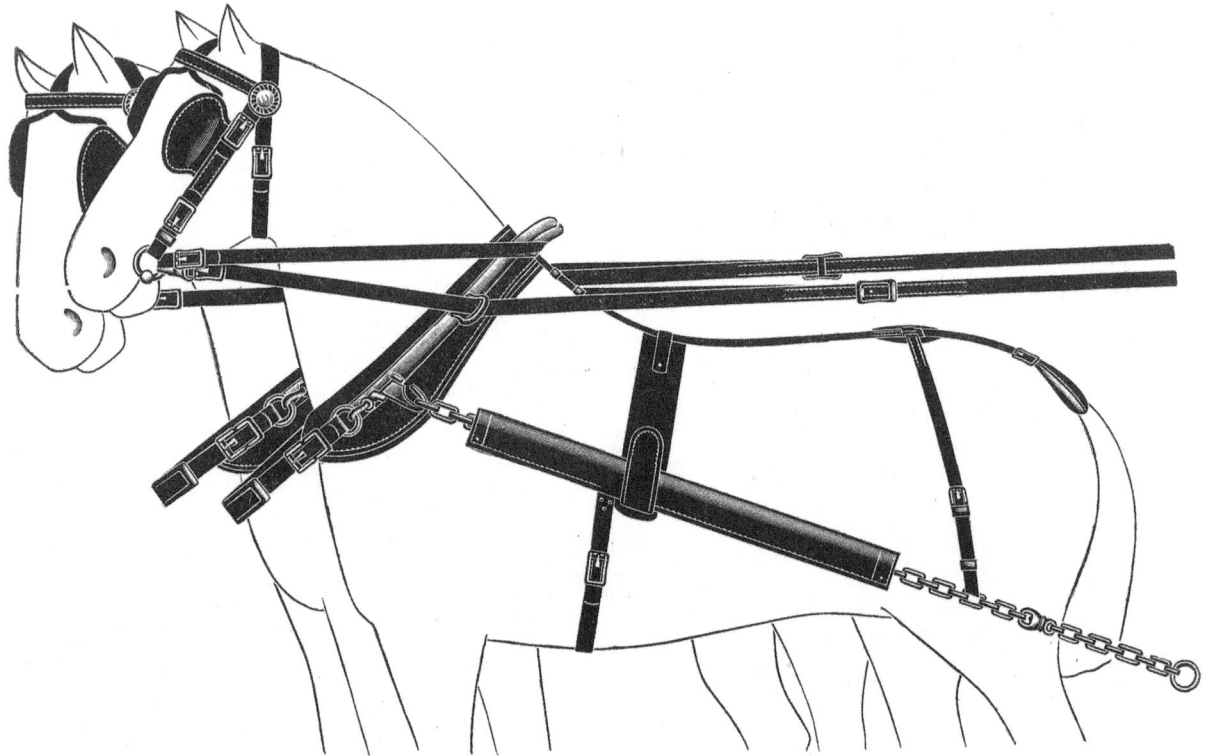

PIPE CHAIN HARNESS, No. 1290
Smooth Finish.

Bridles .... ¾ inch, cup shape harness leather blinds, flat side checks, round winker stays.

Lines .... 1 inch, 18 foot.

Hames .... No. 5 Concord, cart hook.

Trace Chains .... 7 foot, 8 link No. 2, polished.

Chain Pipes .... 36 inch.

Breast Straps .... 1½ inch, twin loop.

Back Bands .... Single strap, 3½ inch.

Belly Bands .... Single strap, 1¼ inch.

Back Straps .... 1¼ inch, running to hames, with folded crupper.

Hip Straps .... 1 inch, to reverse.

The pipe chain harness shown above and common in hotter climes, is so named because the chain tugs pass through leather tubes or 'pipes.' Though often used as such, this particular design of harness is not best suited for wagon use, or where a brichen is part of a backing and braking system. The breast strap, tied in line through the hames and collars to the back strap all the way back to the crupper, would have this team stopping a load with its tail.

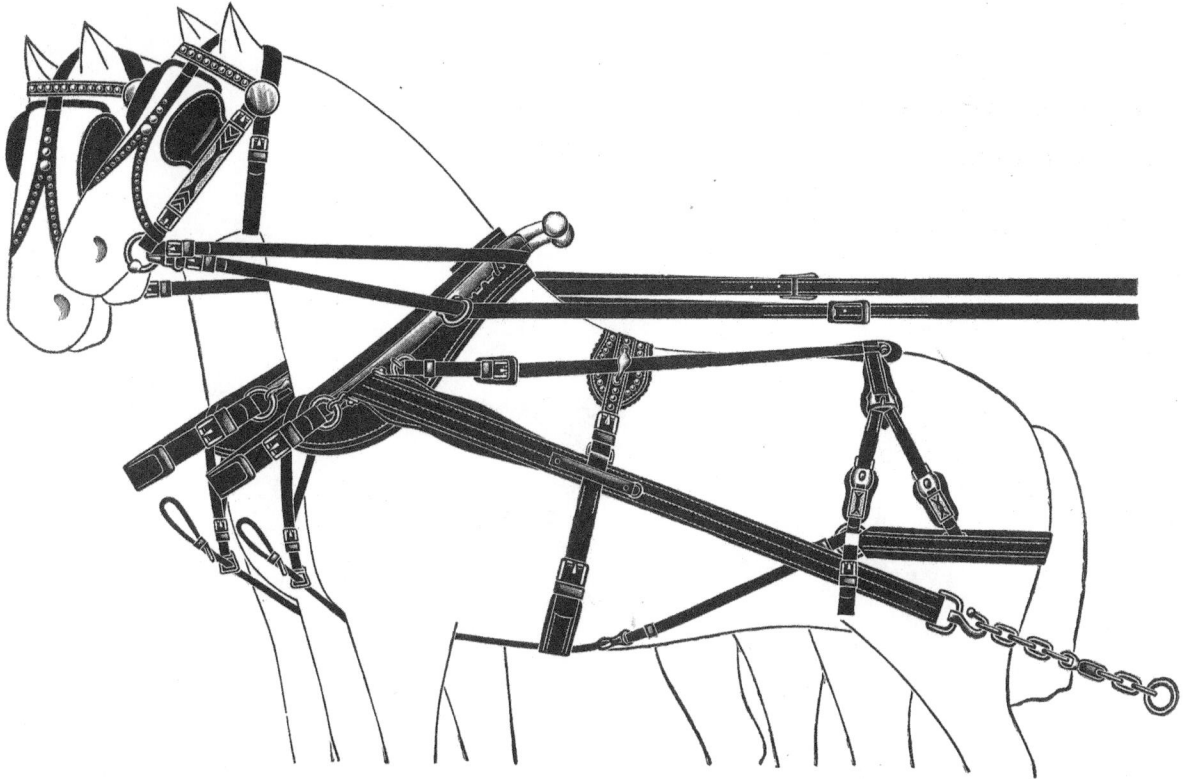

FARM CHAIN TEAM HARNESS, No. 1208

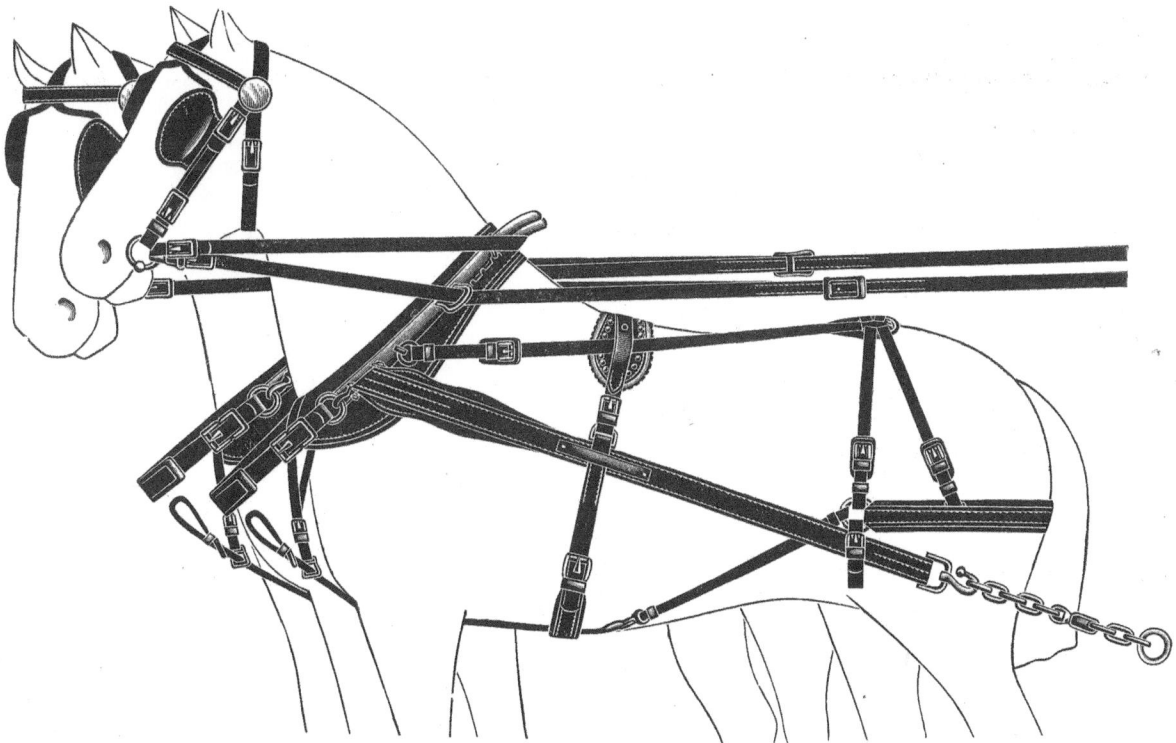

FARM CHAIN TEAM HARNESS, No. 1200

THE WALLACE & SMITH CO

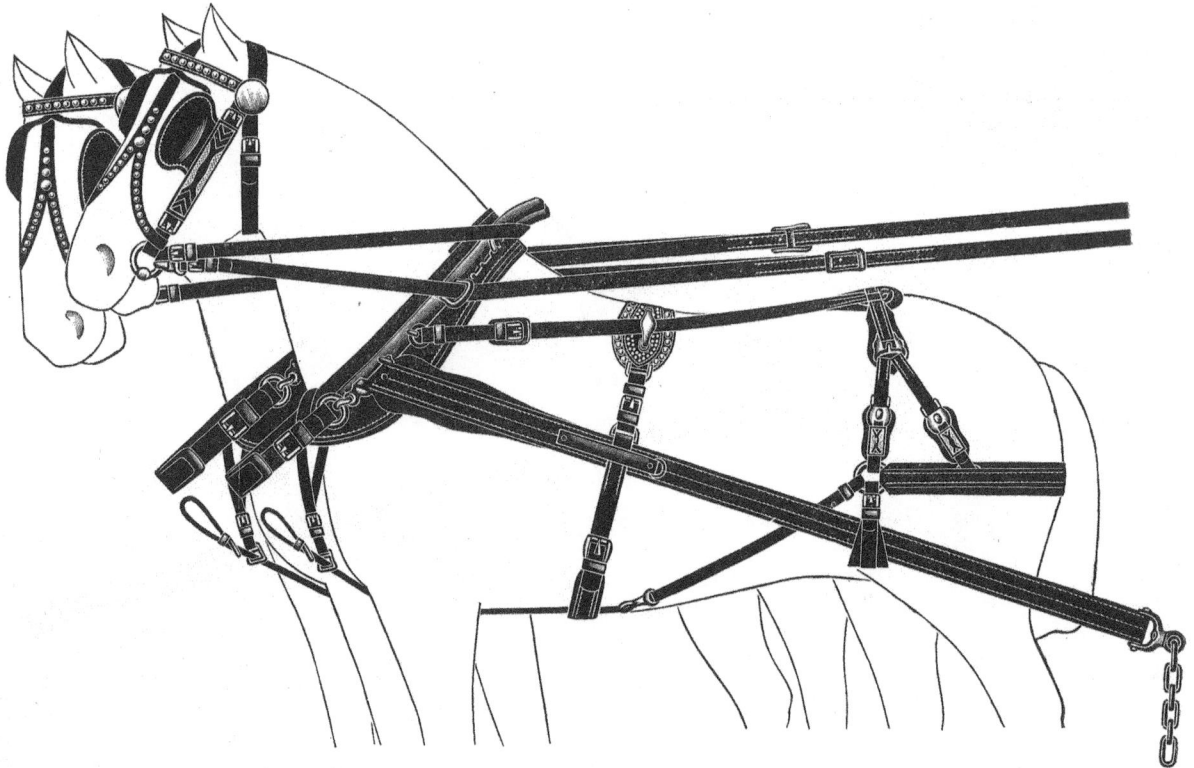

PINERY STYLE BREECHING HARNESS, No. 1179
Smooth Finish.

Bridles . . . . ⅞ inch, long box loop cheeks, spotted fronts, cup shape harness leather blinds, spotted face piece, flat side checks, round winker stays.

Lines . . . . 1⅛ inch, 20 foot.

Traces . . . . 2 inch, 3-row. doubled and stitched, with safes and heel chains, No. 10 oiled Concord bolt hames.

Martingales . . . . 2 inch, with safe and collar straps.

Breast Straps . . . . 2 inch, 5 foot.

Pads . . . . No. 153, swell shaped harness leather, leather lined, spotted, with Colorado bridges and 1½ inch market straps.

Belly Bands . . . . Folded, with layer.

Rib Straps . . . . 1⅛ inch, to hames.

Hip Straps . . . . 1⅛ inch, 3-ring, with buckle shields.

Breechings . . . . Folded, with layer, 1¼ inch side straps sewed in and 1 inch lazy straps, with safes.

It should be obvious, from all the parts, pieces and harness variables you see in this book, the intelligent teamster can customize his or her harness for the work and the individual animals. Rather than purchasing a new harness, and taking what comes 'out of the box' or 'off the rack' so to speak, think about making specific requests of your harness maker for what YOU want and need. Perhaps a wider belly band or narrower lines or pigeon-wing bridles or no crupper. Perhaps buckles on line cross-checks rather than Conways.

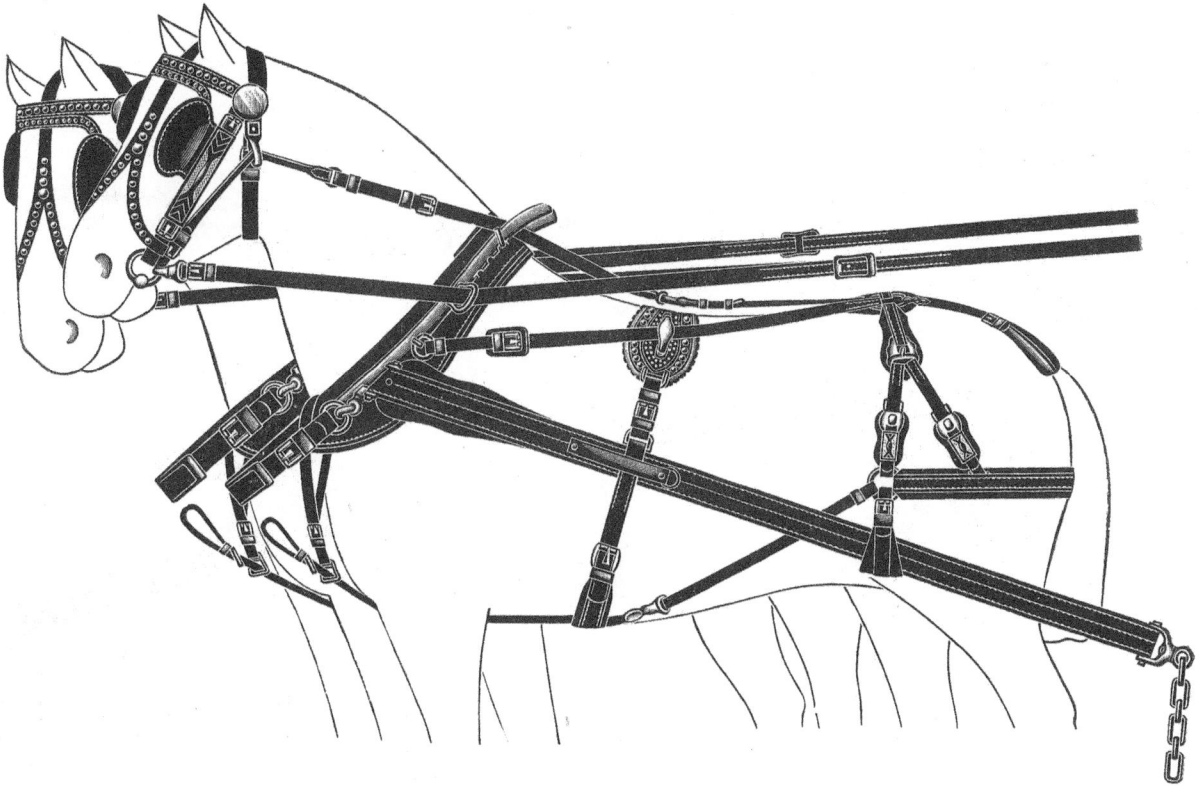

PINERY STYLE BREECHING HARNESS, No. 1166

Above, the check reins run through bridle-mounted swivel loops and back to spider then on to crupper around the tail. All this is meant to hold the horses' heads up. While below the check rein has a straight path back to loop around the top of the hame.

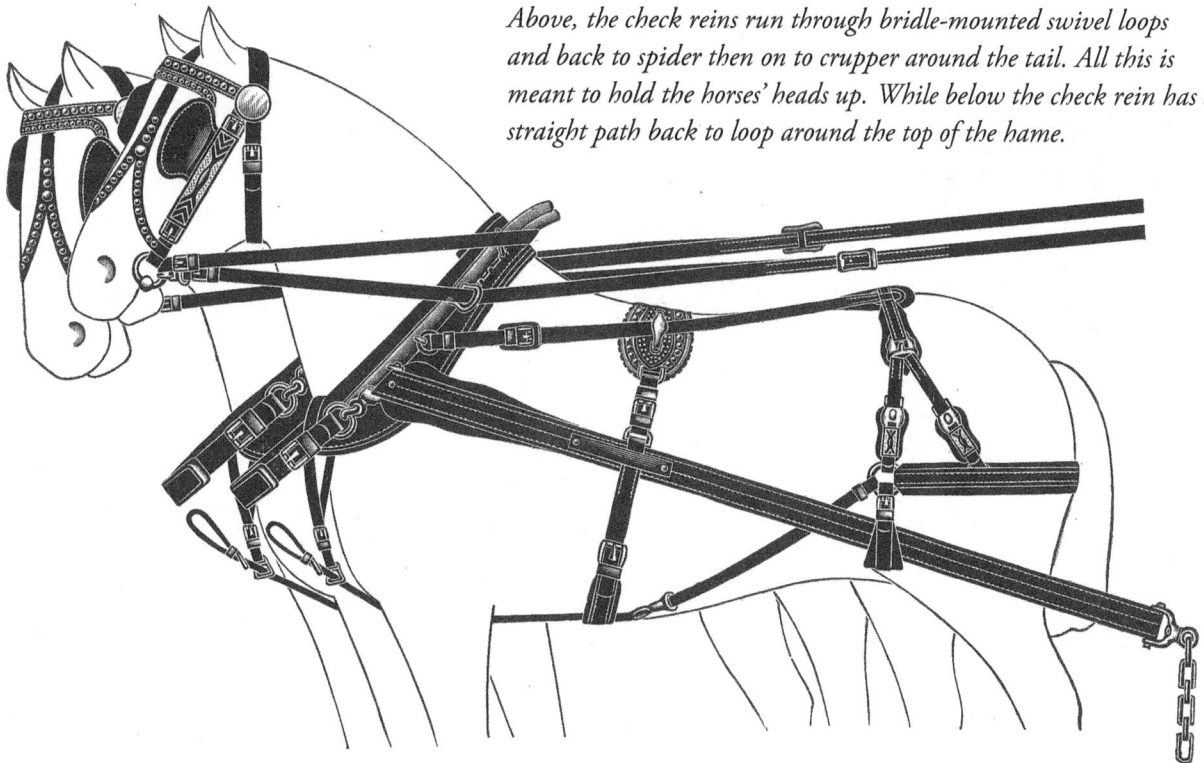

PINERY STYLE BREECHING HARNESS, No. 1162

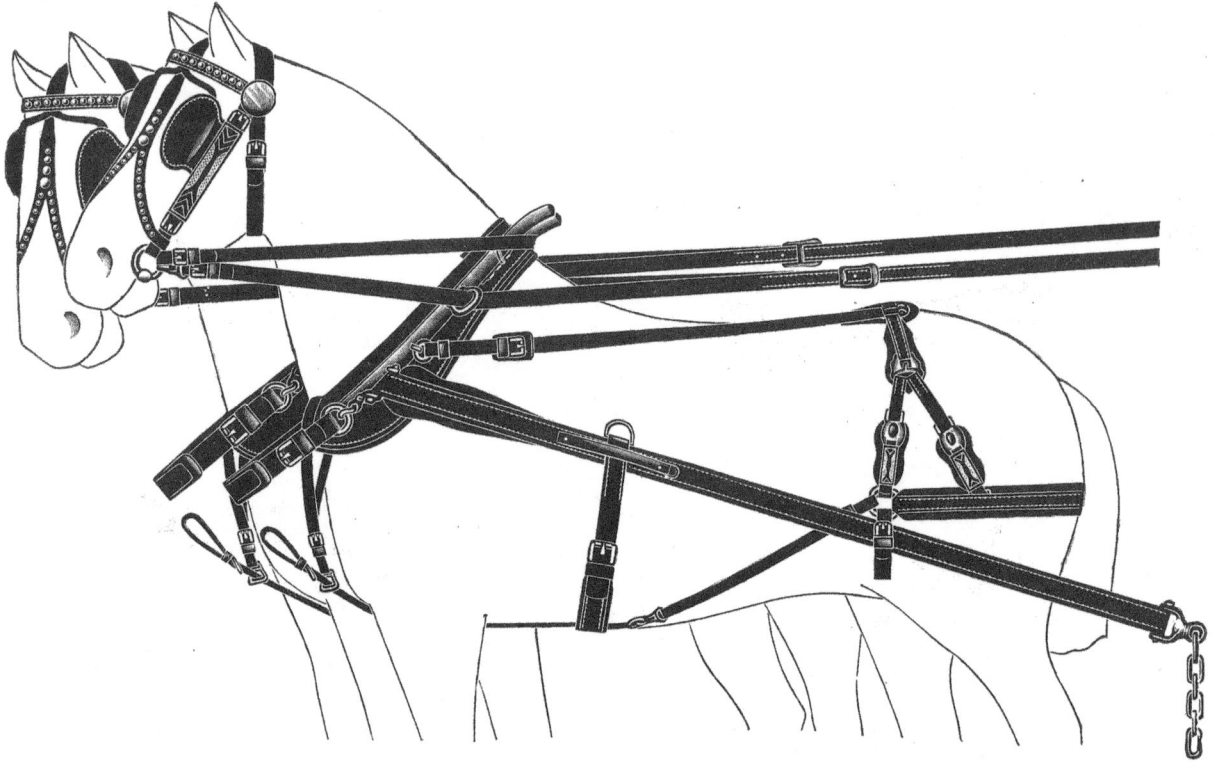

PINERY STYLE BREECHING HARNESS, No. 1130

Above, the absence of saddle or back pad frees the horses of weight and pressure on their backs. Below, this 'plow' style harness has no carrier or restraint on the rear portion of the traces. From the belly band billets back, the angle of draft is free to be whatever the load requires. This can be problematic in situations where there is a lot of hooking and unhooking, such as logging, because it would be easy for the animals to step sideways over a tug or two.

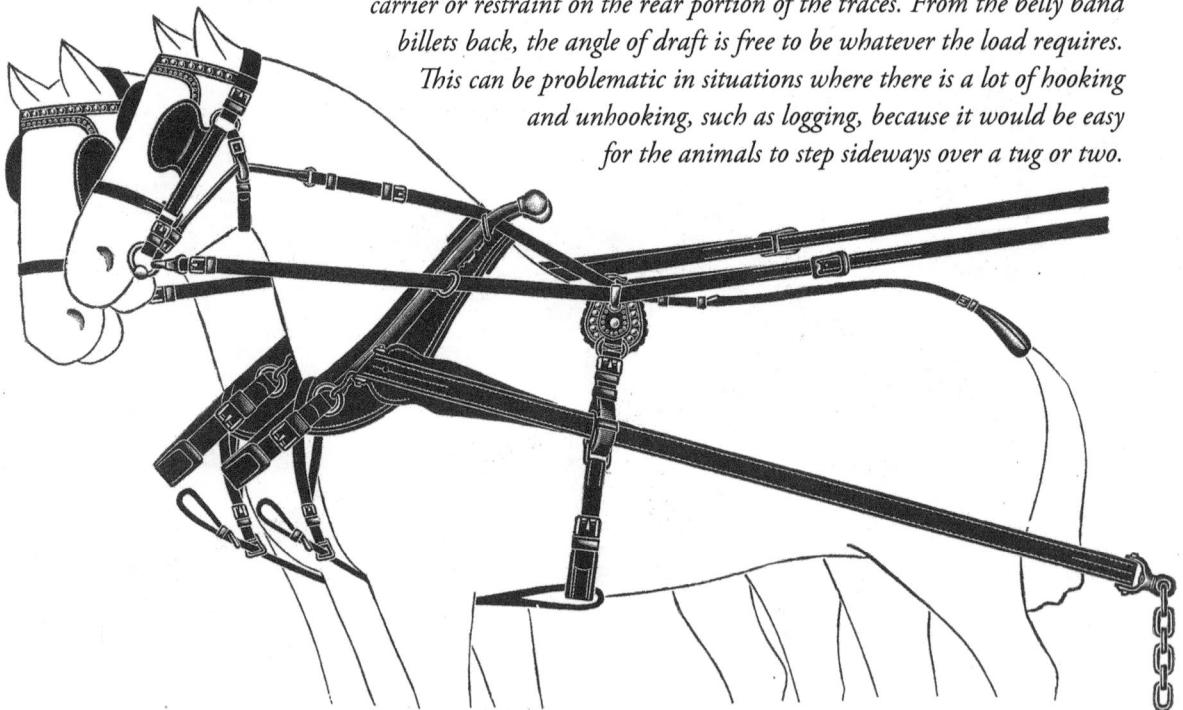

PINERY FARM TEAM HARNESS, No. 1114

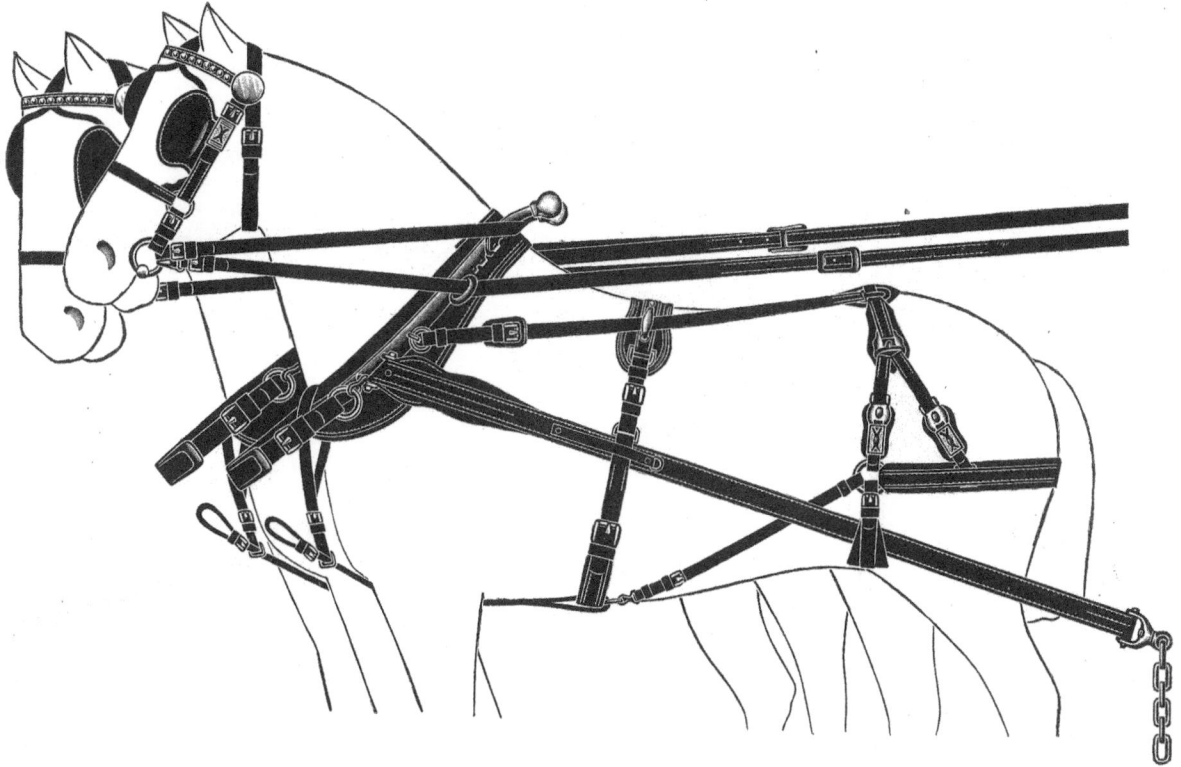

"GLADIATOR" PINERY STYLE BREECHING HARNESS

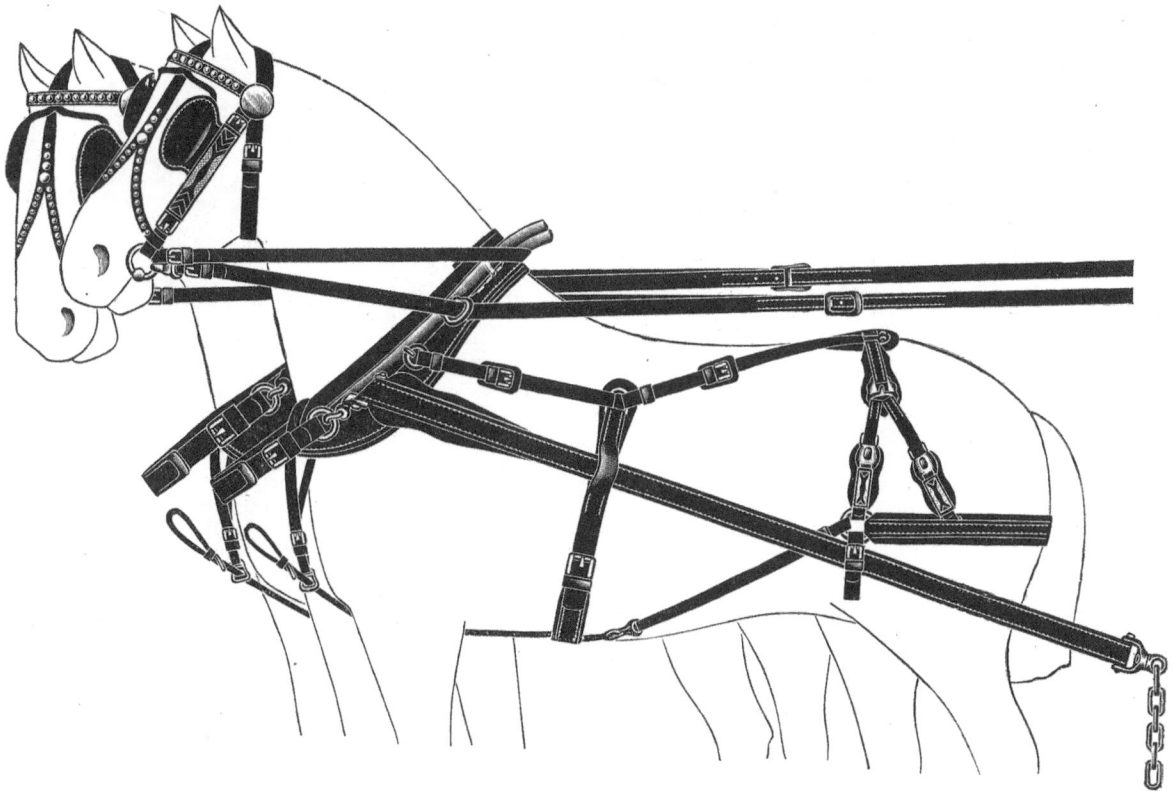

PINERY STYLE BREECHING HARNESS, No. 1135

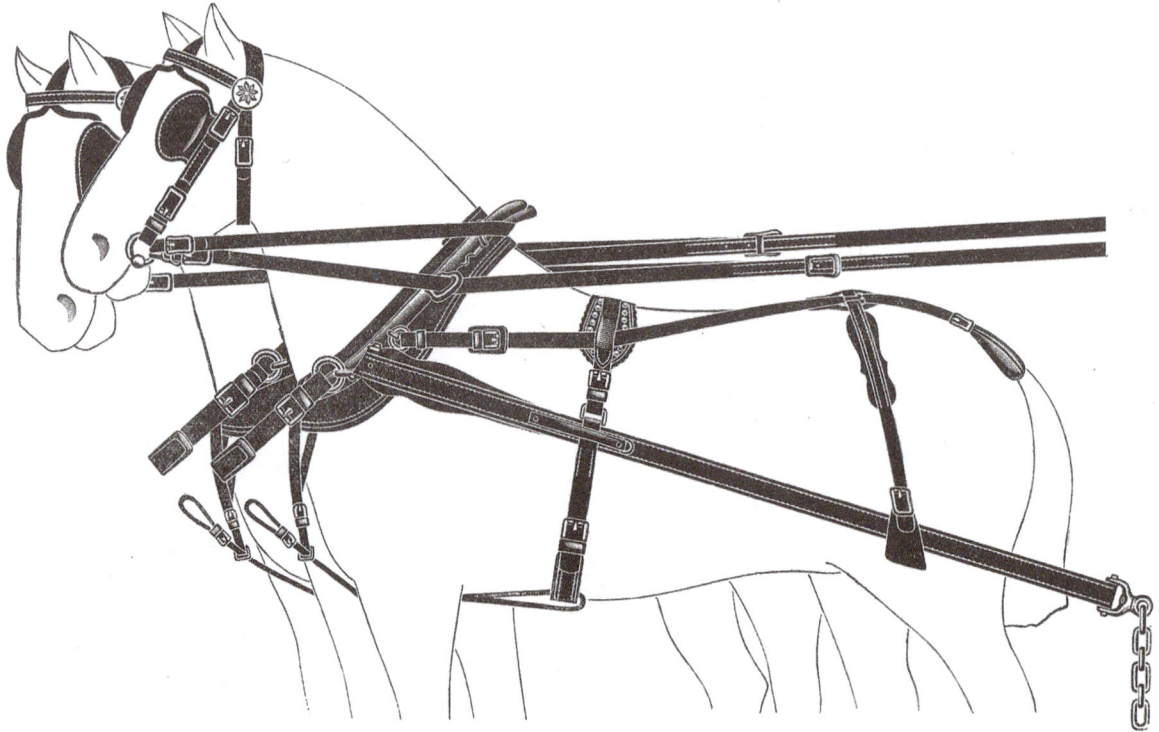

PINERY TEAM PLOW HARNESS, No. 1111

In this scene, captured by Jerry Hunter at 2022 Horse Progress Days, we see a new biothane harness.
Substantially lighter than leather, it cleans easily and, if properly made, withstands hard work. It
most often comes down to personal preference. I like a well-built harness of the best leathers.

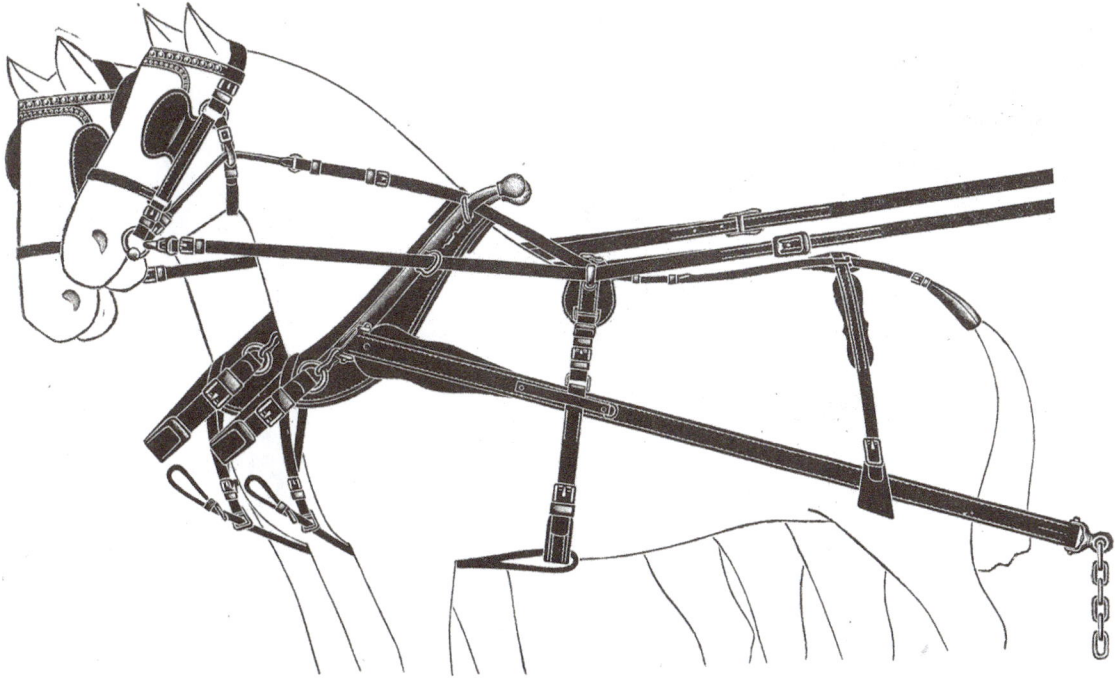

PINERY FARM TEAM HARNESS, No. 1115

Jonathon's team in a biothane side-backer harness.

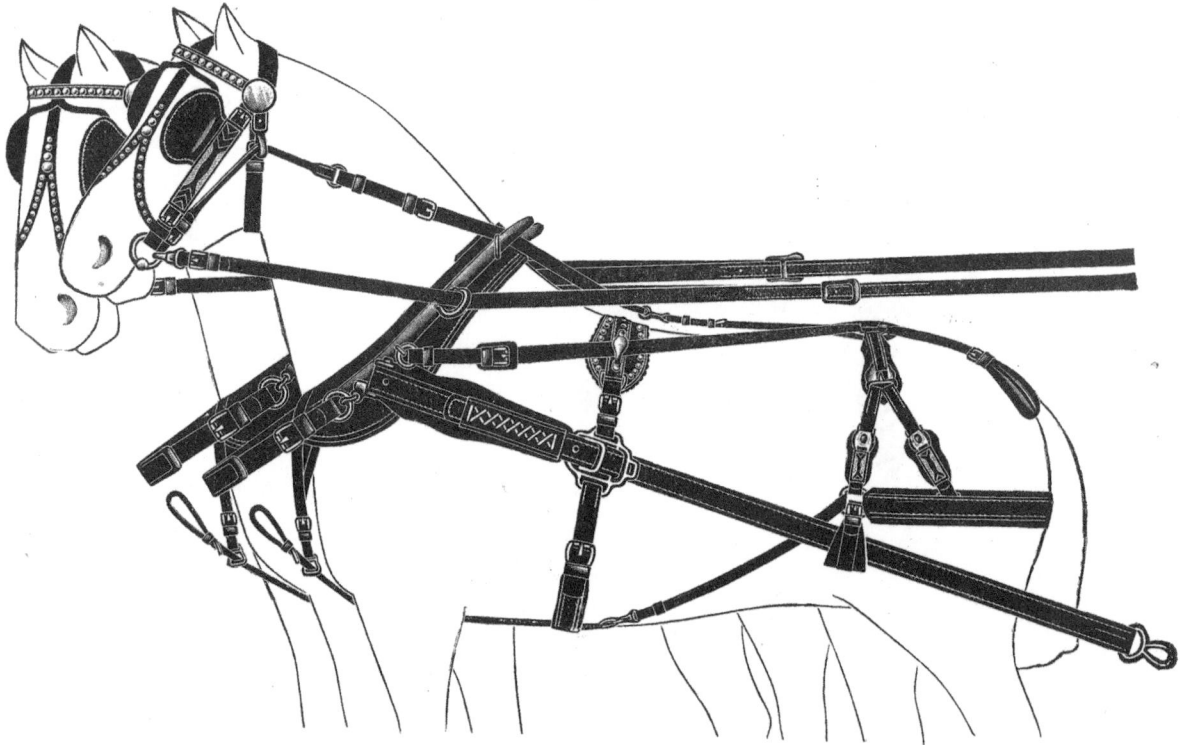

FARM BREECHING HARNESS, No. 1042
Smooth Finish.

Bridles . . . . ⅞ inch, long box loop cheek, cup shape harness leather blinds, spotted fronts, spotted face piece, round side check and round winker stays.

Lines . . . . 1 inch, 20 foot.

Hame Tugs . . . . 1½ inch, hard, doubled and stitched, scalloped safes, laced box loops, No. 5 oiled Concord bolt hames.

Traces . . . . 1½ inch, doubled and stitched, triangular cockeyes.

Martingales . . . . 1½ inch, with collar straps.

Breast Straps . . . . 1½ inch.

Pads . . . . No. 188, swell harness leather, felt lined, fancy spotted, with Colorado bridges and 1½ inch market strap and layer.

Belly Bands . . . . Folded, with layer.

Hip Straps . . . . 1 inch, double, 3-ring, with trace carriers and buckle shields, crupper forks, cruppers to buckle and check up straps.

Rib Straps . . . . 1⅛ inch to hames.

Breeching . . . . Folded, with layer.

Side Straps . . . . 1 inch.

Note that the harness above features 'hame tugs,' a short portion of the overall trace which stays connected to the hames (and at the belly band billet) and features hardware that allows the trace portion may be disconnected. The end of the trace has Cockeye hardware. This style of work harness is no longer commonly available.

T H E   W A L L A C E   &   S M I T H   C O

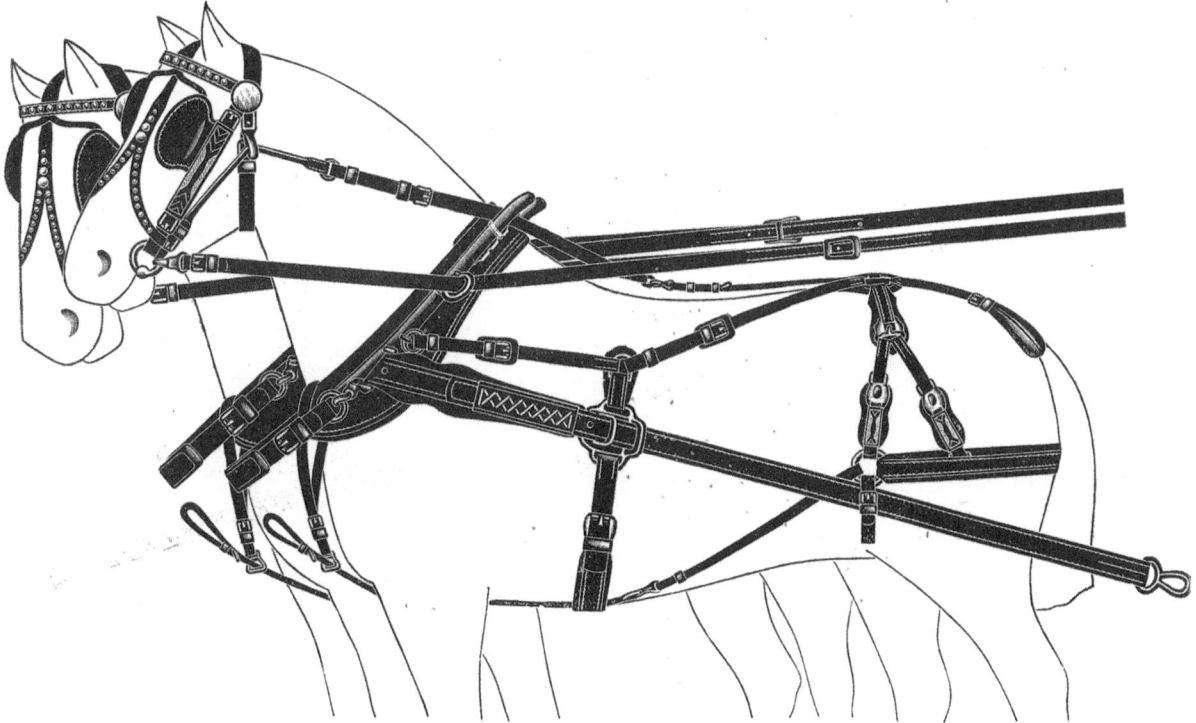

FARM BREECHING HARNESS, No. 1038
Smooth Finish.

Bridles . . . . ¾ inch, long box loop cheek, spotted face piece, harness leather cup shape blinds, round side check and round winker stays, spotted fronts.

Lines . . . . 1 inch, 20 foot.

Hame Tugs . . . . 1½ inch, hard, doubled and stitched, scalloped safes, hand laced box loops, No. 5 oiled Concord bolt hames.

Traces . . . . 1½ inch, 6 foot, doubled and stitched, with triangular cockeyes.

Martingales . . . . 1½ inch, with collar straps.

Breast Straps . . . . 1½ inch.

Belly Bands . . . . Folded, with layer.

Hip Straps . . . . 1 inch, double, 3-ring, spotted with trace carriers and buckle shields, cruppers to buckle, with crupper forks and check up straps.

Rib Straps . . . . 1 inch, extending to lead-up on trace buckle, with quarter straps to hames.

Breeching . . . . Folded, with layer.

Side Straps . . . . 1 inch.

These W&S harness illustrations are outstanding in their correctness and clarity, but I find the earlier 1904 red-colored illustrations the pinnacle. Notice that the horses above were traced from those old original colored plates you see on coming pages.

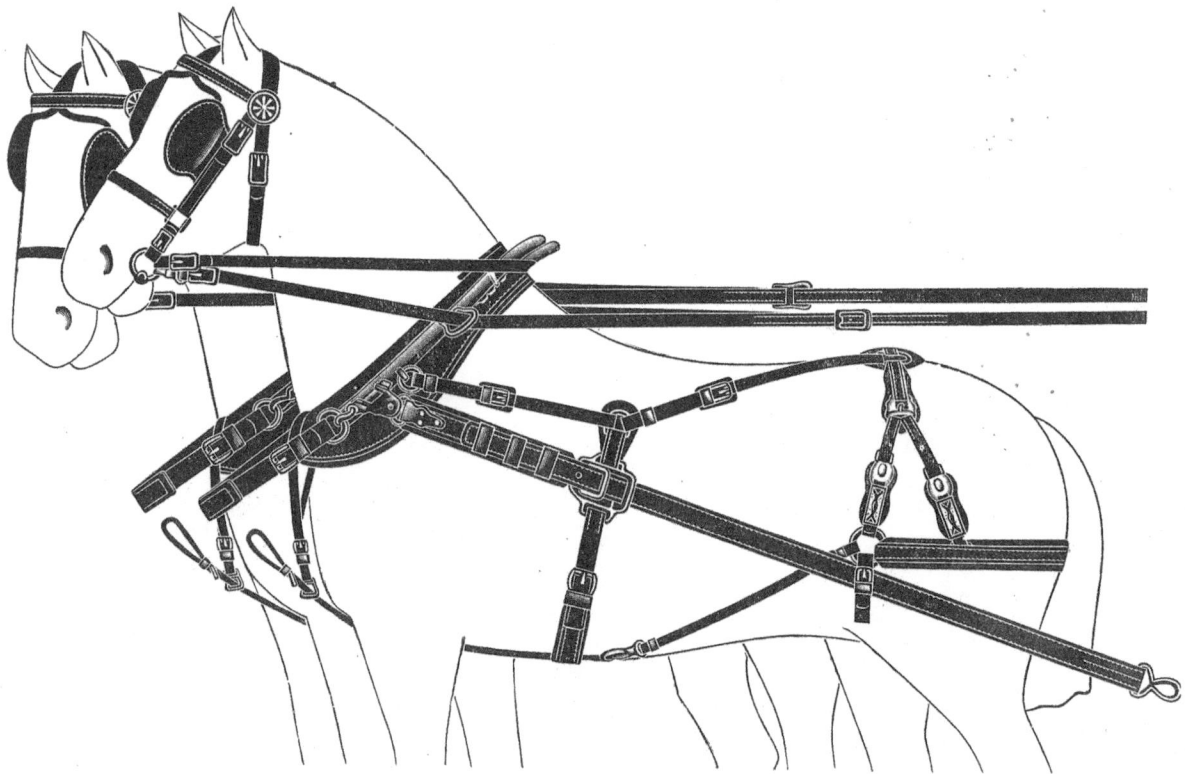

FARM BREECHING HARNESS, No. 1009
Smooth Finish.

Bridles . . . . ⅞ inch, short cheek cup shape harness leather blinds, flat nose band, flat side checks, round winker stays.

Lines . . . . 1 inch, 18 foot.

Hame Tugs . . . . 1½ inch, folded, with jointed Concord clips and No. 5 oiled Concord bolt hames.

Traces . . . . 1½ inch, doubled and stitched, with triangular cockeyes.

INTERESTING SIDE NOTES:

1. Nowhere in the extensive Wallace & Smith Harness catalog do they exhibit or offer any New England, Yankee, Boston, New York sidebacker or D-Ring style harness. Regional preferences amount to a great deal. The heat of the south, the rugged mountains of the west, the tight city streets of Boston, New York, etc., these had great bearing on harness design evolutions.

2. In 1915 there were more than 25 million horses and mules in the U.S. How many more in Canada and Mexico? We guess approximately one million harnesses were sold that year. We have not been able to determine how many mid to large harness making concerns existed in 1915 but we do know that major cities had at least two.

3. While quite popular in the mountain west, we found few examples of the Yankee Brichen style, which features the reinforced brichen riding above the tail head of the animal, with crupper passing through a loop. Some are forgiven for mistaking a plow harness' hip straps for the Yankee Brichen.

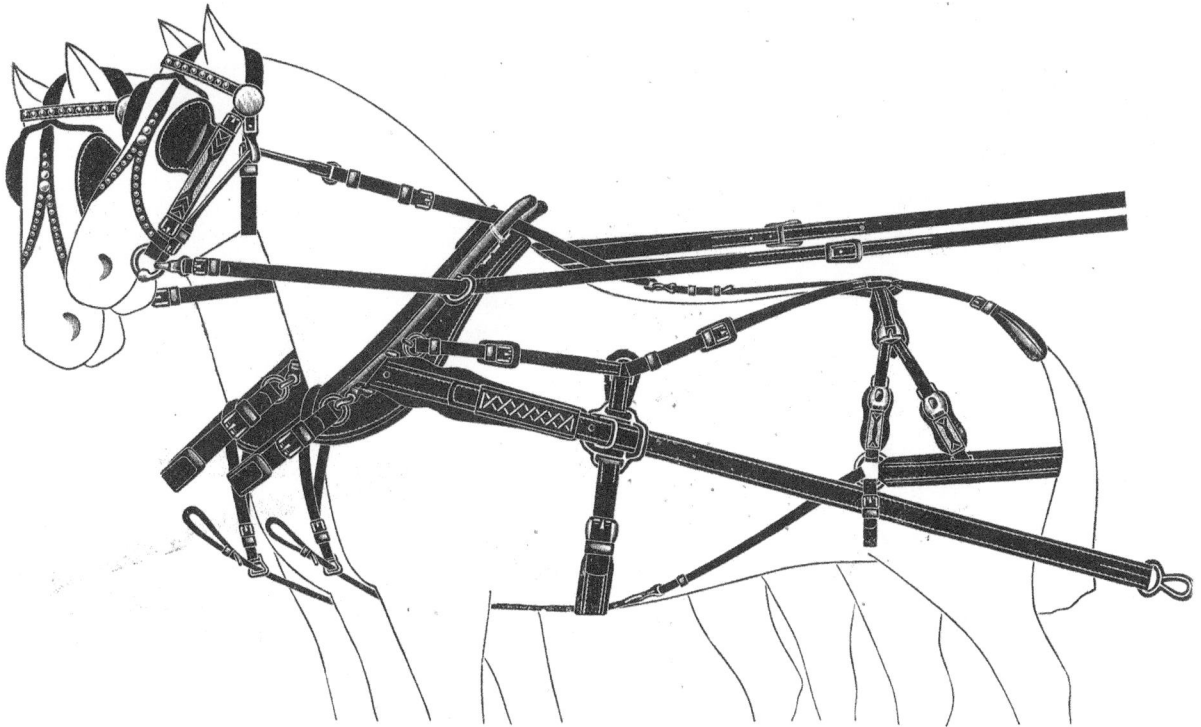

FARM BREECHING HARNESS, No. 1038
Smooth Finish.

Bridles . . . . ¾ inch, long box loop cheek, spotted face piece, harness leather cup shape blinds, round side check and round winker stays, spotted fronts.

Lines . . . . 1 inch, 20 foot.

Hame Tugs . . . . 1½ inch, hard, doubled and stitched, scalloped safes, hand laced box loops, No. 5 oiled Concord bolt hames.

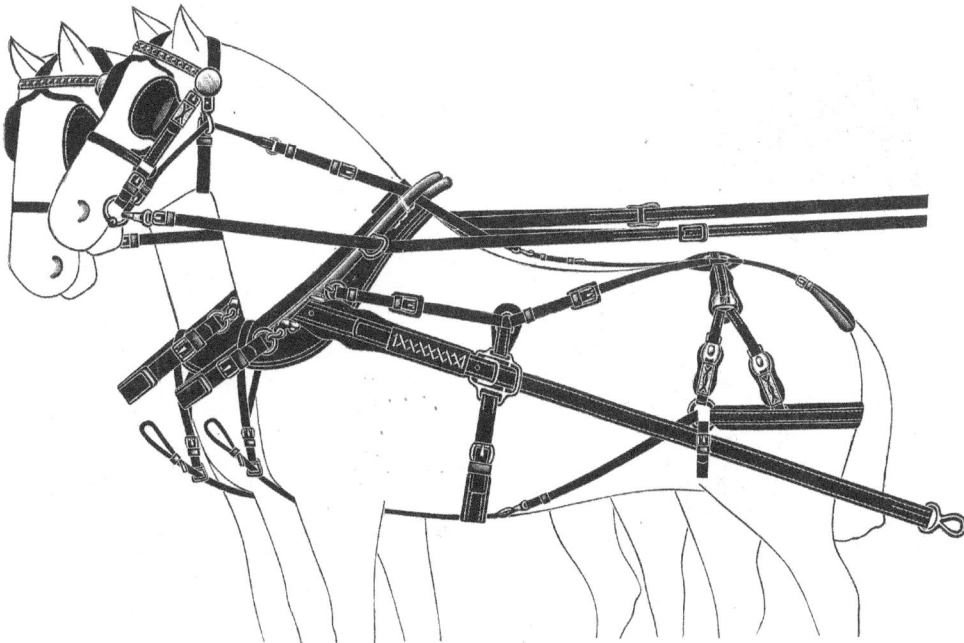

FARM BREECHING HARNESS, No. 1032

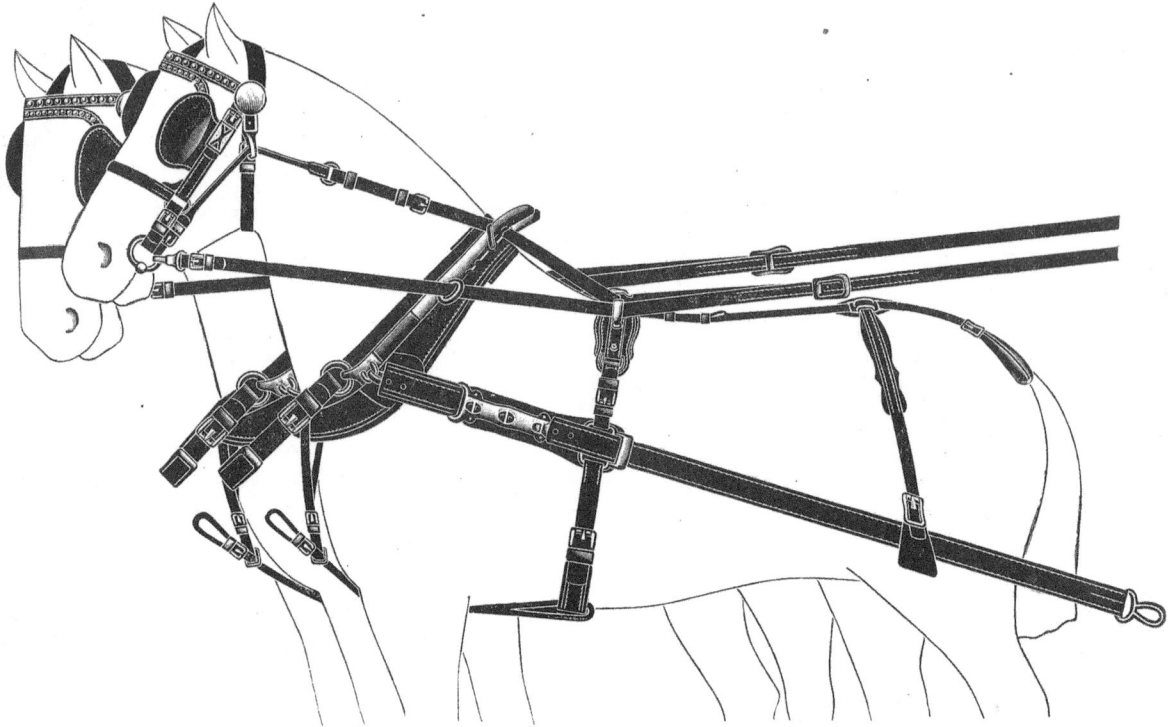

TEAM FARM HARNESS, No. 982
Smooth Finish.

Bridles . . . . ¾ inch, short cheek, harness leather cup shape blinds, round side check, spotted combination fronts and winker stays, flat nose bands.

Lines . . . . 1 inch, 20 foot.

Hame Tugs . . . . 1½ inch, with Moeller hame tug plates, No. 450 or 460 clip hames.

Traces . . . . 1½ inch, doubled and stitched, with triangular cockeyes.

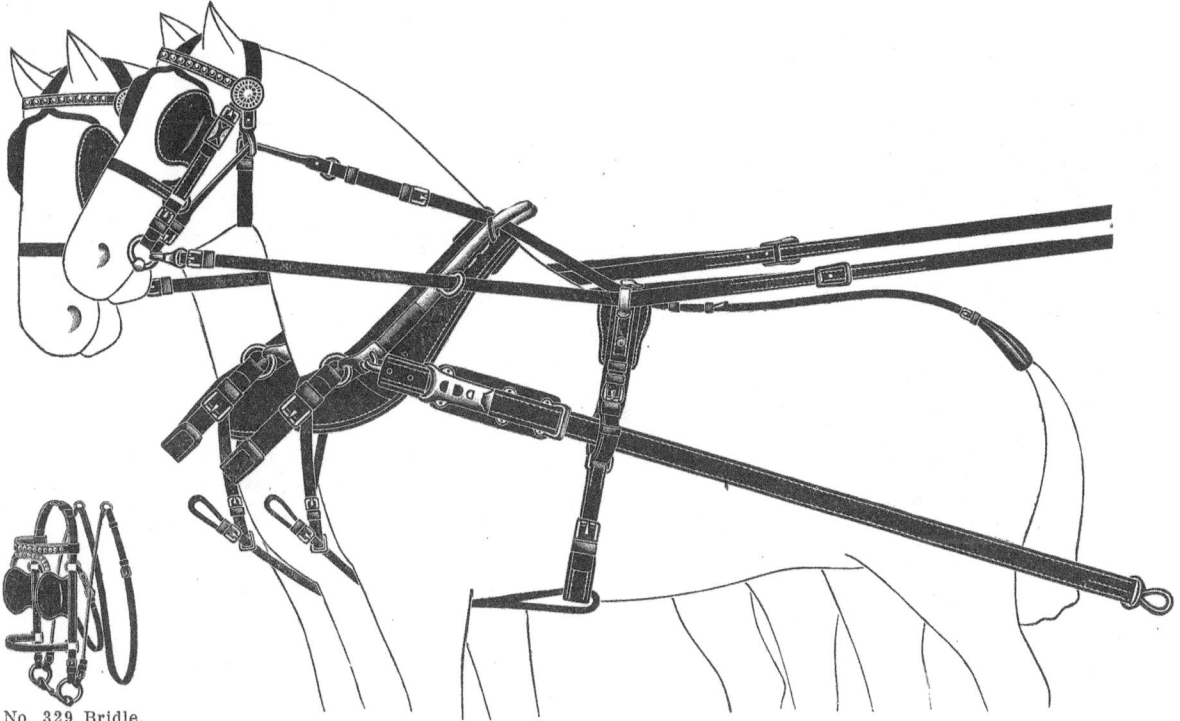

No. 329 Bridle.

TEAM FARM SLIP TUG HARNESS, No. 969
Smooth Finish.

Bridles .... ⅞ inch, short cheek, harness leather cup shape blinds, round side checks, round winker stays, flat nose bands, spotted fronts.

Lines .... 1 inch, 20 foot.

Hame Tugs .... 1½ inch, Moeller hame tugs, with No. 450 or 460 clip hames.

Traces .... 1½ inch, 6 foot, doubled and stitched, triangular cockeyes.

T H E   W A L L A C E   &   S M I T H   C O

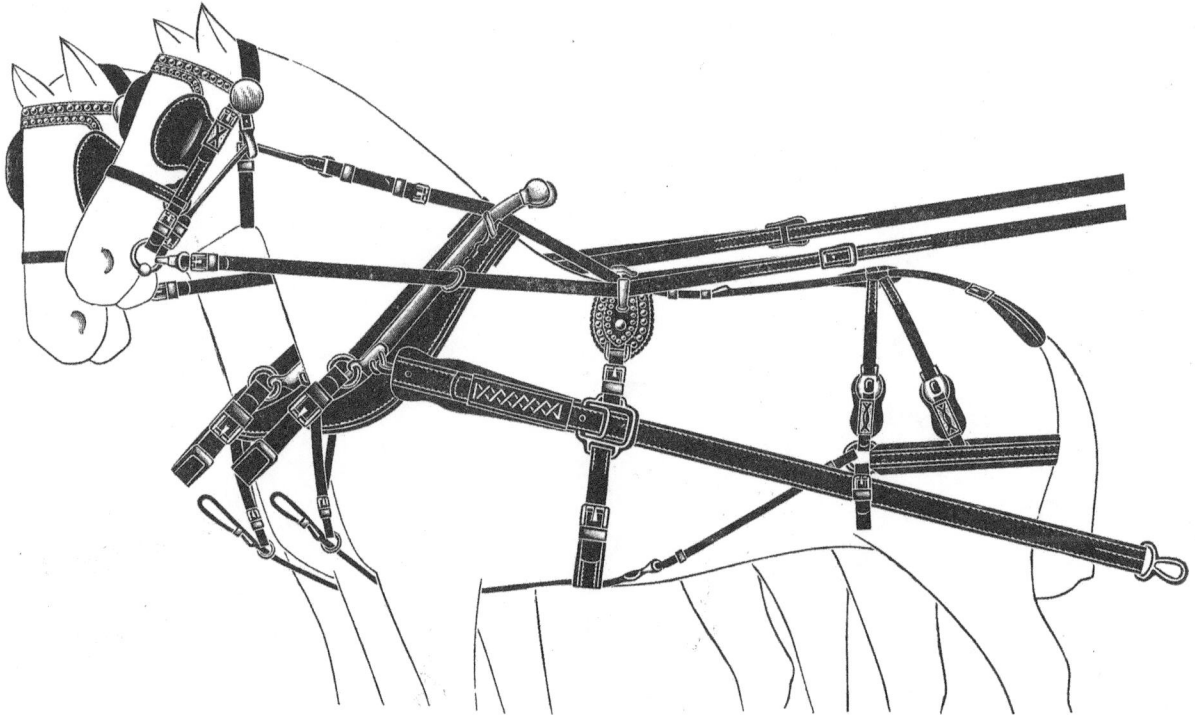

FARM BREECHING HARNESS, No. 0964

Smooth Finish.

Bridles . . . . ⅞ inch, short cheek cup shape harness leather blinds, flat nose bands, spotted combination front and winker stays, round side checks.

Lines . . . . 1⅛ inch, 20 foot.

Hame Tugs . . . . 1⅛ inch, doubled and stitched, scalloped safe hand laced box loops, No. 500 nickel or brass ball top clip hames.

Traces . . . . 1½ inch, doubled and stitched, triangular cockeyes.

Martingales . . . . 1½ inch, with collar straps.

Breast Straps . . . . 1½ inch.

Pads . . . . No. 182, swell shape, harness leather, felt lined, German silver spotted, Dwight hook and terrets, with market straps.

Belly Bands . . . . Folded, with layer.

Back Straps . . . . 1⅛ inch, with folded cruppers buckled on.

Hip Straps . . . . 1⅛ inch, double.

Breeching . . . . Folded, with layer.

Side Straps . . . . 1 inch.

T H E   W A L L A C E   &   S M I T H   C O

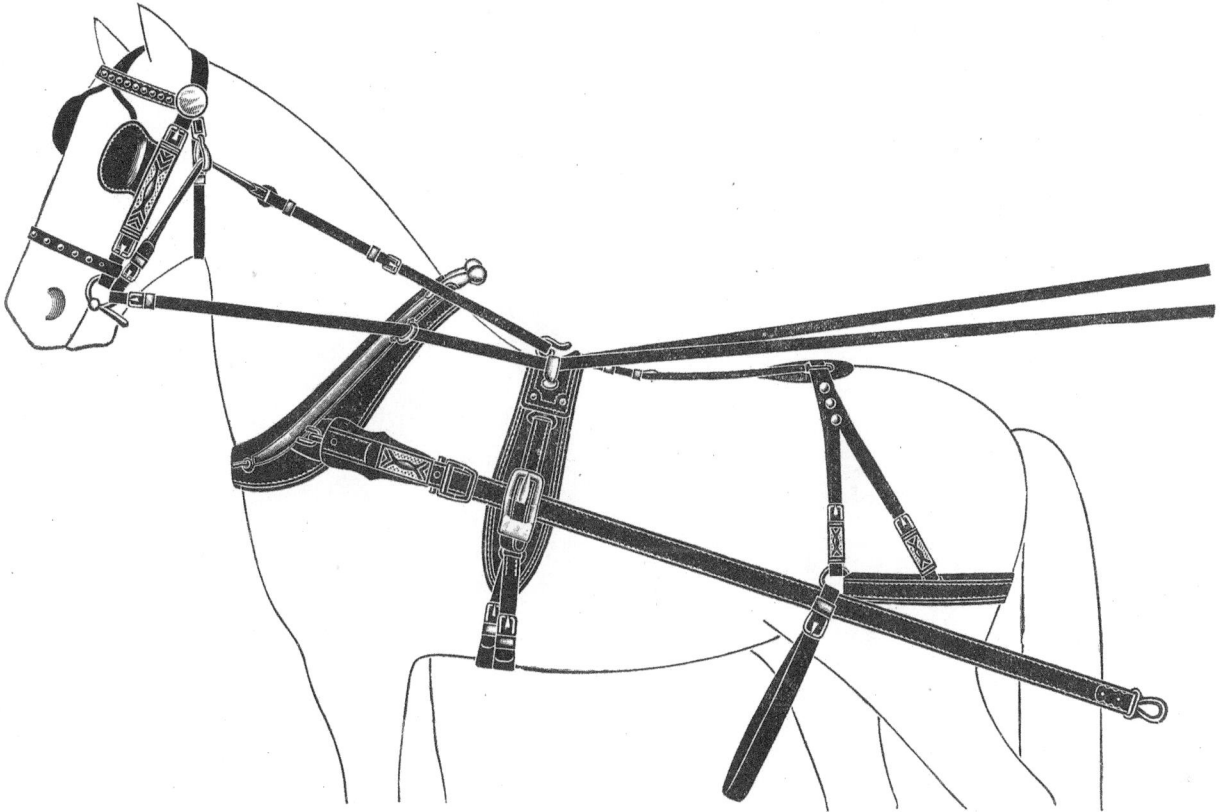

EXPRESS HARNESS, No. 540
Smooth Finish.

Bridle . . . . ¾ inch, box loops cup shaped blinds, round side check, spotted front and nose band, round winker stay.

Lines . . . . 1 inch throughout, with buckles and billets.

Hame Tugs . . . . 1½ inch, box loops, with safe, No. 5 Red Concord hame with Dandy balls.

Traces . . . . 1½ inch, doubled and stitched, clip cockeyes.

Saddle . . . . No. 415, 5 inch, full pad, Kersey lined, express.

Shaft Tugs . . . . 1½ inch, with dee and billets and brass oval express shaft tug buckles.

Belly Bands . . . . One folded and one flat.

Turnback . . . . 1 inch, with ¾ inch split hip straps, spotted.

Breeching . . . . Folded, with layer and box loop tugs.

Side Straps . . . . 1 inch.

Tie Straps . . . . ⅝ inch.

Here's the tattered cover of my copy of that 1906 catalog previously mentioned. It is very thick and includes several hundred illustrations many of which I have borrowed for this volume.

You will see, with the harness style variations, minuscule differences in the offering. So prevalent was horse power during the golden age of agriculture (1910-1913) that competition between harness companies demanded that every conceivable teamster interest and need be satisfied.

WALLACE, SMITH & CO., MILWAUKEE, WIS.

SINGLE SURREY HARNESS No. 123.

IMITATION HAND-SEWED.

Bridle—⅝-inch, box loops, 3-buckle overcheck with nose band, round winker stay.

Lines—1-inch, buckle and billet.

Hame Tugs—1⅛-inch, with safe, 3½-pound iron hames.

Traces — 1⅛ - inch, doubled and stitched, solid raised.

Saddle—No. 87, 3-inch patent leather jockey.

Shaft Tugs—1-inch, with box loops and Dee billets.

Belly Band—Folded.

Breeching—Folded waved layer.

Side Straps—⅞-inch.

Hip Strap—⅝-inch, split.

Tie Strap—⅝-inch.

Full Nickel French Wire Trimmed.
Full Brass French Wire Trimmed.
The above Harness with No. 440, 3½-inch Surrey Saddle and 1¼-inch Hame Tugs and Traces.

SINGLE SURREY HARNESS No. 124.

IMITATION HAND-SEWED.

Bridle—⅝-inch, box loops, nose band, round side check.

Lines—1-inch throughout, buckle and billet.

Hame Tugs—1¼-inch, with safe, 3½-pound iron hames, fancy draft eye.

Traces — 1¼ - inch, doubled and stitched.

Martingale—⅞-inch.

Saddle—No. 447, 3½-inch surrey.

Shaft Tugs—1⅛-inch Dee billets with box loops.

Belly Bands—One single strap, one doubled and stitched.

Turnback—⅞-inch, scalloped, with round crupper to buckle.

Breeching — Folded with waved layer.

Side Straps—⅞-inch.

Hip Strap—⅝-inch, split.

No Tie Strap.

Full Nickel French Wire Trimmed.
Full Brass French Wire Trimmed.

Lynn and Justin Miller on the hay wagon, Polly and Anna in the hitch, headed to the barn.

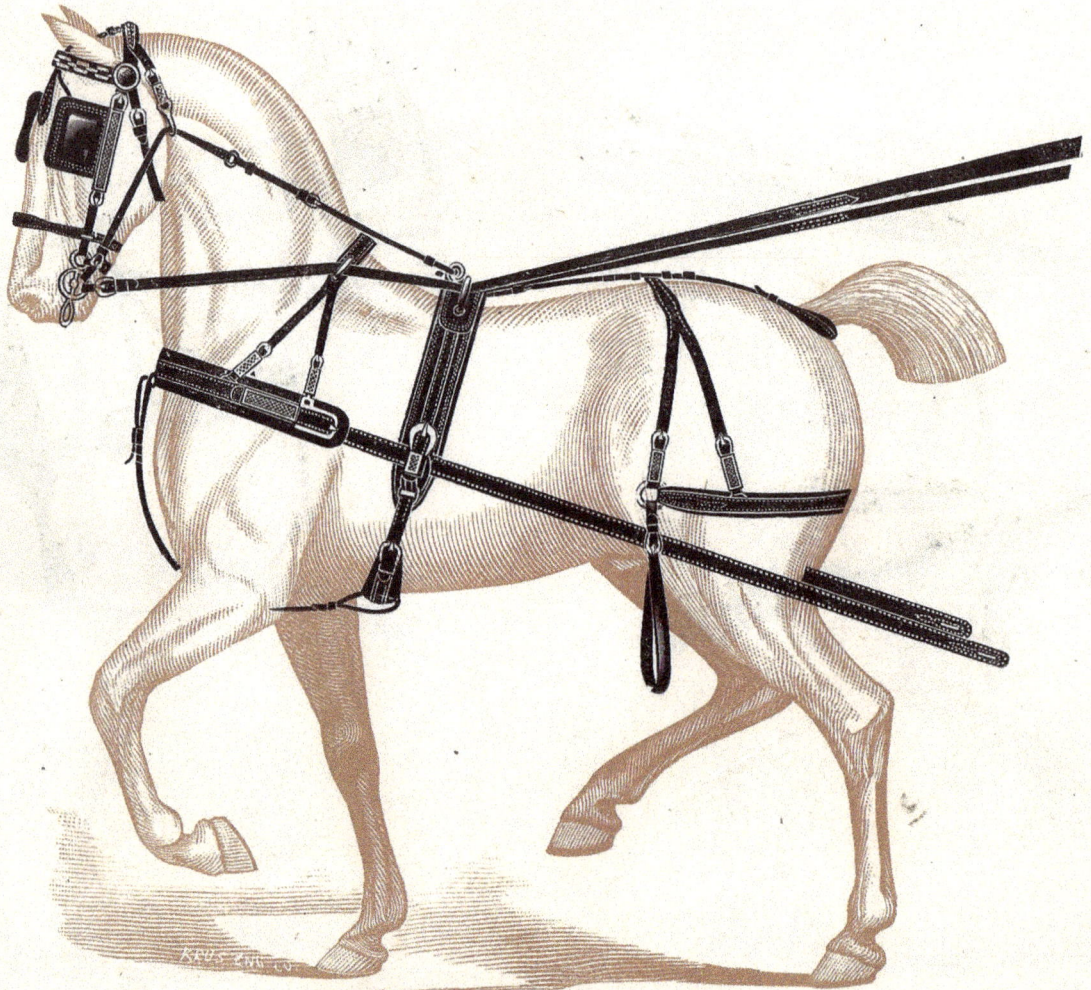

SINGLE SURREY BREAST COLLAR HARNESS No. 1125.

IMITATION HAND-SEWED.

Bridle—⅝-inch, box loops, nose band, round side check, flat doubled and stitched winker stay and face drop.

Lines—1-inch, with buckle and billet.

Breast Collar—Folded, with box loop tugs.

Neck Strap—⅝-inch, split, with line rings.

Traces—1¼-inch, doubled and stitched, raised round edge.

Martingale—⅞-inch.

Saddle—No. 447, 3½-inch surrey.

Shaft Tugs—1⅛-inch, with Dee billets.

Belly Bands—One doubled and stitched, one single strap.

Turnback—⅞-inch, long reversed with narrow loops, round crupper to buckle.

Breeching—Folded, raised layer with ⅝-inch box loop tugs.

Hip Strap—⅝-inch, split.

Side Straps—⅞-inch.

No Tie Strap.

Full Nickel French Wire Trimmed.
Full Brass French Wire Trimmed.

DOUBLE SURREY HARNESS No. 263.

IMITATION HAND-SEWED.

Bridles—⅝-inch, box loops, round side check, coach blinds, gag swivels, plain leather nose bands and chin strap.

Lines—1-inch, with 1¼-inch hand parts, box loops.

Hame Tugs—1⅜-inch, plain leather ends, box loops, with 8-lb. full spot fancy draft eye iron coach hames.

Traces—1⅜-inch, doubled and stitched, 4-row, round edge finish.

Martingales—1-inch, with patent leather drops.

Pole Straps—1½-inch, with box loops.

Coach Pads—No. 104, Newport, with fine chain underhousing.

Skirts—Doubled and stitched, 4-row, with waved, doubled and stitched bearers.

Belly Bands—Folded with layer and box loops.

Turnback—⅞-inch, doubled reversed with heavy coach crupper, 1-inch hip straps, with patent leather trace and hip drops.

DOUBLE SURREY HARNESS No. 265.

IMITATION HAND-SEWED.

Bridles—⅝-inch, narrow fancy creased loops, round side check, coach blinds, gag swivels, plain leather nose band and chin straps.

Lines—1-inch, with 1¼-inch hand parts, narrow creased loops.

Hame Tugs—1⅜-inch, plain leather ends, narrow fancy creased loops, with 8-lb. full spot fancy draft eye coach hames.

Traces—1⅜-inch, doubled and stitched; 4-row, round edge finish.

Martingales—1-inch, with patent leather drops.

Pole Straps—1½-inch, with narrow creased loops.

Coach Pads—No. 104, Newport, with fine chain underhousing.

Skirts—Doubled and stitched, 4-row, with waved doubled and stitched bearers.

Belly Bands—Folded with layer, narrow edge creased loops.

Turnbacks—⅞-inch, long reversed, with narrow edge creased loops, with heavy coach cruppers, 1-inch hip straps with patent leather trace and hip drops.

LIGHT DRAUGHT WAGON HARNESS No. 312.

IMITATION HAND-SEWED.

Bridles—¾-inch, short cheeks, cup shape harness leather blinds, round side check and winker stay.

Lines—1-inch, 20-foot.

Hame Tugs—1¼-inch, with scalloped safes, box loops, slip tug, oval iron wood coach hames.

Traces—1¼-inch, doubled and stitched, raised, smooth finish, with triangular cockeyes.

Martingales—1¼-inch, smooth finish.

Breast Straps—1¼-inch, smooth finish.

Pads—No. 3½ Superba.

Skirts—Single strap, round bearers, trace wear loops, with 1¼-inch billet.

Belly Bands—Folded.

Turnbacks—1-inch, scalloped, with round cruppers buckled on.

Breechings—Single strap, with 1⅛-inch side strap sewed in ring, ¾-inch double hip straps.

Tie Straps—⅝-inch.

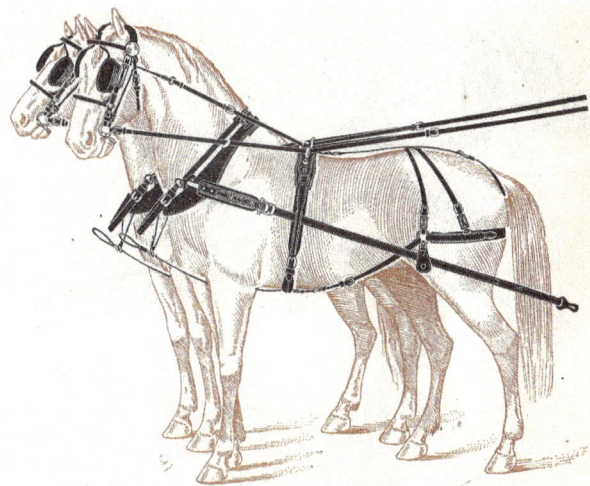

LIGHT DRAUGHT WAGON HARNESS No. 314.

IMITATION HAND-SEWED.

Bridles—¾-inch, short cheeks, cup shape harness leather blinds, round side check and winker stay.

Lines—1-inch, 20-foot.

Hame Tugs—1⅜-inch, with scalloped safes, box loops, slip tug, oval iron wood coach hames.

Traces—1⅜-inch, doubled and stitched, smooth finish, with triangular cockeyes.

Martingales—1¼-inch, smooth finish.

Breast Straps—1¼-inch, smooth finish.

Pads—No. 3½ Superba.

Skirts—Single strap, round bearers, with trace wear loops, with 1¼-inch billet.

Belly Bands—Folded.

Breechings—Single strap, with 1⅛-inch side straps sewed in ring, ¾-inch double hip straps.

Turnbacks—1-inch, scalloped, with round cruppers buckled on.

Tie Straps—⅝-inch.

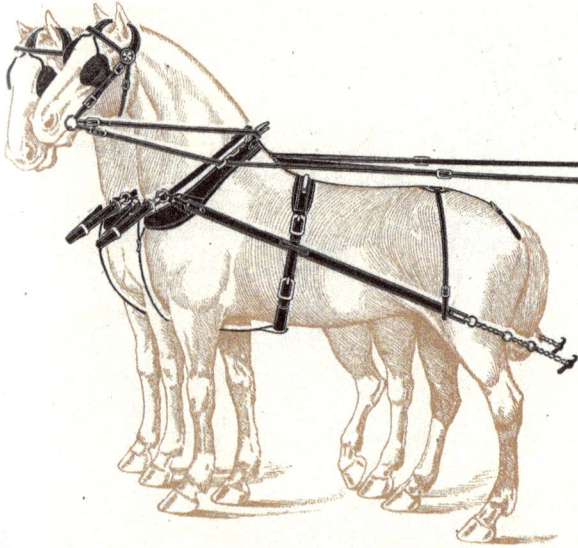

FARM PLOW HARNESS No. 400.

IMITATION HAND-SEWED.

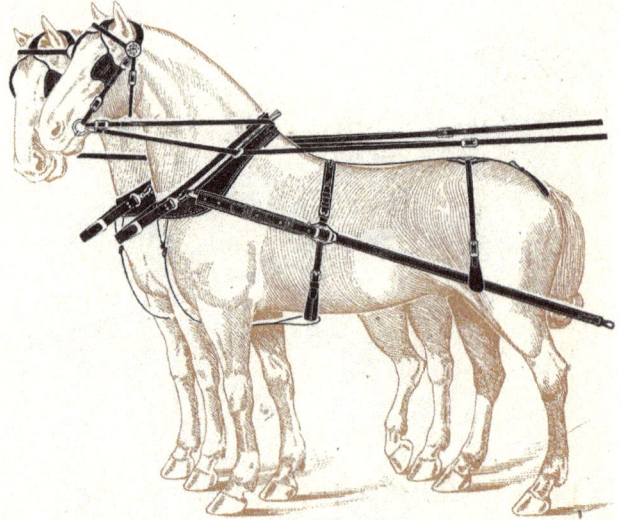

FARM PLOW HARNESS No. 401.

IMITATION HAND-SEWED.

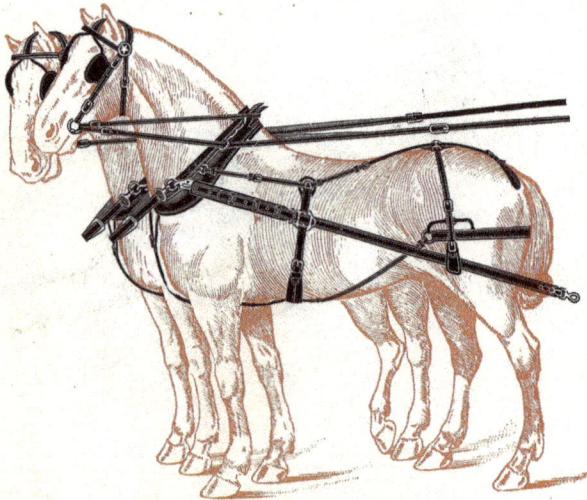

FARM BREECHING HARNESS No. 450.

IMITATION HAND-SEWED.

Bridles—¾-inch, cup shape harness leather blinds, flat side check, round winker stay.

Lines—⅞-inch, 18-foot.

Hame Tugs—1½-inch, folded, 450 I. O. T. hames.

Traces — 1½ - inch, doubled and stitched, with clip cockeyes.

Martingales—1½-inch, with collar straps.

Breast Straps—1½-inch, doubled and stitched.

Belly Bands—Folded.

Rib Straps—⅞-inch to lead up, thence to hames.

Hip Straps—1-inch, with Cooper's trace carriers and cruppers to buckle.

Breeching—Folded with layer, 3-ring brace.

Side Straps—⅞-inch.
No Tie Straps.

Full XC Trimmed.

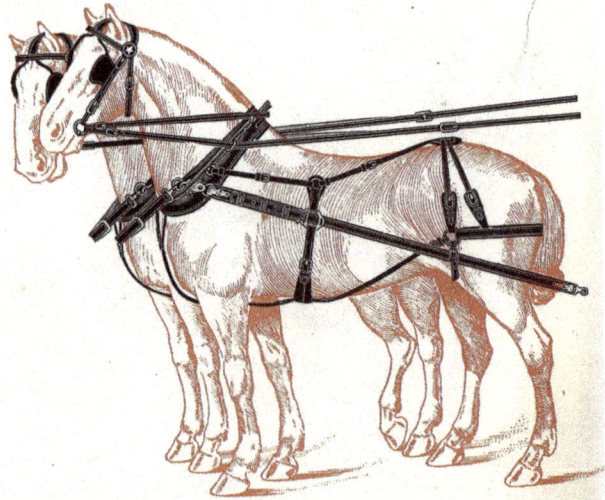

FARM BREECHING HARNESS No. 454.

IMITATION HAND-SEWED.

Bridles—⅞-inch, cup shape harness leather blinds, flat side check, round winker stay.

Lines—1-inch, 18-foot.

Hame Tugs—1½-inch, folded with Cooper's joint clips and No. 5 oiled Concord bolt hames.

Traces — 1½ - inch, doubled and stitched, clip cockeyes.

Martingales—1½-inch, with collar straps.

Breast Straps—1½-inch, doubled and stitched.

Belly Bands—Folded.

Rib Straps—1-inch to lead up, thence to hames.

Hip Straps—⅞-inch, double.

Breeching—Folded with layer.

Side Straps—⅞-inch.
No Tie Straps.

Full XC Trimmed.
No. 1454 Harness, same as 454, with Rib Straps to Hame.
No. 2454 Harness, same as 454, with Laced Box Loop Hame Tugs.

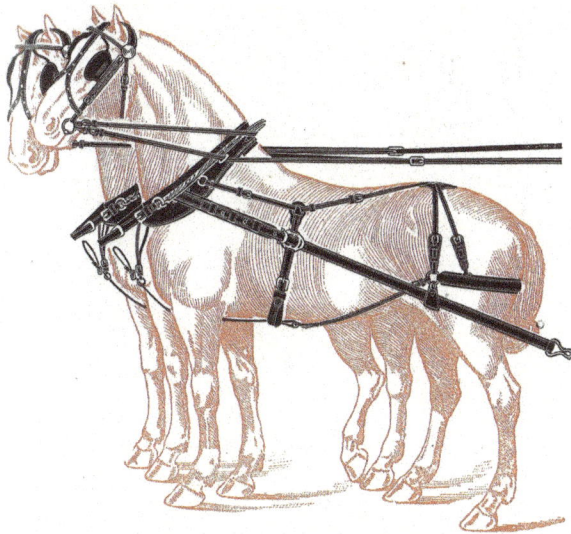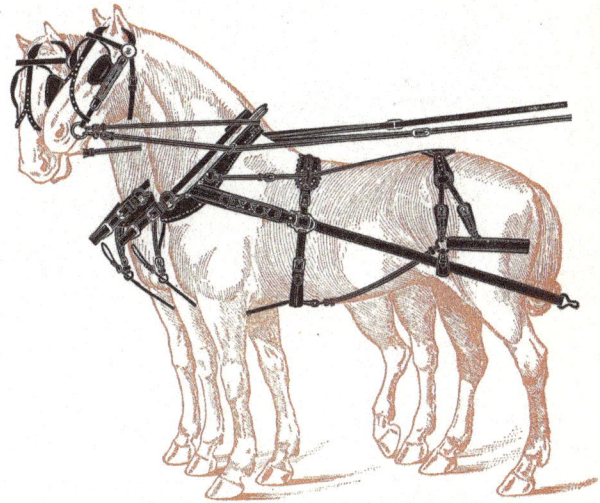

FARM BREECHING HARNESS No. 456.
IMITATION HAND-SEWED.

Bridles—¾-inch, long box loop cheek, spotted face piece, harness leather Concord blinds, flat side check and round winker stay.

Lines—1-inch, 20-foot.

Hame Tugs—1½-inch, folded, doubled and stitched, No. 5 oiled Concord bolt hames.

Traces—1½-inch, 6-foot, doubled and stitched, with triangular cockeyes.

Martingales—1½-inch, with collar straps.

Breast Straps—1½-inch.

Belly Bands—Folded.

Hip Straps—1-inch, double.

Rib Straps—1-inch, extending to lead up on trace buckle, thence to hames.

Breeching—Folded with layer.

Side Straps—1-inch.

Tie Straps—¾-inch.

FARM BREECHING HARNESS No. 1456.
IMITATION HAND-SEWED.

Bridle—⅞-inch, long box loop cheek, spotted face piece, flat side rein and round winker stay.

Lines—1-inch, 20-foot.

Hame Tugs—1½-inch, folded laced box loop, No. 5 oiled Concord bolt hames.

Traces—1½-inch, doubled and stitched, triangular cockeyes.

Martingales—1½-inch, with collar straps.

Breast Straps—1½-inch.

Pads—Swell harness leather, felt lined, spotted, with market strap skirt.

Belly Bands—Folded.

Hip Straps—1-inch, double, 3-ring spotted, rib strap 1 inch to hame.

Breeching—Folded with layer.

Side Straps—1-inch.

Tie Straps—¾-inch.

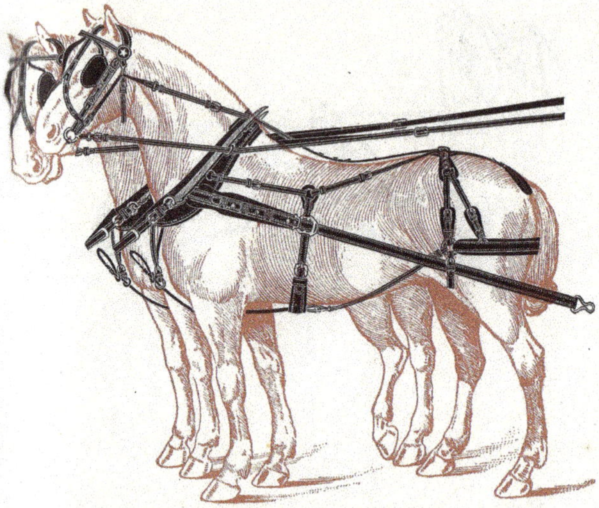

FARM BREECHING HARNESS No. 462.

IMITATION HAND-SEWED.

Bridles—⅞-inch, long box loop cheeks, face piece with spots, harness leather Concord blinds, round side check, round winker stay.

Lines—1-inch, 20-foot.

Hame Tugs—1½-inch, with safes, doubled and stitched, No. 5 oiled Concord screw bolt hames.

Traces—1½-inch, 6-foot, doubled and stitched, with triangular cockeyes.

Martingales—1½-inch, with collar straps.

Breast Straps—1½-inch.

Belly Bands—Folded with layer.

Hip Straps—3-ring, spotted, with trace carriers and crupper fork with cruppers to buckle and check-up strap.

Breeching—Folded, with layer.

Rib Straps—1-inch to lead up, thence to hames.

Side Straps—1-inch.

Tie Straps—¾-inch.

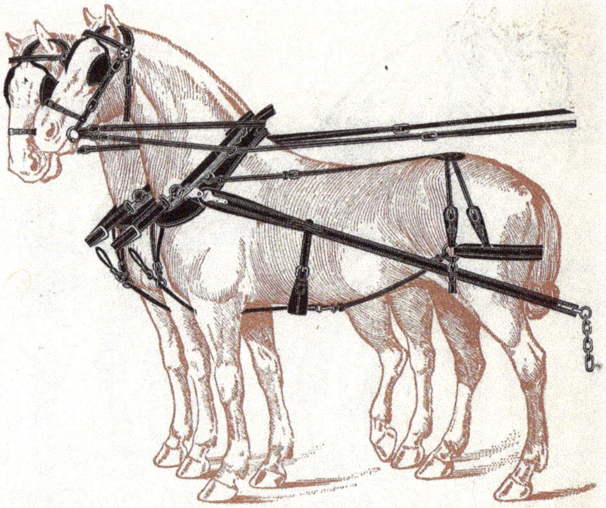

PINERY STYLE BREECHING HARNESS No. 470.

IMITATION HAND-SEWED.

Bridles—⅞-inch, long cheek round winker stay, flat side rein, spotted nose band.

Lines—1-inch, 18-foot.

Traces—1½-inch, heel chain, 15 red brass ball bolt hames, with joint Concord clips.

Martingales—1½-inch, with collar straps.

Breast Straps—1½-inch, doubled and stitched.

Belly Bands—Folded.

Rib Straps—1-inch, to hames.

Hip Straps—⅞-inch, double.

Breeching—Folded with layer.

Side Straps—⅞-inch.

No Tie Straps.

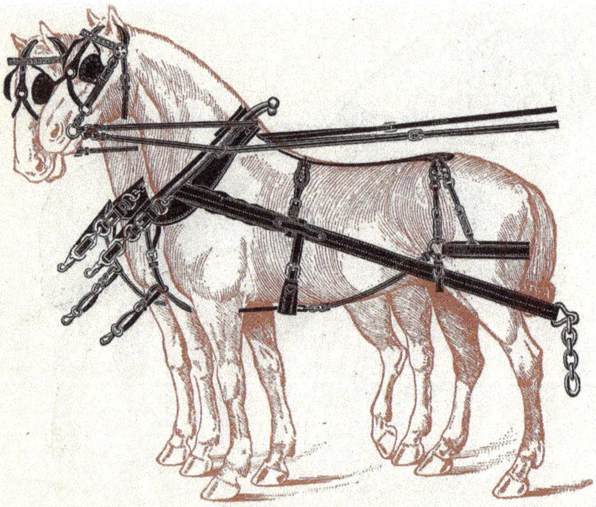

PINERY STYLE FARM BREECHING HARNESS No. 484.

IMITATION HAND-SEWED.

Bridle—⅞-inch, short cheek with bit straps, ring face piece fancy spotted, round winker stay, spotted Concord harness leather blinds, flat side check.

Lines—1⅛-inch, 18-foot.

Traces—1¾-inch, 3-row stitched, scalloped safes with heel chains, No. 10 brass dandy bolt hames, with fancy spotted collar housings.

Martingales—1¾-inch, truck style.

Breast Straps—1¾-inch.

Breeching—Folded with layer, box loop tugs.

Side Straps—1⅛-inch.

Back Straps—1½-inch, to hames.

Hip Straps—1-inch, double, 5-ring, scalloped hip safe fancy spotted.

Belly Bands—Folded with layer.

Pads—Concord, spotted.

Tie Straps—⅞-inch.

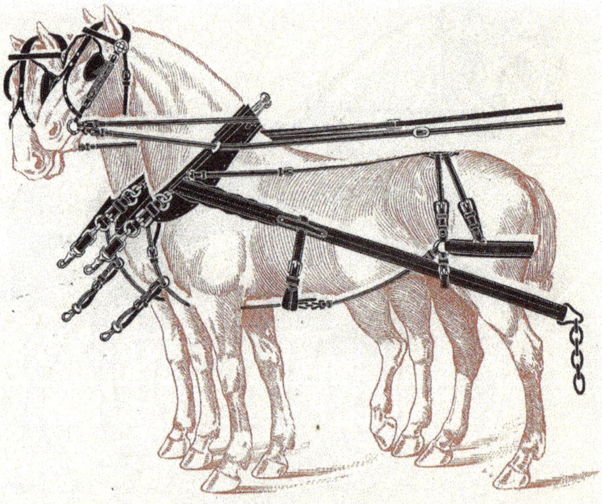

PINERY STYLE FARM BREECHING HARNESS No. 487.

IMITATION HAND-SEWED.

Bridles—1-inch, Concord blinds, long box loop cheeks and spotted face piece, round winker stay, 1-inch flat side check, brass fronts and rosettes.

Lines—1⅛-inch, 18-foot.

Traces—2-inch, 3-row stitched, large scalloped safes with heel chains, No. 10 red Concord bolt hames.

Martingales—1¾-inch, doubled and stitched, pinery style.

Breast Straps—1¾-inch, with extension straps.

Breeching—Folded, with 1½-inch layer, 1¼-inch side straps, 1⅛-inch double hip straps, 1⅛-inch rib straps to hames.

Belly Bands—Folded with layer.

Tie Straps—⅞-inch.

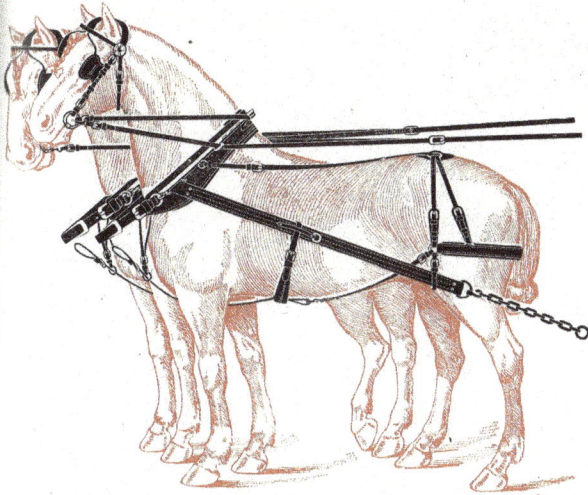

BUTT CHAIN TRACE HARNESS No. 490.

IMITATION HAND-SEWED.

Bridles — ¾-inch, long cheek, double and stitched winker stay, flat side check.

Lines — 1-inch, 20-foot.

Traces — 1½-inch, 4 feet 8 inches, with pinery hook and Dees, No. 5 oiled Concord bolt hames, joint Concord clips and 3½-foot trace chains.

Martingales — 1½-inch, with collar straps.

Breast Straps — 1½-inch.

Belly Bands — Folded.

Hip Straps — 1-inch, double, with 1-inch rib straps to hames.

Breeching — Folded with layer.

Side Straps — 1-inch.

No Tie Straps.

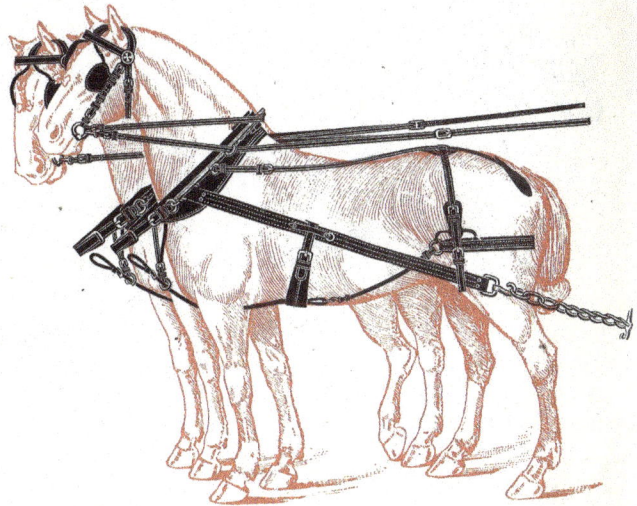

BUTT CHAIN TRACE HARNESS No. 491.

IMITATION HAND-SEWED.

Bridles — ⅞-inch, long box loop cheeks, Concord harness leather blinds, flat side check and round winker stay.

Lines — 1-inch, 20-foot.

Traces — 2-inch, 4 foot 8 inches, doubled and stitched, 3-row with pinery hook and Dee, and 3½-foot grab link chains, oiled Concord bolt hames.

Martingales — 1¾-inch.

Breast Straps — 1¾-inch, with snaps and slides.

Belly Bands — Folded.

Hip Straps — 1¼-inch, with folded crupper to snap.

Rib Straps — 1⅛-inch to hames.

Breeching — Shifting, folded with 1½-inch layer.

Side Straps — 1¼-inch, with snaps.

No Tie Straps.

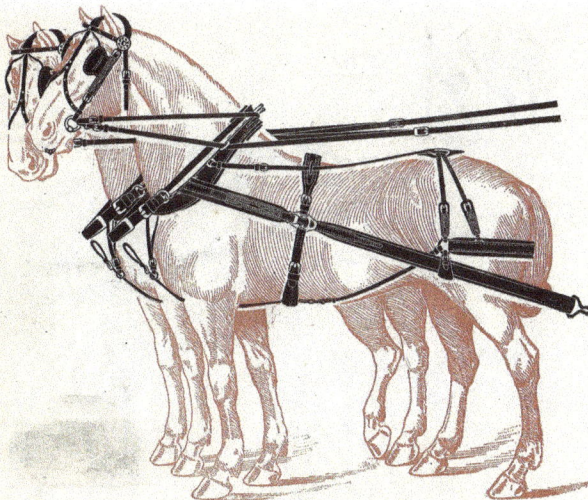

CONCORD TEAM HARNESS No. 611.

IMITATION HAND-SEWED.

Bridles — ⅞-inch, long box loop cheeks, Concord blinds, spotted face piece, round winker stay and flat side check.

Lines — 1-inch, 20-foot.

Hame Tugs — 2-inch, long box loop, No. 5 Concord bolt hames.

Traces — 2½-inch single strap, 2-inch points, with triangular cockeyes.

Martingales — 1¾-inch.

Breast Straps — 1¾-inch.

Pads — Concord style, spotted, for crotch back straps.

Belly Bands — Folded.

Rib Straps — 1-inch to hame.

Hip Straps — 1-inch, double.

Breeching — Folded with layer and box loop tugs.

Side Straps — 1-inch.

No Tie Straps.

CONCORD TEAM HARNESS No. 616.

IMITATION HAND-SEWED.

Bridles — ¾-inch, long box loop cheeks, Concord blinds, spotted face piece, round winker stay, flat side check.

Lines — 1-inch, 20-foot.

Hame Tugs — 1¾-inch, long box loop, with No. 91 XC bolt hames.

Traces — 2½-inch single strap, with 1¾-inch points, triangular cockeyes.

Martingales — 1½-inch.

Breast Straps — 1½-inch.

Belly Bands — Folded.

Rib Straps — 1-inch to lead up, thence to hames.

Hip Straps — 1-inch, double, 3-ring style, spotted.

Breeching — Folded with layer and box loop tugs.

Side Straps — 1-inch.

No Tie Straps.

PIPE CHAIN HARNESS No. 640.

IMITATION HAND-SEWED.

Bridles—¾-inch, cup shape harness leather blinds, flat side check.

Lines—⅞-inch, 18-foot.

Hames—No. 6 Concord, cart hook.

Trace Chains—7-foot, 8-2 polished.

Chain Pipes—36-inch.

Breast Straps—1½-inch, doubled and stitched.

Back Bands—Single strap, 3½-inch.

Belly Bands—Single strap, 1¼-inch.

Back Straps—1¼-inch, running to hames, with folded crupper.

Hip Straps—1-inch.

No Tie Straps.

PIPE CHAIN HARNESS No. 642.

IMITATION HAND-SEWED.

Bridles—¾-inch, cup shape harness leather blinds, flat side check.

Lines—⅞-inch, 18-foot.

Hames—No. 6 Concord, cart hook.

Trace Chains—7-foot, 8-2 polished.

Chain Pipes—36-inch.

Breast Straps—1½-inch, doubled and stitched.

Back Bands—Single strap, 3½-inch.

Belly Bands—Single strap, 1¼-inch.

Back Straps—1¼-inch, running to hames, with folded crupper.

Hip Straps—1-inch.

Breechings—Folded, 3-ring brace.

Side Straps—⅞-inch.

No Tie Straps.

Oregon's Jacob McIntosh at a plowing match with his Percherons in their three-strap brichen parade harness.

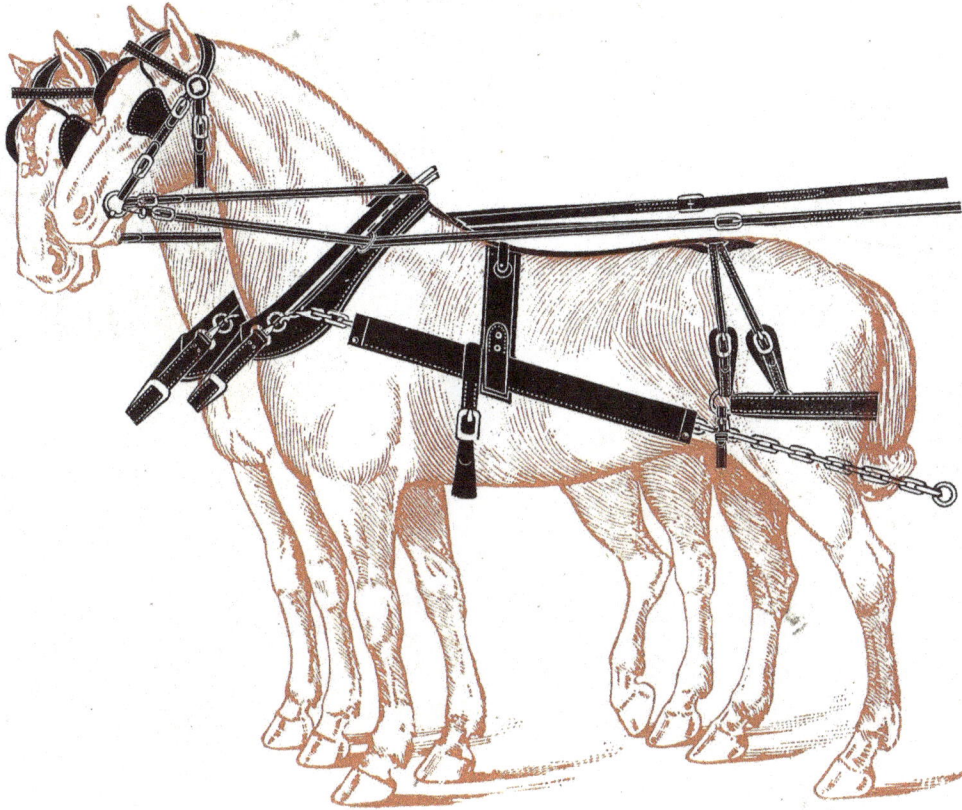

GRADERS HARNESS No. 643.

IMITATION HAND-SEWED.

Bridles—⅞-inch, cup shape harness leather blinds, flat side check.

Lines—1-inch, 20-foot.

Hames—No. 6 Concord cart hook trace chains, 7-foot polished, extra heavy.

Chain Pipes—36-inch.

Breast Straps—2-inch, double and stitched, with snaps and slides.

Martingales—2-inch.

Back Bands—Single strap, 3½-inch.

Belly Bands—Single Strap, 1½-inch.

Back Straps—1¼-inch, running to hames with 1-inch double hip straps.

Breeching—Folded, with layer.

Side Straps—1-inch.

No Tie Straps.

Full XC Trimmed.

Even the road-building teamsters of 1906 needed their own designated harness style, this one with the extra strong, gloved, or leather-tubed, chain tugs.

*Lise Hubbe's magnificent team of working mares staring down the
next path for the side delivery rake on the farm in Scio, Oregon.*

Thrashing Oats In Whatcom County

DODSON-FISHER CO.   7

No. 2768

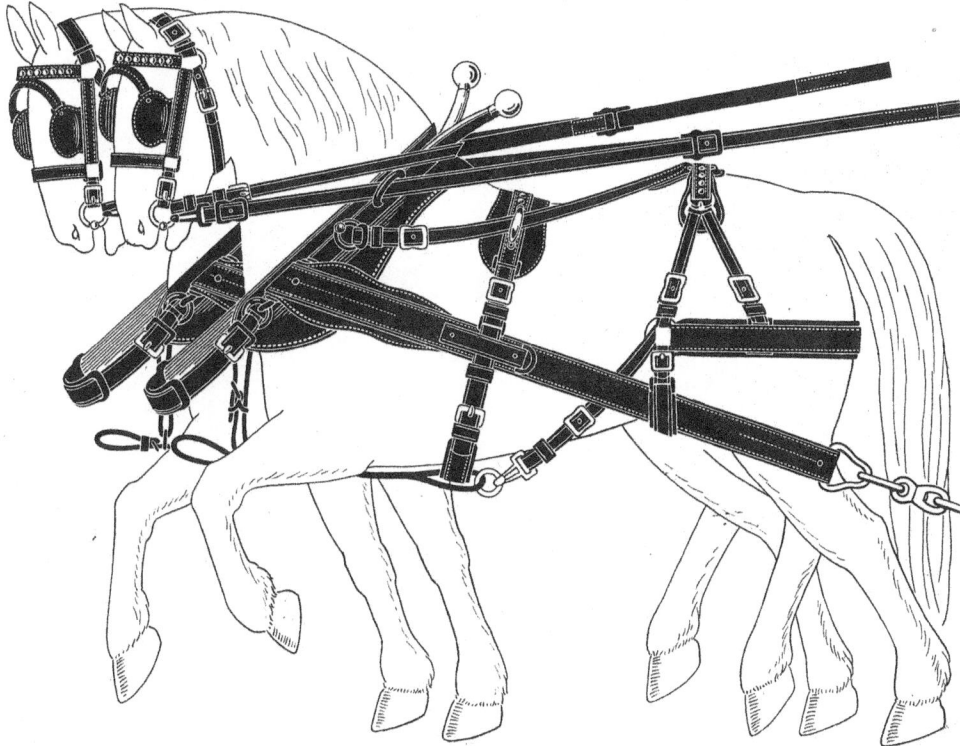

No. 2768

X C TRIMMED. CREASED.

No. 2768A—1½-inch Harness with No. 5 Oiled Bolt Hames. Per Set $110.00
No. 2768C—1½-inch Harness with Nickel Ball Steel Hames. Per Set 112.00
If wanted with 1¾-inch Traces, Breast Straps and Martingales, add 4.70

BRIDLES—Per set $13.10. Ring crown, 1-inch cheeks, Concord blinds, folded crown, spotted front.
LINES—1⅛-inch by 20 feet with Conway loop and snap.
TRACES—1½-inch by 6 feet with 6-link heel chain. Sewed bolt ends, 1½-inch billets.
BACK BANDS—Per set $9.20. 5½-inch Swell harness leather housings, leather lined, 1½-inch layer with dee, 1½-inch market straps to reverse. X C Colorado bridges, spotted between bridges.
BREECHING—Per set $29.50. 2¼-inch single strap with 1½-inch full length layer, 1½-inch split turnbacks, 1⅛-inch reverse hipstraps, 1¼-inch reverse sidestraps, 1-inch lazy strap with wear leather and dee. Spotted hell diver.
BELLY BANDS—2-inch single strap with 1½-inch layer—1½-inch buckles.
BREAST STRAPS—1½-inch with snaps and slides.
MARTINGALES—1½-inch with ring.
COLLAR STRAPS—⅞-inch.

No. 2769A—1½-inch Harness with No. 5 Oiled Bolt Hames. Per Set $114.70
No. 2769C—1½-inch Harness with Nickel Ball Tubular Steel Hames. Per Set 116.80

The Number 2769 harness is same as Number 2768, only has long round side check bridles and third turnbacks.

No. 2769—X C Bridles. Per Set...$16.50

A good heavy serviceable harness where one is wanted at a lower price than our Gopher Brand

PRICES SUBJECT TO CHANGE WITHOUT NOTICE

No. 2565

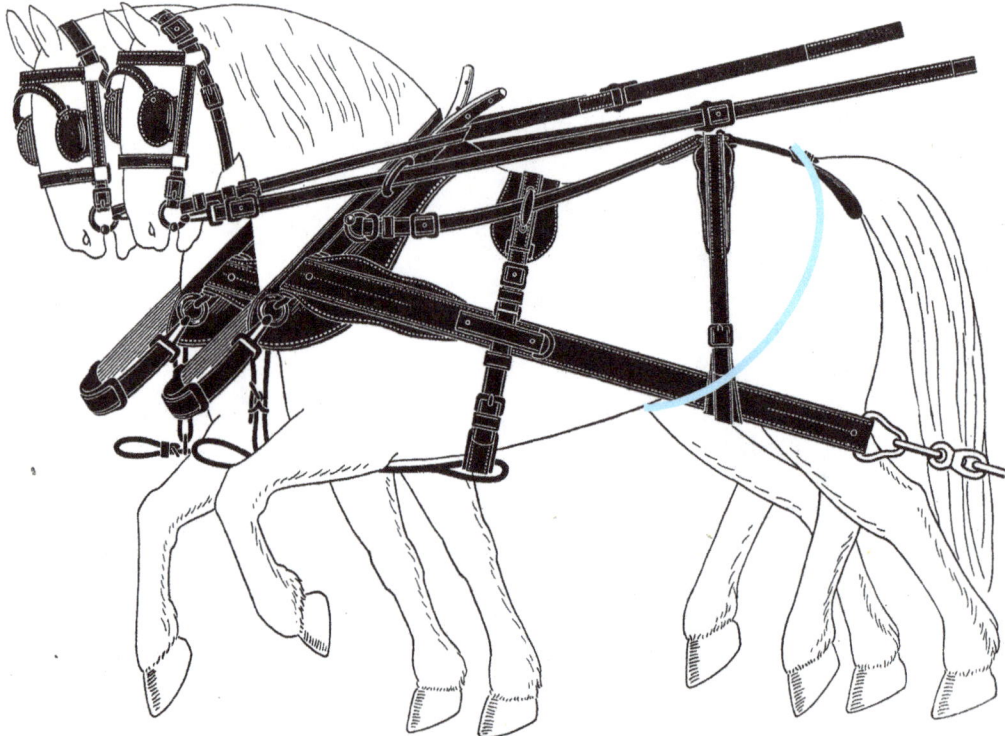

No. 2565

FULL X C TRIMMED. CREASED.

No. 2565A—With No. 5 Oiled Bolt Hames. Per Set $96.00

No. 2565C—With Nickel Ball Tubular Steel Bolt Hames. Per Set 98.20

BRIDLES—Ring crown, 1-inch cheeks.

LINE—1⅛-inch x 20 feet with Conway loops and snaps.

TRACES—1½-inch x 6 feet, sewed bolt end, scalloped wear leather, 1½-inch belly band billets, 6-inch link chain.

BELLY BANDS—2-inch single strap with 1½-inch layer and buckle.

BACK BANDS—4¾-inch swell harness leather housings, leather lined, 1½-inch layer with dee, 1½-inch market straps to reverse X C Colorado bridges.

BACK AND HIPSTRAPS—Per set $15.70. 1⅛-inch, scalloped wear leather, wide mud carrier, ¾-inch folded crupper to buckle on.

BREAST STRAPS—1½-inch, Martingales 1½-inch, collar straps, ⅞-inch.

For 1¾-inch traces, breast straps and Martingale's add $4.70 per set.

A good heavy serviceable harness where one is wanted at a lower price than our Gopher Brand.

Above is one variation on a 'plow' harness, sometimes referred to as a 'lead' harness (as in for the 'lead' team of a four or six or eight up). The primary distinction of this design is that there's no brichen, which means it is lighter weight and subsequently cheaper. Should this harness design be employed to hold back or back up or stop a load on a tongue it will possibly cause discomfort to the animals as it transfers the weight to the top of the neck and applies forward pinching motion to the belly band. This harness type is sometimes confused with a Yankee Brichen which has a reinforced wide strap (see blue line above) which rides across the top of the tail and passes beneath the traces to connect to the pole strap.

No. 2765

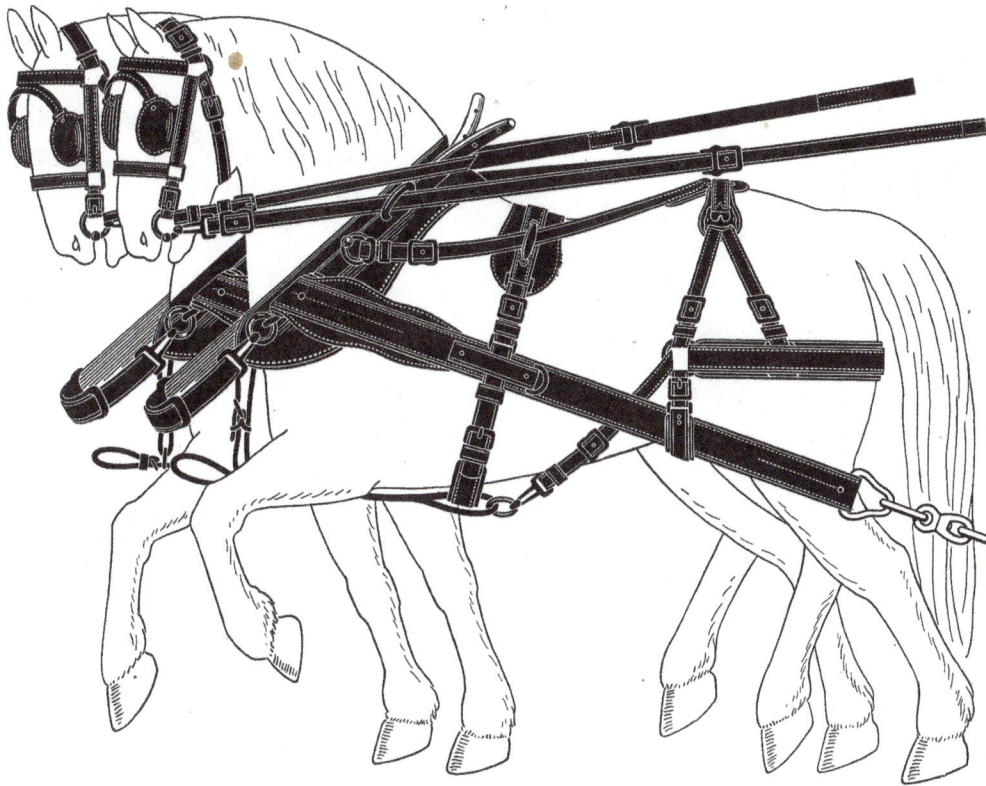

No. 2765

X C TRIMMED. CREASED.

No. 2765A—1½-inch Harness with No. 5 Oiled Bolt Hames. Per Set $107.90
No. 2765C—1½-inch Harness with Nic. Ball Steel Hames. Per Set 110.00
If wanted with 1¾-inch Traces, Breast Straps and Martingales, add 4.70

BRIDLES—Per set $11.70. Ring crown, 1-inch cheeks, Concord blinds.
LINES—1⅛-inch by 20 feet with Conway loop and snap.
TRACES—1½-inch by 6 feet with 6-link heel chain. Sewed bolt ends, 1½-inch billets.
BACK BANDS—4¾-inch Swell harness leather housings, leather lined, 1½-inch layer with dee, 1½-inch market straps to reverse. Colorado bridges.
BREECHING—Per set $28.70. 2¼-inch single strap with 1½-inch full length layer, 1½-inch split turnbacks, 1⅛-inch reverse hipstraps, 1¼-inch reverse sidestraps, 1-inch lazy strap with wear leather and dee.
BELLY BANDS—2-inch single strap with 1½-inch layer—1½-inch buckles.
BREAST STRAPS—1½-inch with snaps and slides.
MARTINGALES—1½-inch with ring.
COLLAR STRAPS—⅞-inch.

No. 2766A—1½-inch with No. 5 Oiled Bolt Hames. Per Set $114.70
No. 2766C—1½-inch with Nickel Ball Tubular Steel Hames. Per Set 116.80
The Number 2766 harness is the same as Number 2765, only has long round side check bridles and third turnbacks.
No. 2766—Bridles. Per Set $13.90

A good heavy serviceable harness where one is wanted at a lower price than our Gopher Brand.

Most of the older brichen harnesses were sold with collar straps that buckled around the collar throat and into a square D on the front of the pole strap (or martingale). Combination snaps were later employed to attach the pole strap and breast straps negating the need for the collar strap.

No. 2604

(GOPHER BRAND)

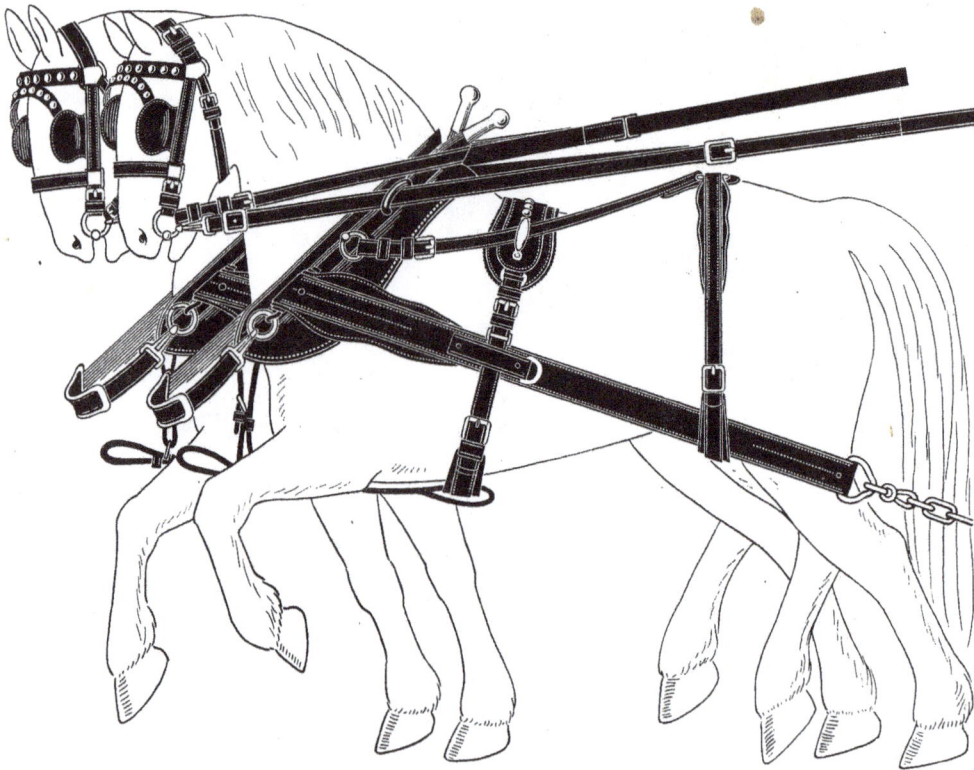

No. 2604

FARM HARNESS

FULL X C TRIMMED. LOCK STITCHED. BLACKED ON THE FLESH.   MADE SMOOTH, NO CREASE.

No. 2604A—With No. 5 Concord Bolt Hames. No Balls $110.50
No. 2604C—With Nickel or Brass Ball Top Tubular Steel Bolt Hames.......................... 112.60

BRIDLES—⅞-inch, short cheeks, ring crown, spotted combination fronts and plain braces, short flat side checks. Folded crown.

LINES—1-inch x 20 feet with buckle and billet, doubled and stitched coupling.

TRACES—1½-inch, heavy solid three ply, eight link screw dee chain, 1½-inch belly band billets with liner, scalloped wear leathers.

BACK BANDS—5-inch swell harness leather housings, leather lined with scalloped and stitched layer, 1½-inch market straps, nickel Colorado bridges.

BELLY BANDS—Folded with 1½-inch layer and buckles.

BACK AND HIP STRAPS—Per set $15.80. 1⅛-inch turnbacks, 1⅛-inch hip straps, scalloped wear leather, folded cruppers to buckle on.

BREAST STRAPS—1½-inch sewed with snaps and slides.

MARTINGALES—1½-inch, with ring.

COLLAR STRAPS—1-inch.

For 1¾-inch traces, breast straps and Martingales add $4.70 per set.

Popular with many modern Amish harness shops are detachable brichens (page 134) which snap on and convert a plow harness such as this to a full two-strap brichen style.

No. 2765

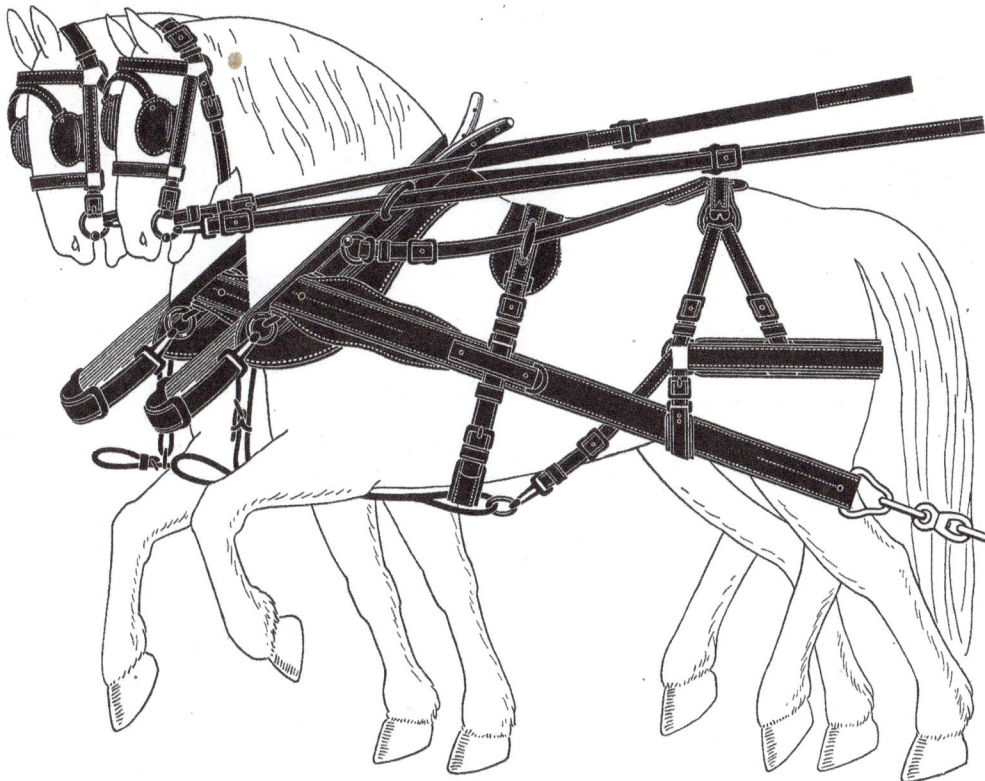

No. 2765

X C TRIMMED. CREASED.

No. 2765A—1½-inch Harness with No. 5 Oiled Bolt Hames. Per Set $107.90
No. 2765C—1½-inch Harness with Nic. Ball Steel Hames. Per Set 110.00
If wanted with 1¾-inch Traces, Breast Straps and Martingales, add 4.70

BRIDLES—Per set $11.70. Ring crown, 1-inch cheeks, Concord blinds.
LINES—1⅛-inch by 20 feet with Conway loop and snap.
TRACES—1½-inch by 6 feet with 6-link heel chain. Sewed bolt ends, 1½-inch billets.
BACK BANDS—4¾-inch Swell harness leather housings, leather lined, 1½-inch layer with dee, 1½-inch market straps to reverse. Colorado bridges.
BREECHING—Per set $28.70. 2¼-inch single strap with 1½-inch full length layer, 1½-inch split turnbacks, 1⅛-inch reverse hipstraps, 1¼-inch reverse sidestraps, 1-inch lazy strap with wear leather and dee.
BELLY BANDS—2-inch single strap with 1½-inch layer—1½-inch buckles.
BREAST STRAPS—1½-inch with snaps and slides.
MARTINGALES—1½-inch with ring.
COLLAR STRAPS—⅞-inch.

No. 2766A—1½-inch with No. 5 Oiled Bolt Hames. Per Set $114.70
No. 2766C—1½-inch with Nickel Ball Tubular Steel Hames. Per Set 116.80

The Number 2766 harness is the same as Number 2765, only has long round side check bridles and third turn-backs.

No. 2766—Bridles. Per Set $13.90

A good heavy serviceable harness where one is wanted at a lower price than our Gopher Brand.

This configuration of harness was less costly as it featured, among other discounts, blackened hardware as opposed to nickel or brass. No ornamentation. Each little feature, such as balls on the hame tops and spotting on the bridles, added expense.

DODSON-FISHER CO.

CUSTOM HARNESS
No. 2962

(GOPHER BRAND)

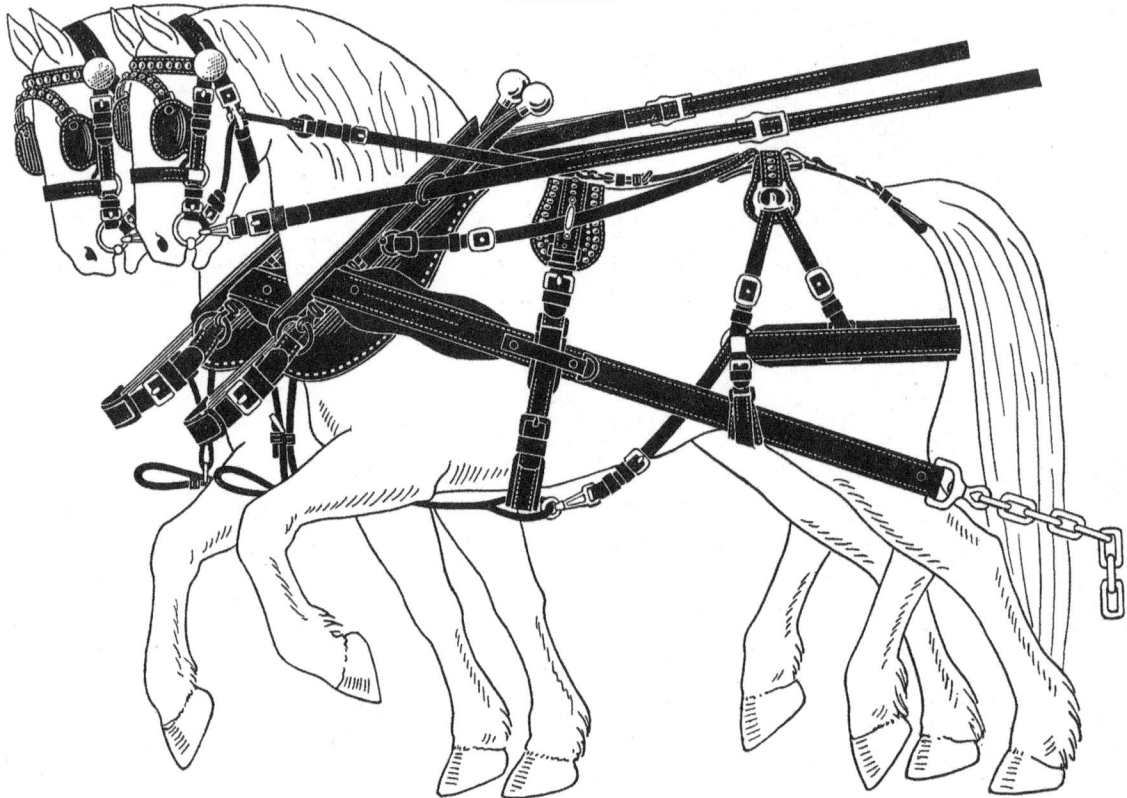

No. 2962

TRUCK FARM HARNESS

No. 2962A—With No. 5 Concord Bolt Hames. No Balls. Per Set $130.50
No. 2962C—With Nickel Ball Top Tubular Steel Bolt Hames. Per Set 132.60

BRIDLES—¾-inch cheeks, combination spotted fronts and winker braces, long round side checks.

LINES—1-inch by 20 feet, buckle and billet, doubled and stitched coupling.

TRACES—1⅜-inch heavy solid three ply, 6 feet 3-inch long, scalloped wear leathers, eight link screw dee heel chain, 1¼-inch belly band billets with 1¾-inch liner.

BACK BANDS—4¾-inch swell harness leather housings, leather lined, 1¼-inch layer, 1¼-inch market straps to buckle into housings, nickel spotted housings with layer, nickel Colorado bridges.

BELLY BANDS—1¼-inch folded, with 1¼-inch layer and buckles.

BREECHING—2-inch single strap with 1¼-inch layer, 1-inch turnbacks, 1-inch hip straps to reverse with nickel shield buckles, 1⅛-inch side straps to reverse, ⅞-inch lazy straps with dee and wide wear leather, crupper forks with snap, folded cruppers, ⅞-inch checkup strap, spotted hell diver, nickel trace carrier.

BREAST STRAPS—1½-inch, sewed with snaps and slides.

MARTINGALES—1½-inch, with doubled and stitched neck yoke ends.

COLLAR STRAPS—⅞-inch.

An exceptionally good grade where a light harness is wanted.

No. 2824

(GOPHER BRAND)

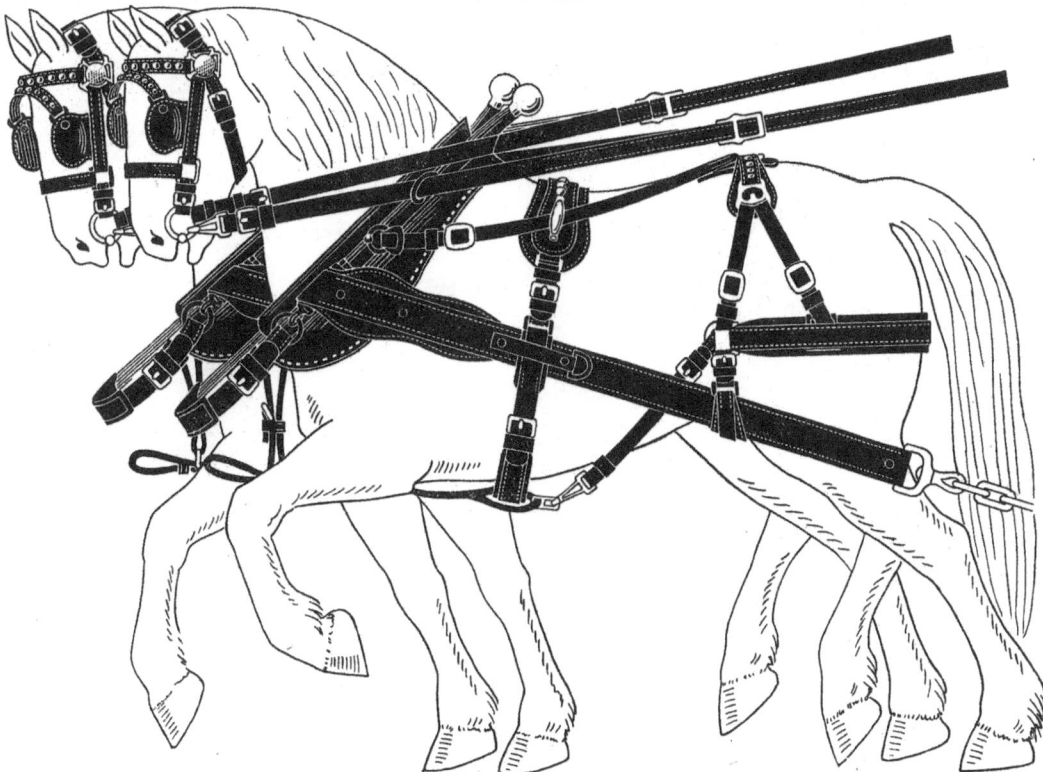

No. 2824

TRUCK FARM HARNESS

FULL X C PLATE. LOCK STITCHED. BLACKED ON THE FLESH. MADE SMOOTH.

No. 2824A—1½-inch with No. 5 Concord Bolt Hames. No Balls. Per Set......................$133.10
No. 2824C—1½-inch with Nickel Top Tubular Steel Bolt Hames. Per Set...................... 135.30
No. 2814A—1¾-inch with No. 5 Concord Bolt Hames. No Balls. Per Set...................... 138.40
No. 2814C—1¾-inch with Nickel Top Tubular Steel Bolt Hames. Per Set...................... 140.50

BRIDLES—Per set $.... No. 2814, ⅞-inch ring crown, Concord harness leather blinds, nickel spotted combination fronts and plain winker braces, short flat side checks, folded crown, ring bits.

LINES—Per set $.... No. 1425, 1⅛-inch x 20 ft. buckle and billet.

TRACES—Per Set $.... No. 2832X, 1½-inch x 6 ft. 2-inch long, solid three ply, eight link screw dee heel chain, wide scalloped wear leather, 1½-inch belly band billets with a 2-inch liner.

BACK BANDS—Per set $.... No. 2824, 5-inch swell harness leather housings, leather lined with scalloped and stitched layer, nickel or brass Colorado bridges and spots, 1½-inch market straps.

BELLY BANDS—Per set $.... No. 1200, folded with 1½-inch full length layers, 1½-inch buckles.

BREECHING—Per set $.... No. 2814, folded with 1¼-inch full length layers, 1⅛-inch turnbacks, 1⅛-inch hip straps to reverse with nickel or brass shield buckles, 1⅛-inch lazy straps with dee and wide wear leather, 1¼-inch side straps, single to reverse.

BREAST STRAPS—No. 4, 1½-inch.

MARTINGALES—No. 4, 1½-inch.

COLLAR STRAPS—No. 5, ⅞-inch.

No. 2826A—1½-inch Harness with No. 5 Oiled Bolt Hames. Per Set $137.90
No. 2826C—1½-inch Harness with Nickel Ball Tubular Steel Hames. Per Set 140.00

The Number 2826 Harness is same as Number 2824, only bridles have long round side checks, and third turnbacks.

GOPHER BRAND HARNESS "WEARS LIKE A PIG'S NOSE"

No. 2530

(GOPHER BRAND)

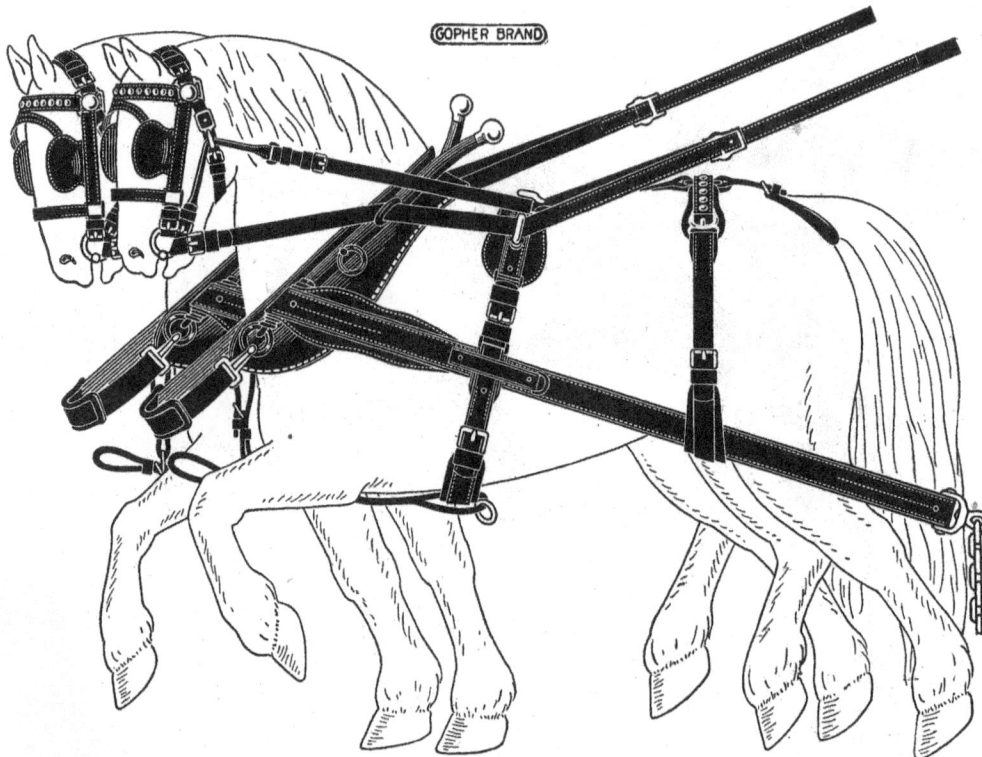

No. 2530

FULL X C TRIMMED. MADE SMOOTH.

Hook and Terret Farm Harness

No. 2530A—With No. 5 Concord Bolt Hames, no Balls. Per Set $116.30

No. 2530C—With Nickel or Brass Ball Top Tubular Steel Bolt Hames. Per Set.................118.40

BRIDLES—Per set $18.70. ⅞-inch folded ring crown, combination fronts and winker braces, spotted front, long round side checks.

LINES—1⅛-inch by 20 feet buckle and billet, doubled and stitched coupling.

TRACES—1½-inch, heavy solid three-ply, 8 link screw dee chain, 1½-inch belly band billets with liner.

PADS—5¼-inch swell, doubled and stitched leather housings, felt interlined, 1½-inch layer, 1½-inch market straps, dwight hook and terrets.

BELLY BANDS—Folded with 1½-inch layer and buckles.

BACK AND HIP STRAPS—Per set $14.40. 1⅛-inch turnback, 1⅛-inch hip straps with spotted hell diver and 352 X C trace carriers, mud carrier with liner and roller buckles.

BREAST STRAPS—1½-inch, sewed with snaps and slides.

MARTINGALES—1½-inch, with doubled and stitched neck yoke ends.

COLLAR STRAPS—1-inch.

SPREADERS—¾-inch with celulloid rings.

On these two pages we see variations of a plow harness. What that choice of name connotes is that, in the 'beginning,' **plow** *harness and plow horses were employed hitched to walking plows, which, being drug along (without tongue) and often into the ground, were never going to 'run up' on a horse's heels. And with walking plows there was absolutely no way to have horses back up the plow. Ergo there was no need for brichen. When poor farmers realized they could purchase new plow harness for less money than brichen harness, they tinkered with ways to use the harness, on mostly level ground, to brake and back up rolling implements. The result is pole strap and breast strap assemblies depending on collars and belly bands.*

No. 2510

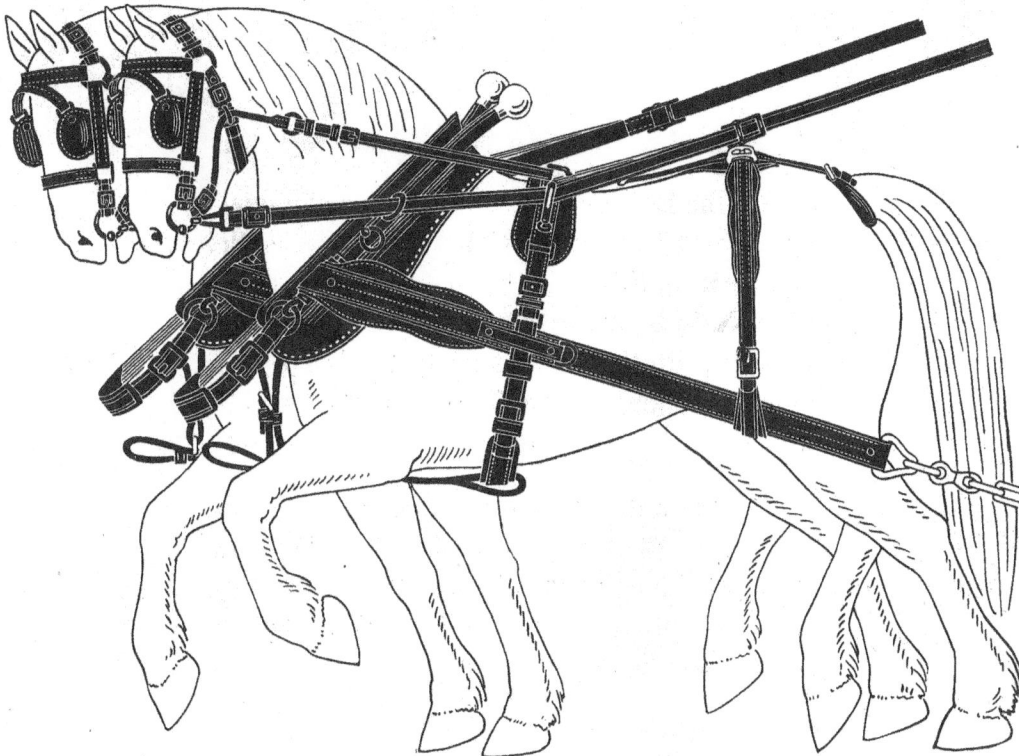

No. 2510

X C OR JAPAN. CREASED.

No. 2510A—With No. 5 Concord Bolt Hames, no Balls. Per Set ....$ 97.90

No. 2510C—With Nickel or Brass Ball Top, Tubular, Steel Bolt Hames. Per Set............... 100.00

BRIDLES—Per set $16.00. Ring crown, 1¼-inch crown, 1-inch cheeks, plain combination fronts and winker braces, long round side checks.

LINES—1⅛-inch by 20 feet, with Conway loop and snaps.

TRACES—1½-inch, with scalloped wear leathers, 1½-inch billets, six link swivel heel chains.

PADS—Per set $9.90. 5¼-inch swell doubled and stitched housings, 1½-inch layer, 1½-inch market straps, dwight hook and terrets.

BELLY BANDS—2-inch single strap, with 1½-inch layer and buckles.

BACK AND HIP STRAPS—Per set $11.90. 1⅛-inch back straps, 1⅛-inch hip straps, scalloped wear leather, wide mud carrier, folded cruppers buckled on.

BREAST STRAPS—1½-inch sewed with snaps and slides.

MARTINGALES—1½-inch with ring.

COLLAR STRAPS—⅞-inch.

A good heavy serviceable harness where one is wanted at a lower price than our Gopher Brand.

Notice that the check reins on both these styles are hooked to turrets mounted on the back pad, and that the spider and crupper are held in place with a back strap also fastened to the turret. This creates a direct line from bit to crupper broken only by a limiting connection at the back pad. Neat bit of design, but this author finds this disrespectful of the horse's comfort.

No. 3040

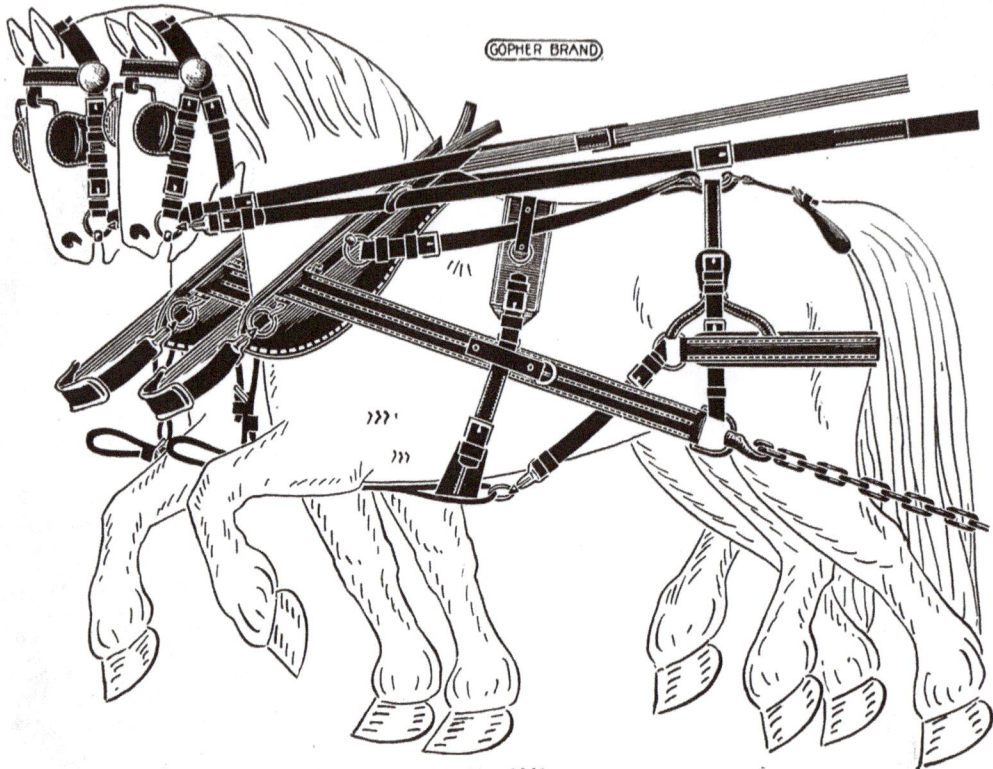

No. 3040

HEAVY MICHIGAN LUMBER HARNESS

A harness seen at a recent Horse Progress Days employing triangulated secondary straps tying saddle and billet forward to hame ring.

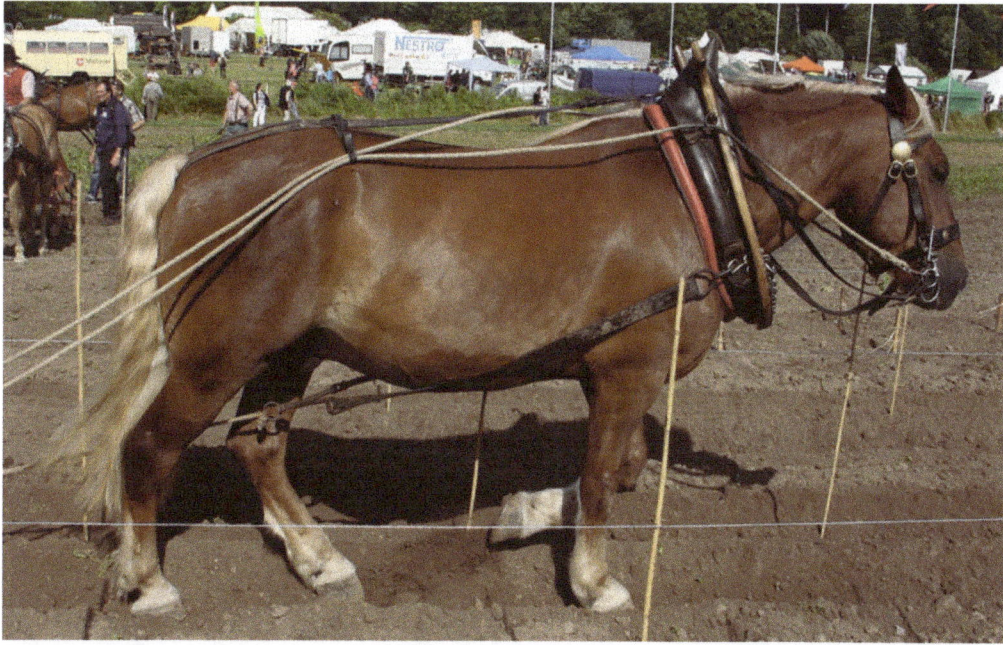

Views of three European variations of harness. There are many dozens of different styles seen across the continent.

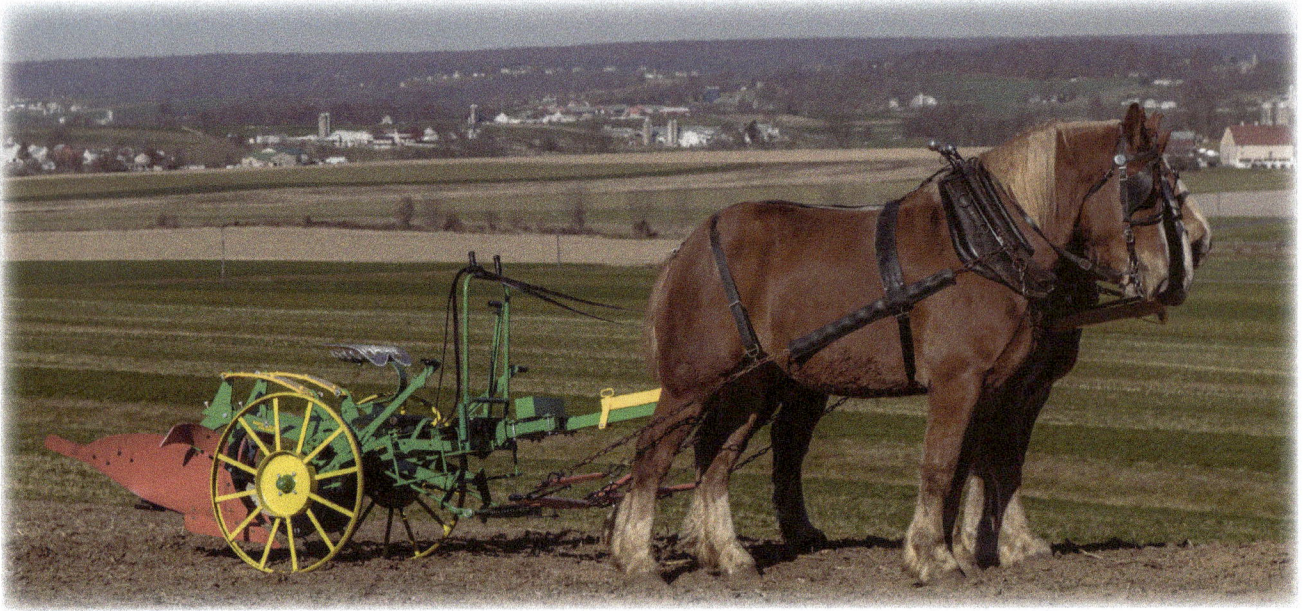

New White Horse spring-loaded, two-way plow. Belgians are employing same harness as pictured below.

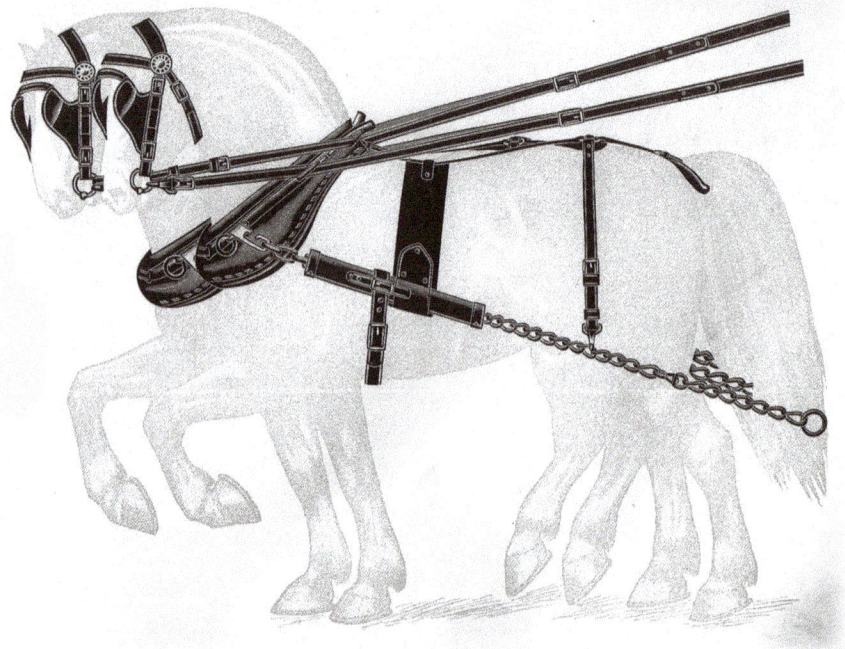

Chain tug plow harness with wide back belt.

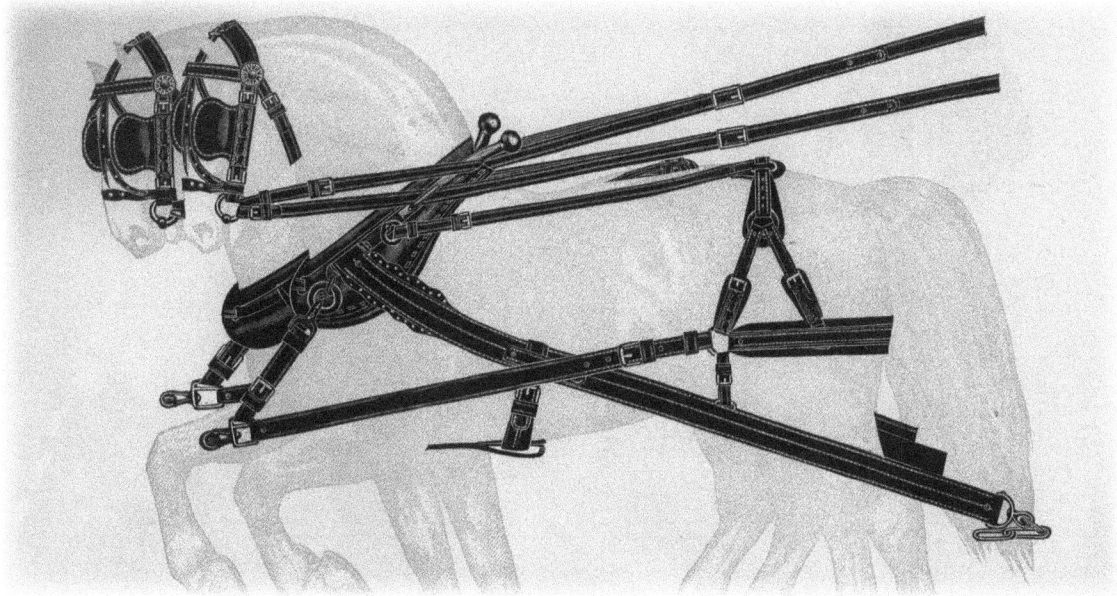

Side-backer harness with no saddle or back pad.

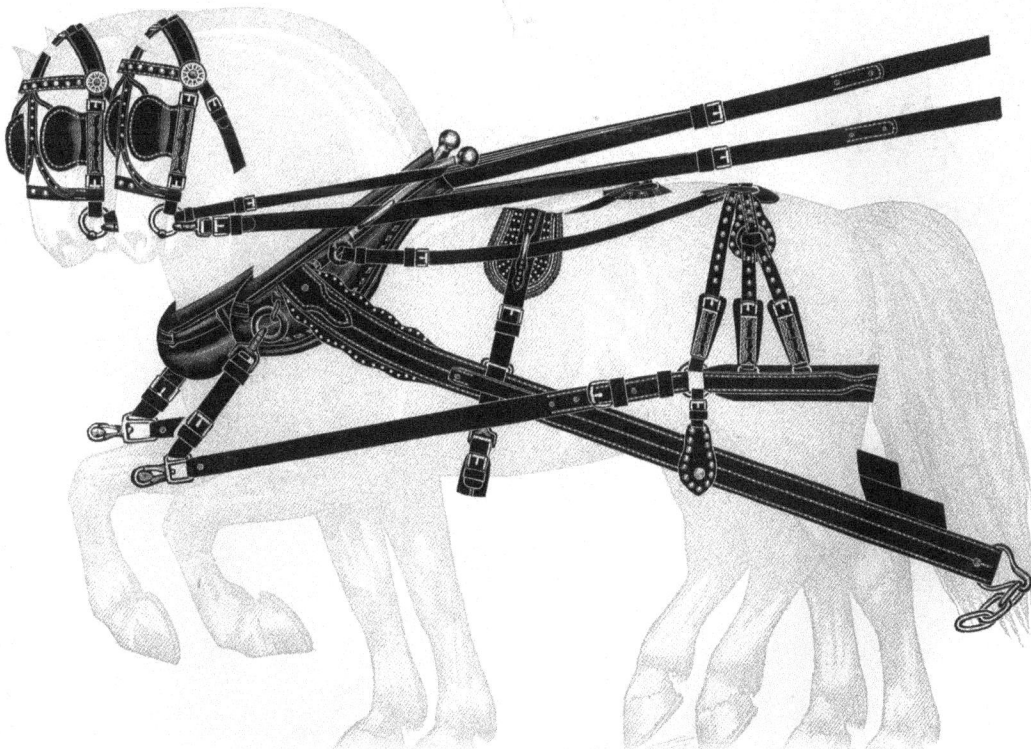

Conventional side-backer harness, different from a D-ring side-backer harness.

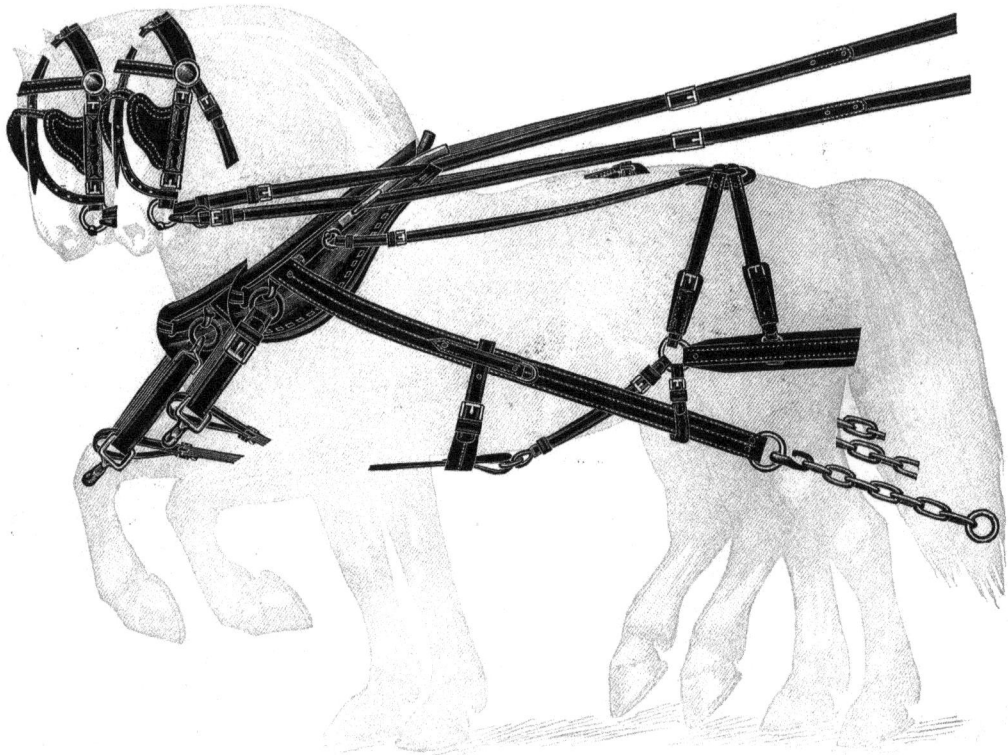

No. 255—Logging Harness

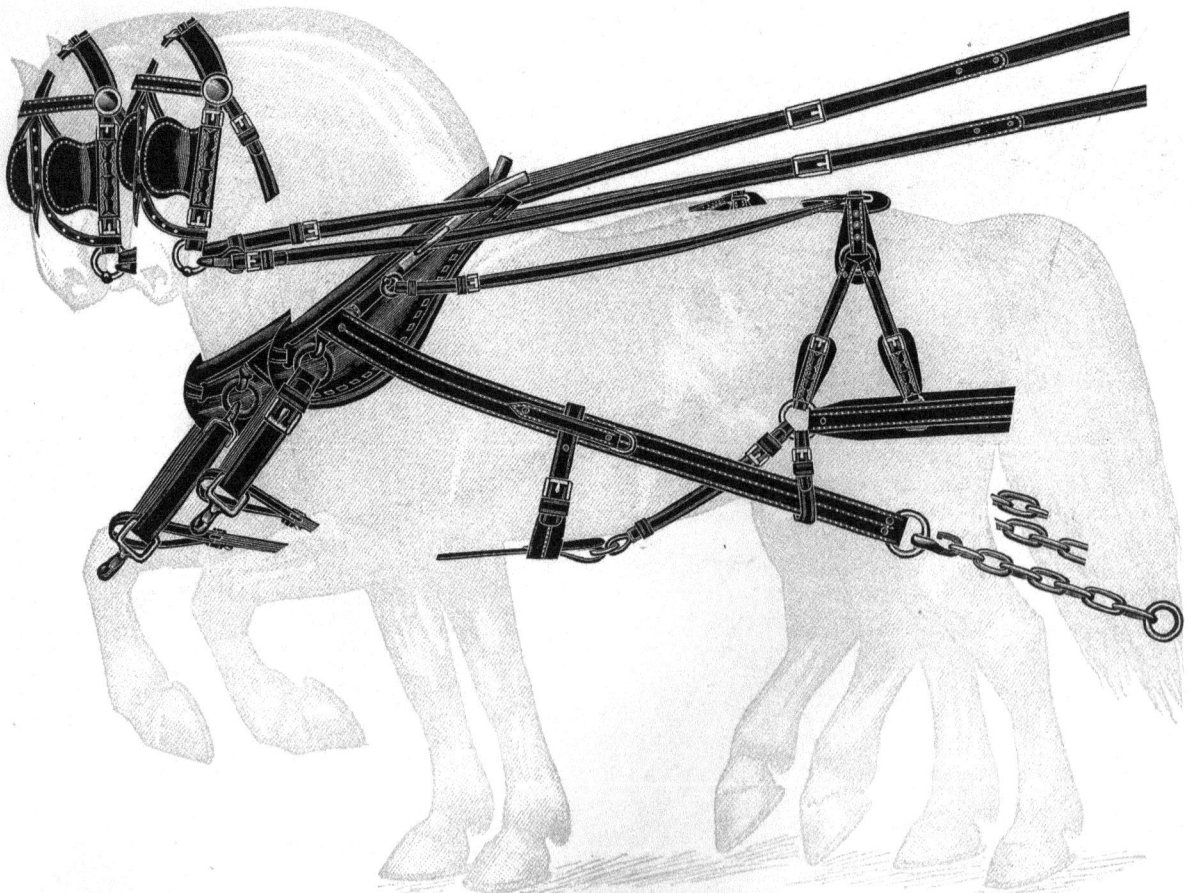

No. 256—Logging Harness

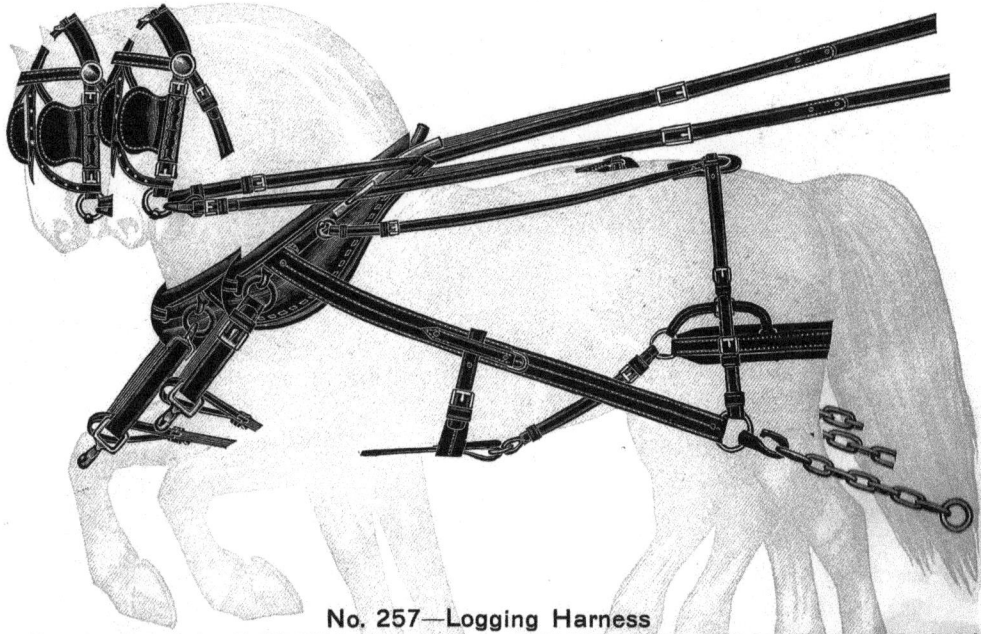

No. 257—Logging Harness

BRIDLES—No. 1; ⅞-inch; Box Loop Cheeks; Sensible Blinds; Spotted Face Pieces.
LINES—No. 1; 1 inch x 18 ft., with Snaps.
HAMES—No. 10; X. C. Black, Concord Bolt.
TRACES—2¼-inch; Three Ply Solid; Three Rows of Stitching; Double and Stitched Billets; Pinery Hooks and Dees; 30-10-1 Butt Chains.
BREAST STRAPS—No. 1; 1½-inch, with Bolt Snaps and Roller Snaps.

POLE STRAPS—No. 1; 1½-inch, with Safe and Ring in end.
BELLY BANDS—No. 2; 1¼-inch; Folded.
BREECHING—Folded; 1½-inch Layers; 1¼-inch Double Turnbacks; 1¼-inch Single Hip Straps; 1½-inch Side Straps.
HAME STRAPS—No. 1; 1 and 1⅛-inch.
SPREADERS—No. 1; ¾-inch; Ring in end.

No. 258—Logging Harness

BRIDLES—No. 1; ⅞-inch; Box Loop Cheeks; Sensible Blinds; Spotted Face Pieces.
LINES—No. 1; 1⅛ inch x 18 ft., with Snaps.
HAMES—No. 114 Lumberman's Bolt.
TRACES—2½-inch; Three Ply, Solid; Three Rows of Stitching; Double and Stitched Billets; Pinery Hooks and Dees; 30-10-0 Butt Chains.
BREAST STRAPS—No. 1; 1¾-inch, with Bolt Snaps and Roller Snaps.

BREAST STRAP EXTENSIONS—No. 1; 1¾-inch.
POLE STRAPS—No. 1; 1¾-inch, with Safe and Ring.
BELLY BANDS—No. 1; 1½-inch; Folded.
BREECHING—Folded; 1½-inch Layers; 1¼-inch Double Turnbacks; 1½-inch Single Hip Straps; 1¼-inch Side Straps.
HAME STRAPS—No. 1; 1 and 1⅛-inch.
SPREADERS—No. 1; ¾-inch, with Rings.

Japan Trimmed   Finish—Double Round Edge Creased

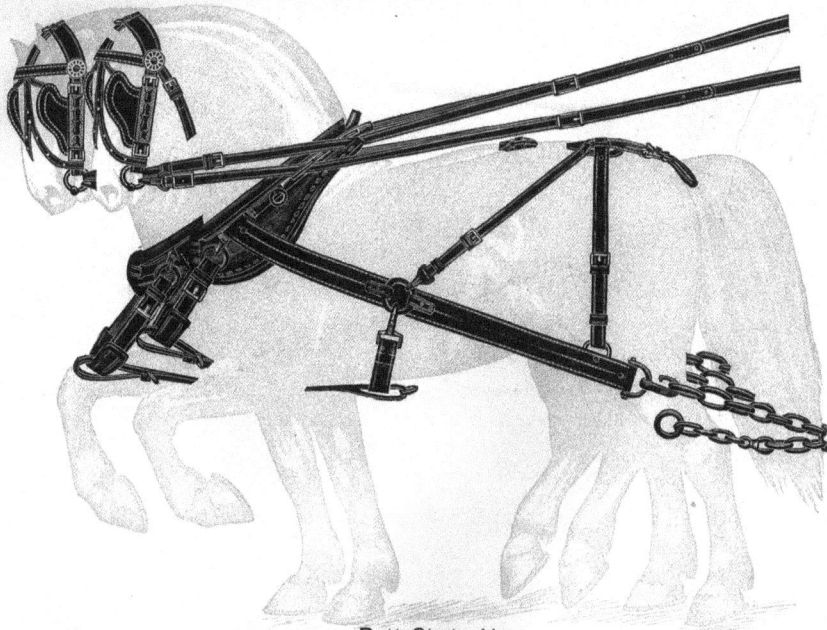

Butt Chain Harness

In Shropshire, England, William Castle, violin maker and farmer, employs Percheron horses to do his field and garden work. He uses a North American two-strap brichen work harness, a padded collar, and an open, blinderless (or blinker-free) bridle over a halter.

However, in this case, he is using a customized cart harness with padded reinforced back pad to carry the weight of the mower shafts. The lines, as is common in his part of the world, are braided cord.

Ears forward, head and neck comfortable, padded collar well fit, this willing mature mare is the very picture of ease, conditioning and good health.

Pokes and Jabs at Harness Futures

It is not within the scope of my knowledge, or intelligence, to begin to imagine what new synthetic materials and redesigned hardware might be brought to use in future harness. On the one hand, I can wish that useful and beneficial innovations continue to be found, while on the other hand, I do lament the meddlesome arrogance which so quickly relegates solid, elegant technologies to obscurity in favor of the untested whimsy of commercially subsidized *resmirch* and *derailment* (R&D).

I like to imagine that plastics, and indiscriminate use of any petroleum products, will be restricted if not outlawed within the next fifty to seventy years.

Cheap and rubbery got us hooked on plastics, who's to say someone won't come up with an entire new kitchen recipe which strengthens and preserves fruit and vegetable leathers, or beef jerky, enough to use it to harness those draft Monitor lizards and genetically modified capons of the future.

Our dear, dear long lost buddy Bulldog Frazier, Montana horselogger and farmer of supreme talents, great good humor and tough allure, once said to the author, "Boss, just smile and wave. They might not understand, but you will and it looks better in the picture."

Sometimes we've got to just for-
get about the theory and let the
job at hand guide the harness.
After all, that hump in the way
may just be the ticket.

Be it dog, goat, turtle, capon,
burro or Bucephalus, it's *the
union of* and *communion with*
which will always give us go.

Well-fitted hames set tight in the perfectly fitted work collar on the Littlefield Farm Fjord horse. Mood is all morning preparation for the working day ahead. Calm, attentive and ready, no teamster could ask for more.

"And indeed, a horse who bears himself proudly is a thing of such beauty and astonishment that he attracts the eyes of all beholders. No one will tire of looking at him as long as he will display himself in his splendor." – Xenophon

"There is a touch of divinity even in brutes, and a special halo about a horse, that should forever exempt him from indignities." – Herman Melville

Notice the stories of the competing lines: The fence line is insistent. The crop rows burgeoning and defining. But it is the curves of the young man's walk, handles in steerage, hat brim capping the perspective, the lift of the trace angle, the ticklish little curve of the tops of the hames, the wave-like flow of the lines, the well fit basket of the harness, and the willing horse picking her steps. I apologize if you wish, but I will never tire of the view. It defines the very best of man's possibility and that given he is but a part of it all.

Photo by Ryan Foxley of his son on Littlefield Farm

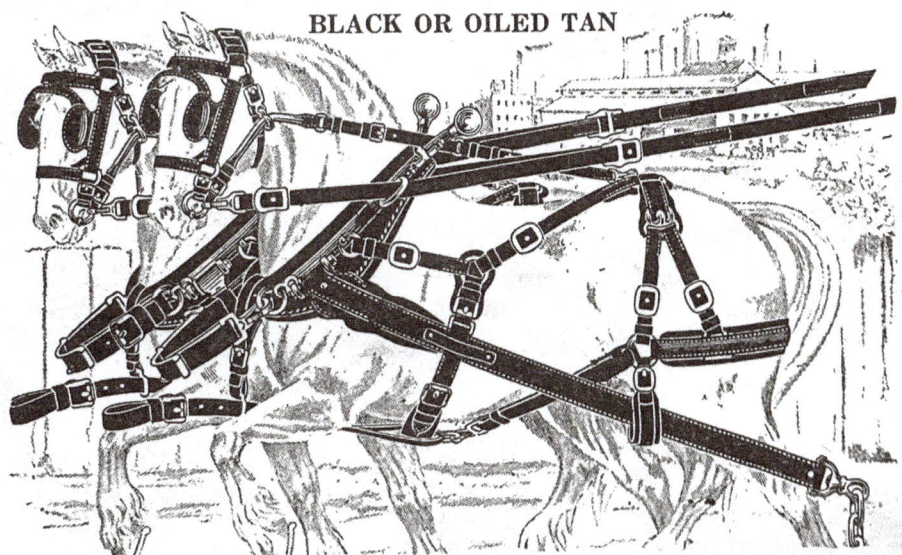

SPRING, 1940

Reproduced thru courtesy "FARM JOURNAL"—The National news magazine for the farm family

Southern Saddlery Company

CHATTANOOGA, TENNESSEE • U.S.A.

BLACK OR OILED TAN

DUNCAN & SONS, Inc., SEATTLE

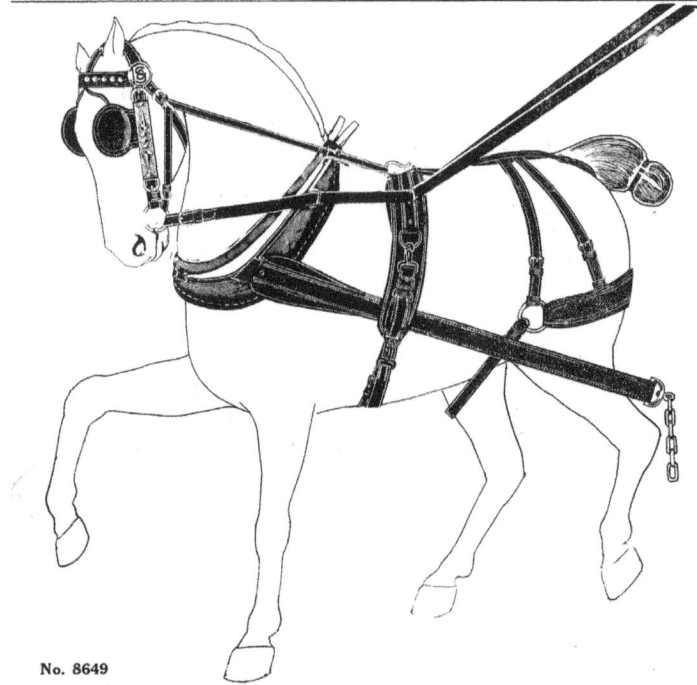

No. 8649

Whether it be Dodson-Fisher or Duncan & Sons or Hobbs or George Lawrence or Southern Harness or Walsh or whomever, most every one of these companies had a 'touch' or 'manner' to their harness design and construction that set them apart. And knowledgeable teamsters developed passionate loyalties for their maker of choice. Way back, before economists stripped Capitalism of its empathy for true craftsmanship and artistry, the elegance of design won the day every day.

REAL QUALITY FARM HARNESS
Round-Edge, Black or "Oiled Tan"

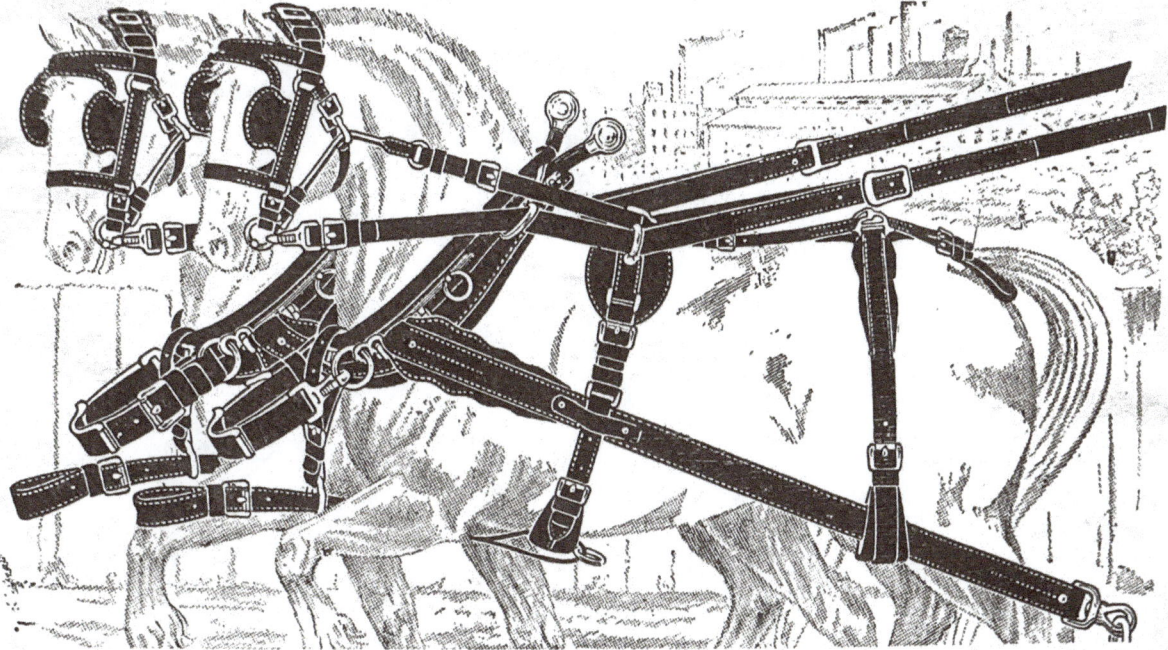

No. 1960

Steve Hagen and his superb Belgians cultivating a Willamette Valley vineyard.

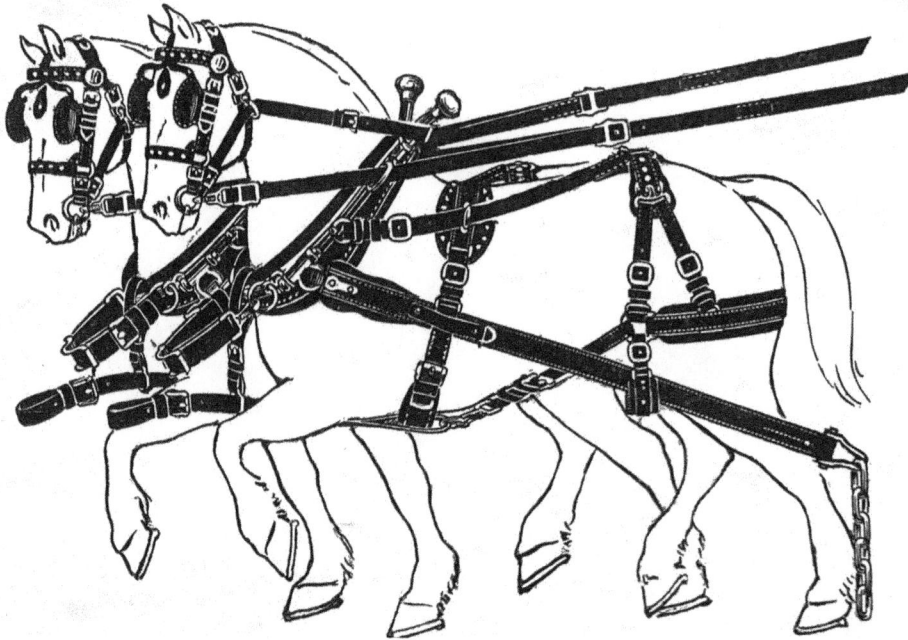

The side-backer harness, such as the one below, gave greater precision to backing and turning, but it required more work hooking and unhooking. In city traffic the precision was rewarded, as with backing manure spreaders into tight barn alleys, but for general field and road work the standard harness picture above worked just fine.

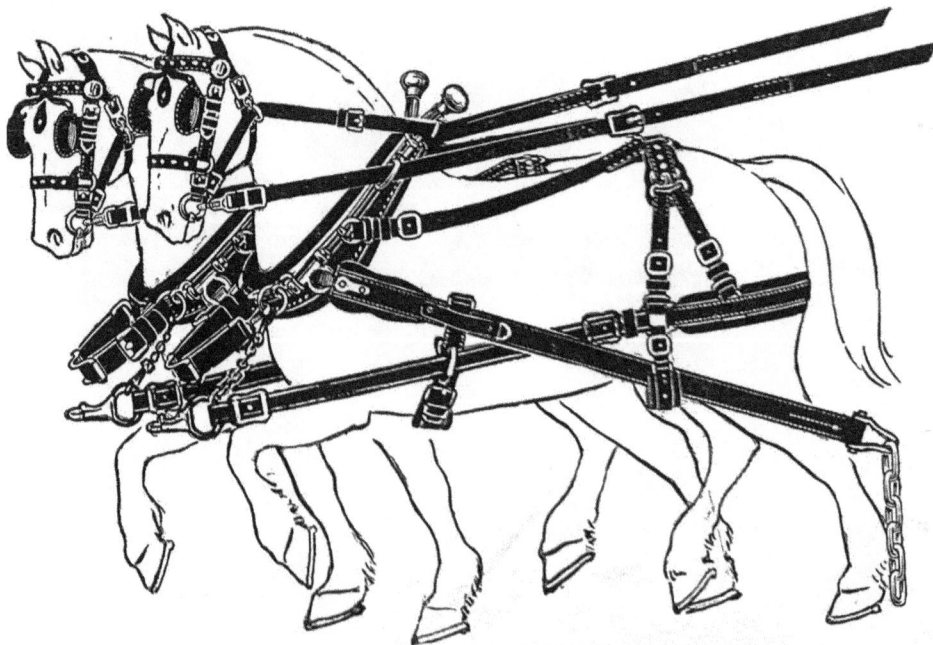

Look, on page 231, at how the trace runs under the side-backer straps, whereas above it is vice versa. Who's right?

No. 1—BOSTON TRUCK HARNESS

A **TRUE** D RING Harness
by another name.

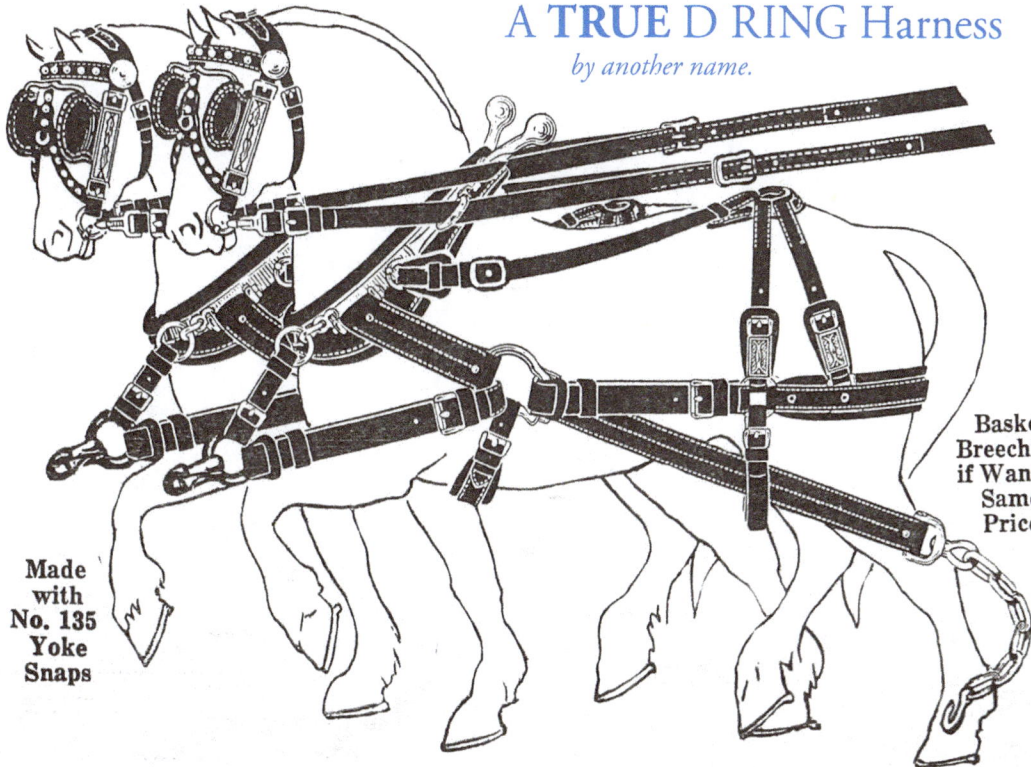

Made
with
No. 135
Yoke
Snaps

Basket
Breeching
if Wanted
Same
Price

No. 1 BOSTON TRUCK

Bridles—Boston Truck ⅞-inch box loop cheeks, ¾-inch throat, 1-inch D & S spotted front, ⅞-inch x 7 ft. flat reins, ⅝-inch spotted ring face piece. Per Pair $13.00

Lines—Boston Truck, 1⅛ x 20 with buckles and billet and Snaps. D. & S. at center check. Per Pair 11.80

Hames—New England, No. 8 mule, bolts, brass balls and long spot. Per Set 15.60

Hame Straps—New York, 1 x 24-inch, roller buckle and leather loops. Per Set 1.40

Spreaders—New York, ¾-inch roller buckle and leather loops, 2-inch brass rings. Per Pair 1.20

Traces—No. 1 Boston Truck, 2-inch front sewed around bolt; No. 331 trace loop sewed and hand copper riveted, 1¾-inch rear, 3-ply, 3-row stitching, No. 86-24-1 screw dee, hook heel chain. Per Set 31.00

Belly Bands—Boston truck, 2½-inch folded, 1½-inch layer and billets, hand copper riveted. Per Set 5.00

Hips and Turnbacks—Boston Truck, 1⅛ crotch turnbacks to hames, heavy padded rump safe, 1⅛-inch Hip straps. Per Set $10.50

Breeching—Indiana, 3 x 42-inch heavy folds, with 1¾-inch full length layer, dees in end, hand copper rivets, 1⅛-inch box loop lead ups with safes. Japan buckles. Per Set 10.50

Lazy Straps—Boston Truck, 1⅛-inch with roller buckle and leather slide loops. Per Set 3.30

Yoke Straps—Boston Truck, 1¾-inch heavy double reverse to go from dee in trace to yoke hooks. Per Set 8.20

Yoke Snaps—No. 135 heavy 1¾-inch snaps. Per Set 3.80

Carrying Straps—1⅛-inch with roller buckles and leather loops. Per Set 3.30

Side Straps—Boston Truck, 1½-inch heavy double reverse to go from dee in trace to breeching rings. Per Set... 8.20

Trimmings—Black harness Japan and brass trimmed; Oiled Tan harness cadmium trimmed.

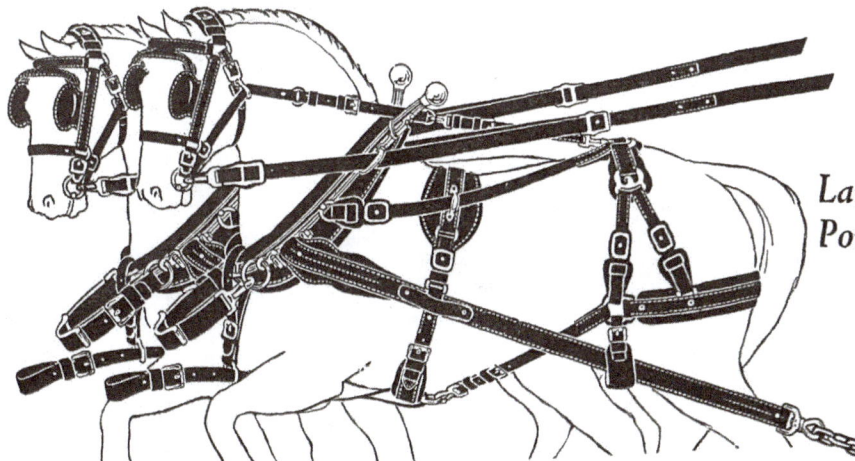

*Large for 1600
Pound Horses*

THESE HARNESSES FIT LARGE PERCHERON HORSES

LARGE TEAM HARNESS FOR 1600-LB. HORSES

QUALITY GUARANTEED—
Black or Oiled Tan

Bridles—No. 1840F, ⅞-inch cheeks, ¾-inch bit straps, ⅞-inch spotted braced nose band, 1-inch spotted front, round winker stay with drop, layer on crown, gag swivels on throat billet, ⅞-inch flat reins. Per Pair. **$10.50**

Lines—No. IE heavy steerhide, 1 inch x 20 feet with 6½-foot cross checks, with No. 150 roller buckle at center and No. 1880 buckles and snaps at front. Per Pair **8.20**

Hames—No. 94, steel, bolt style, brass or nickel ball tops, Per Set **6.20**

Hame Ears—Wear leathers under trace bolt. Per Set... **.50**

Hame Straps—No. 17, 1 inch x 21 inches with roller buckles and twin loops. Per Set **1.30**

Spread Straps—¾-inch with 1¾-inch ring. Per Pair.. **.90**

Traces—1½-inch x 6 feet, 3 ply, 2-row stitched, dee in keeper, advance attachments on front end; No. 488 screw dee and 8-link heel chains on rear end, 1¼-inch belly band billets. Per Set **24.10**

Belly Bands—2-inch double and stitched with 1¼-inch buckles and leather loops. Per Pair **2.20**

Pads—4¾-inch x 21 inches double and stitched, spotted, metal bridges, 1¼-inch layer with thill loops in end and 1¼-inch market straps to reverse. Per Pair **7.30**

Turnback and Hips—1-inch crotch turnbacks, 1¼-inch spotted hip layer, with 1-inch hip straps to reverse. Per Pair ......**$ 9.30**

Breeching Body—2½ x 42-inch single strap body, with 1½-inch full-length layer with wear leather around dees. Per Pair **6.50**

Flank Straps—1⅛-inch with snap one end and Conway buckles on both ends. Per Set **3.30**

Lazy Straps—¾-inch with safe. Per Set **2.20**

Martingales—1½-inch with reverse buckle and slide loop on front end with ring in lower end, loose collar strap slide. Per Pair **2.10**

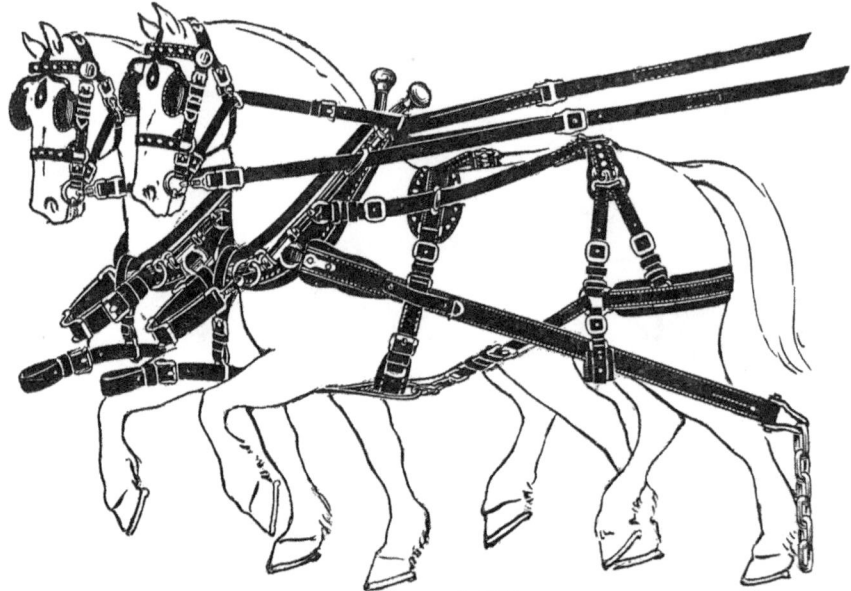

Nos. 570 and 0572

Collar Straps—⅞-inch x 32 inches with roller buckles and twin loops. Per Pair .... **$ .80**

Breast Straps—1½-inch made up 4 feet, 6 inches long with snap and slide. Per Pair **3.20**

Trimmings—Black harness with Japan hardware and brass ball hames. Oiled tan harness with cadmium hardware and nickel ball hames.

Finish—Creased edge.

No. 570 BLACK or 0572 OILED TAN, 1½-inch Standard, creased edge finish, complete less collar. Per Set **$88.00**

No. 571 BLACK or 0573 OILED TAN, 1¾-inch Standard, creased edge finish, as above with following changes—1½-inch lines; 1⅛-inch back strap and hips; 2¾-inch breeching, 1½-inch belly bands, 1¾-inch side straps, 1¾-inch traces, martingales and breast strap, complete less collar **94.00**

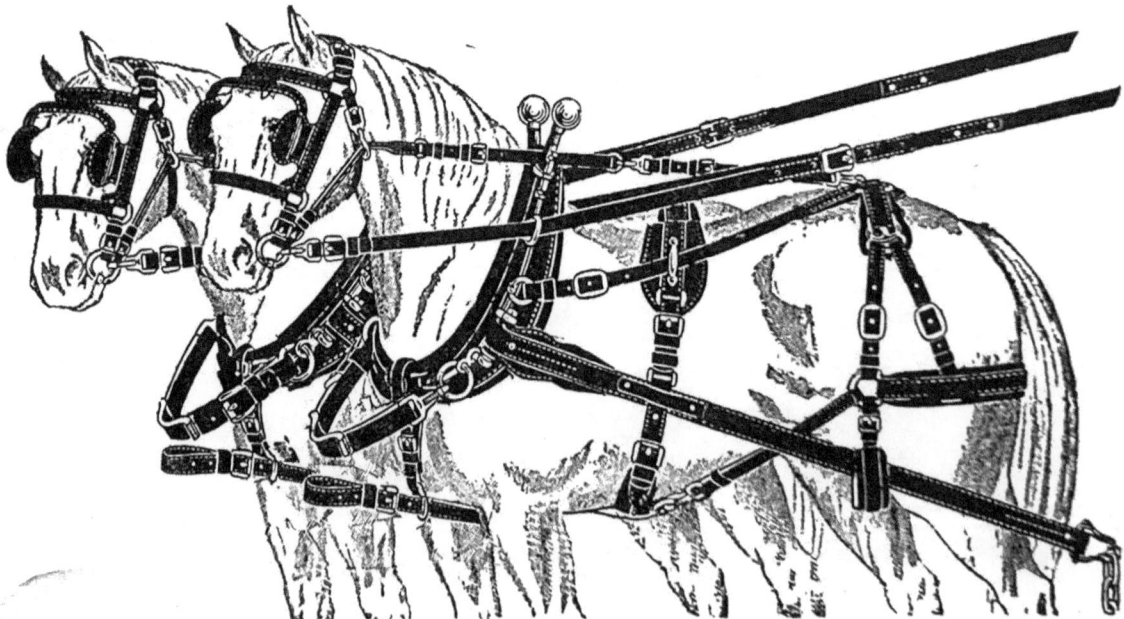

CRAFTSMANSHIP—FACTORY PRICES

BIG BARGAINS IN PLOW HARNESS
A MATCHLESS VALUE, MEDIUM SIZE FOR 1000 LB. HORSES

If wanted
with 26-8-2
Breast Chains
instead of
Straps deduct
$1.00 per set.

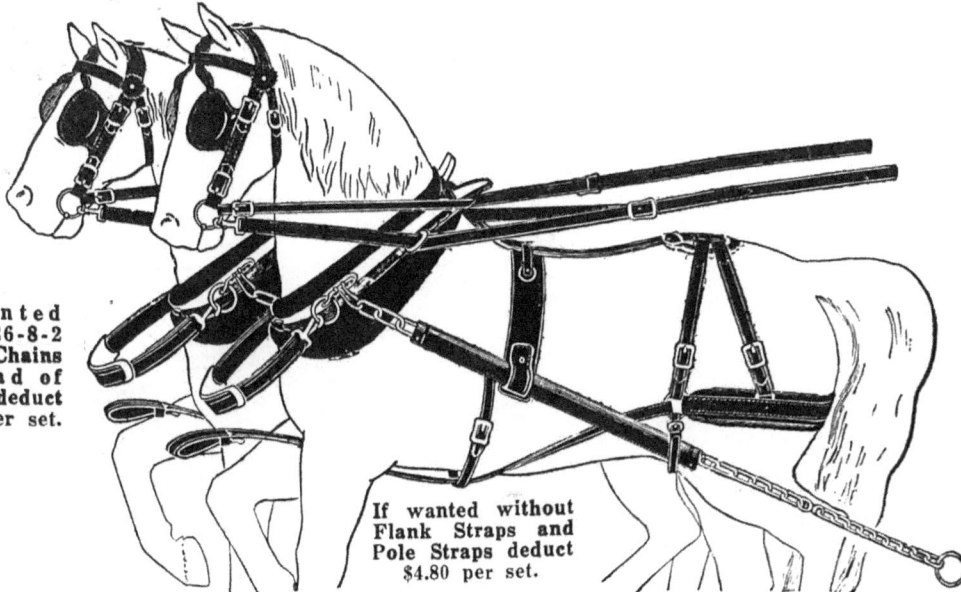

If wanted without
Flank Straps and
Pole Straps deduct
$4.80 per set.

No. 467 LESS COLLARS, PER SET (Shipping Weight about 50 lbs.) ---------------------- $39.70

Bridles—1-inch harness leather, cupped blinds, with check reins.
Lines—No. 11, 1-inch by 14 feet, snaps sewed on.
Hames—No. 1, adjustable draft, varnished, steel over top.
Chains—7 feet long, 8-link, 2-wire.
Piping—No. 50, 36 inches long.
Back Bands—No. 110, 3 inches wide, with loops.
Belly Bands—No. 110, 1¼-inch, single strap.

Breeching—2¼-inch folds, wide layer.
Turnback—1-inch, single to hames.
Hip Straps—1-inch.
Side Straps—1-inch, single.
Breast Straps—1¼-inch, with snap and slides.
Pole Straps—No. 110, 1¼-inch.
Lazy Straps—¾-inch.

If wanted
with 26-8-2
Breast Chains
instead of
Straps deduct
$1.00 per set.

*A page from the depression-era Southern Harness catalog. Please compare to the
Wallace & Smith catalog illustrations and information from twenty years prior.*

6

THESE HARNESSES FIT LARGE PERCHERON HORSES

"INDIANA" TEAM HARNESS

Quality Guaranteed—Black or Oiled Tan

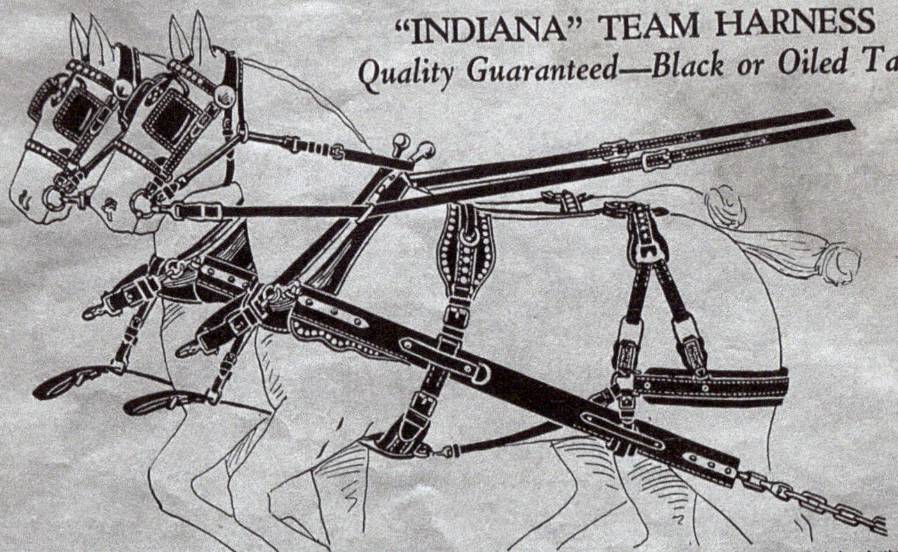

Bridles—Indiana, 7/8-inch short cheeks, box loop, 7/8-inch bit straps, 1-inch doubled and stitched spotted front, 1-inch spotted noseband, round winker with spotted drop, 7/8-inch round reins, 3/4-inch gag sewed over crown. Large square blinds, per pair $15.00

Lines—Indiana, 1 1/8-inch x 20 feet, 1880 buckles with spring snaps, doubled and stitched at check buckle, with 3/4 inch extensions, per pair 12.00

Hames—No. 920 Lone Star, 2-inch brass balls, per set 9.20

Hame Straps—No. 1424—1 x 24-inch roller buckles and leather loops, per set of (4) 1.40

Spreaders—"New York," 3/4-inch with roller buckles and leather loops with 2-inch solid brass rings, per pair 1.20

Traces—"Indiana" folded, 2 1/2 x 5 1/2 feet, 3-hole clips front end, No. 61-24-1 Heel Chains clipped on 3-hole clips, 1 1/2-inch Belly Band billets, per set 29.60

Belly Bands—Folded with 1 1/2-inch layer, hand riveted laps, per pair.. 3.10

Pads—"Indiana," 5 3/4 inches wide, extra long with wide jockey, leather lined, spotted edge and layer, doubled and stitched center loop with brass ring, 1 1/2-inch double market straps, per pair $10.50

Hips and Turnbacks—"Indiana" 1 1/2-inch turnback to buckle at rump, 1 1/2-inch spotted hip layer, 1 1/8-inch hips sewed into brass side trace carriers, per pair 10.50

Breeching—3 x 42-inch folds, dees in ends with 1 3/4-inch layer, 1 1/8-inch box loop tugs with safes and brass buckle shields. Per set.. 12.00

Side Straps—"Indiana," 1 1/4-inch x 5 1/2 feet with liner. Per set 6.00

Lazy Straps—1-inch with wide safe, per set 2.60

Choke Straps—"Indiana," 1 3/4-inch with billet, dee and safe lower end, 1-inch collar straps, per pair 5.60

Breast Straps—New York, 1 3/4-inch with leather loops and roller snaps. Per Pair 4.60

Trimmings—Black Harness, Japan and brass trimmings. "Oiled Tan" Harness cadmium trimmed.

No. "INDIANA"—Breeching style, 2 1/2 in. x 5 1/2 ft. folded traces, less collars, Per Set $123.30

Black or Oiled Tan

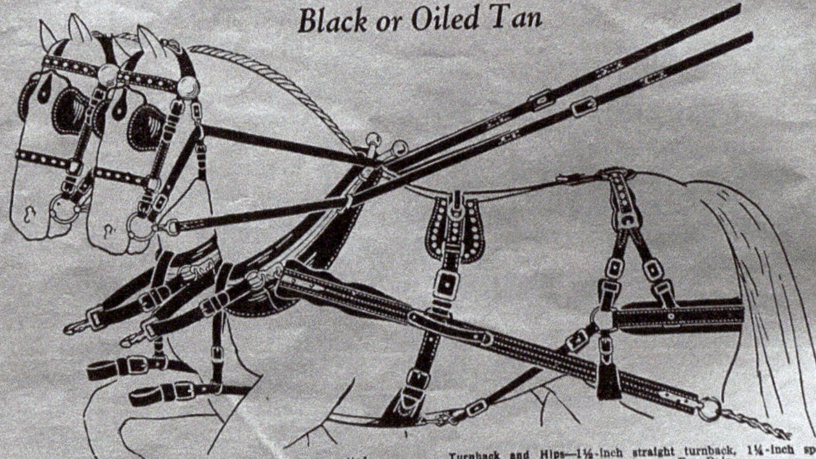

Bridles—7/8-inch bar buckle cheeks, 3/4-inch bit straps, 3/4-inch spotted nose band, 7/8-inch spotted front, round bow stay with billet, gag swivel on 5/8-inch throat, 7/8-inch flat reins. Per Pair .... $ 8.00

Lines—No. 11, 1 1/8 inches x 18 feet with 6-foot cross checks, with No. 121 bar buckle at center and Conway buckles and snaps at front. Per Pair 6.50

Hames—No. 94 steel, adjustable or clip style, brass or nickel ball tops. Per Set 6.20

Hame Straps—1 inch x 22 inches with roller buckles and leather loops. Per Set 1.30

Traces—1 3/4-inch x 6 feet, 3-ply, 3-row stitched, with two-hole clip on front end, and No. 61-24-1 heel chains and three-hole clips on rear end, with 1 1/2-inch belly band billets and with D in keeper. Per Set 24.70

Belly Bands—"Columbus" folded, with layer, 1 1/2-inch buckles and leather loops. Per Pair 2.50

Pads—Columbus, 6 x 20 inches, double and stitched, spotted, with 1 1/2-inch layer, dees in end with 1 1/2-inch market straps double reverse. Per Pair 8.00

Turnback and Hips—1 1/2-inch straight turnback, 1 1/4-inch spotted hip layer, with 1 1/8-inch hip straps. Per Pair .... $ 6.00

Breeching Body—2 1/2 x 42-inch single strap body, with 1 1/2-inch short layer with wear leather around dees, 1 1/8-inch uptugs. Per Pair.. 7.00

Flank Straps—1 inch x 6 feet double reversed with snaps. Per Set.... 5.70

Lazy Straps—7/8-inch with safe. Per Set 2.10

Martingales—1 1/2-inch with reverse buckle and slide loop on front end with rings and safe in lower end and loose collar strap slide. Per Pair 3.20

Collar Straps—7/8 inch x 32 inches with roller buckles and twin loops. Per Pair .... .80

Breast Straps—1 1/2-inch made up 4 feet, 6 inches long with snap and roller snap. Per Pair 3.90

Spread Straps—3/4-inch with 1 3/4-inch ring. Per Pair .... .80

Trimmings—Black harness with Japan hardware and brass ball hames. Oiled tan harness with cadmium hardware and nickel ball hames.

Finish—Creased edge.

No. "J.G.W." Black, complete less collars, Per Set $86.70

No. "O.J.G.W." Oiled Tan, complete less collars, Per Set 86.70

Also made with 2 1/2-inch x 6 foot folded traces with spotted fenders, Add Per Set 5.00

Southern Saddlery Co., LEATHER DEPENDABLE GOODS Chattanooga, Tennessee

No. 1—NEW YORK FARM HARNESS

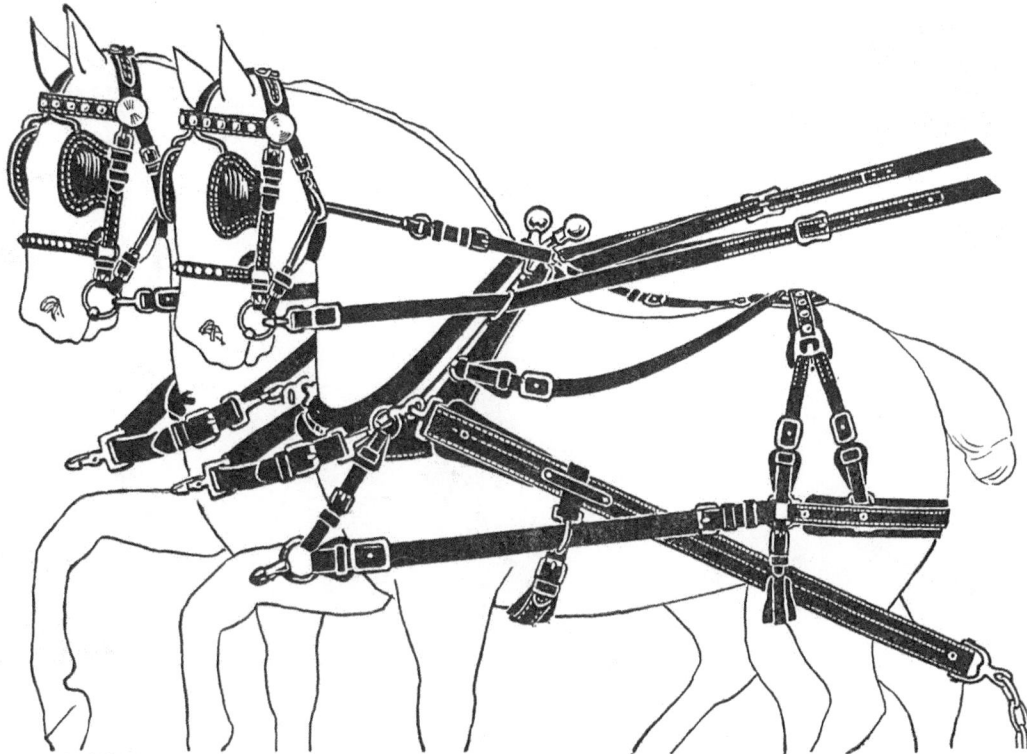

No. 1 NEW YORK

If a novice were to look long and hard at lots of harness catalog drawings and photos, paying attention to angles, adjustments and conformation, it would be difficult to get useful readings. There is so much variation pictured in all these cuts. How's the uninformed to make a judgement?

For example, above, the traces attach to the hames slightly ahead of and below the point of the shoulder which would, after a day of plowing, render these horses less that willing because of the pain. The headset of this team, while upright, tucked and dramatic, is foreign to comfortable work and gives a misleading picture about what a new teamster might expect as ideal. Following the lines of the harness, the suggestion is that the check rein running back to the brichen spider (no crupper employed) suggests the team holds the harness up on the hip by the bits.

If one were to look at this drawing, and many of the others, to determine proper fit, many mistaken conclusions would be arrived at. I believe that harness makers, when designing the catalogs of the future, need to make extra effort that their illustrations do service to their customers, the animals and the craft. In the end loyalty and good business may be expected.

LAWRENCE

1935 HARNESS AND SADDLERY CATALOG
All Prices Subject to Change Without Notice

Honest Horse Power -- Farm Bred -- Farm Fed
Lawrence Harness on Good Horses Will Cut Production Costs

Our Written GUARANTEE of Quality on Each Set

THIS HARNESS IS MADE OF SOLID OAK-TANNED LEATHER. IT IS UNCONDITIONALLY GUARANTEED FOR ONE YEAR AGAINST FLAWS IN MATERIAL OR WORKMANSHIP.

The GEORGE LAWRENCE CO.
PORTLAND, OREGON
Since 1857

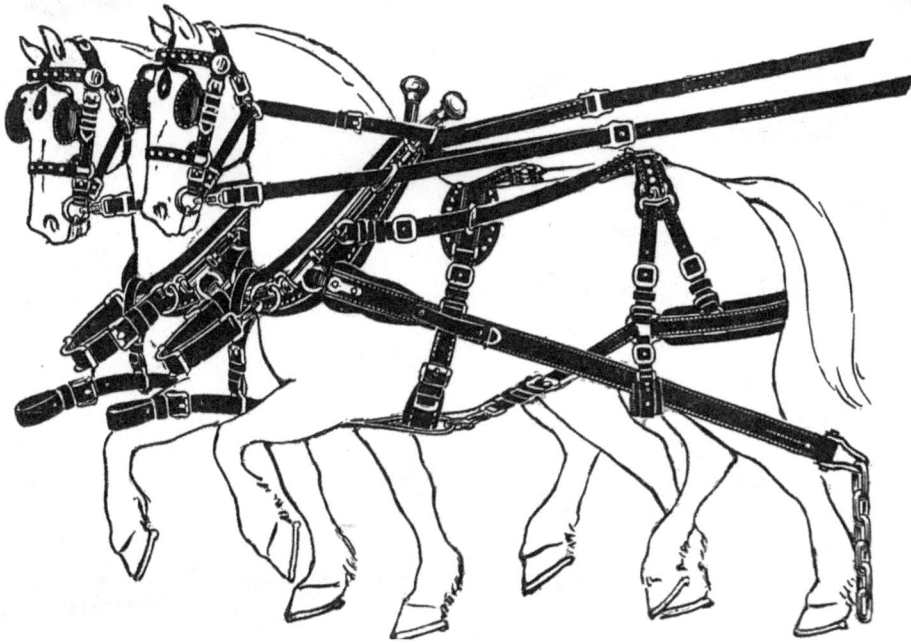

Conventional heavy western two-strap basket brichen work harness, lightly spotted.

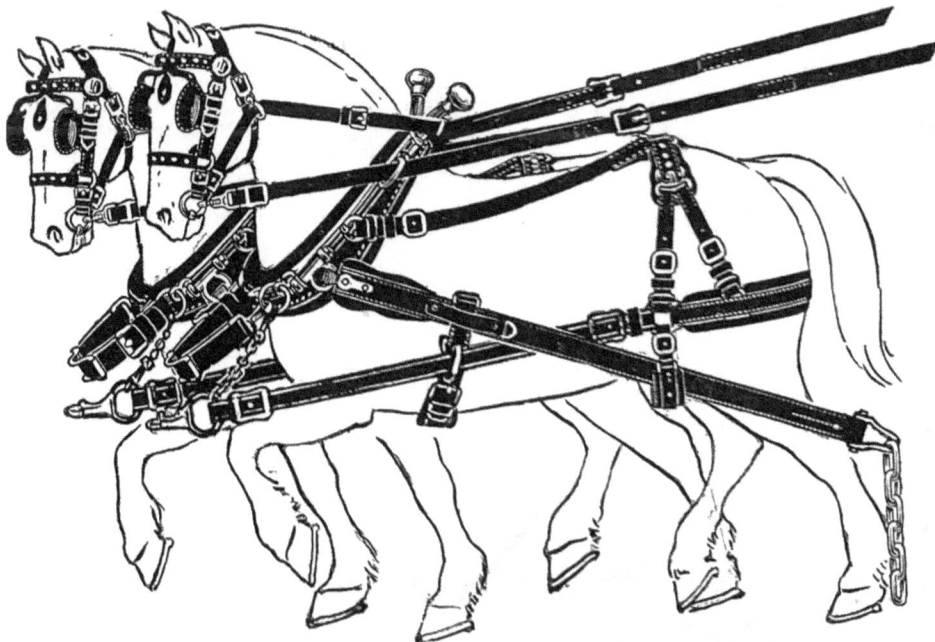

A side-backer harness with no back pad. Side-backer straps threaded through belly band billet slides and running over traces.

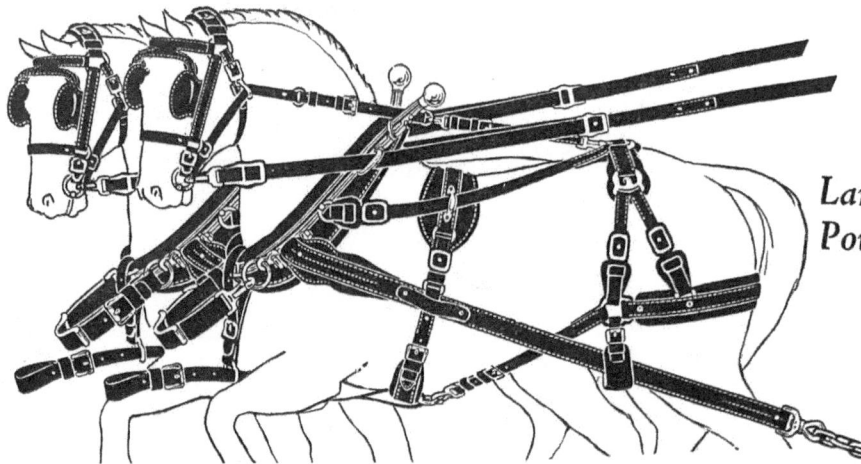

Large for 1600
Pound Horses

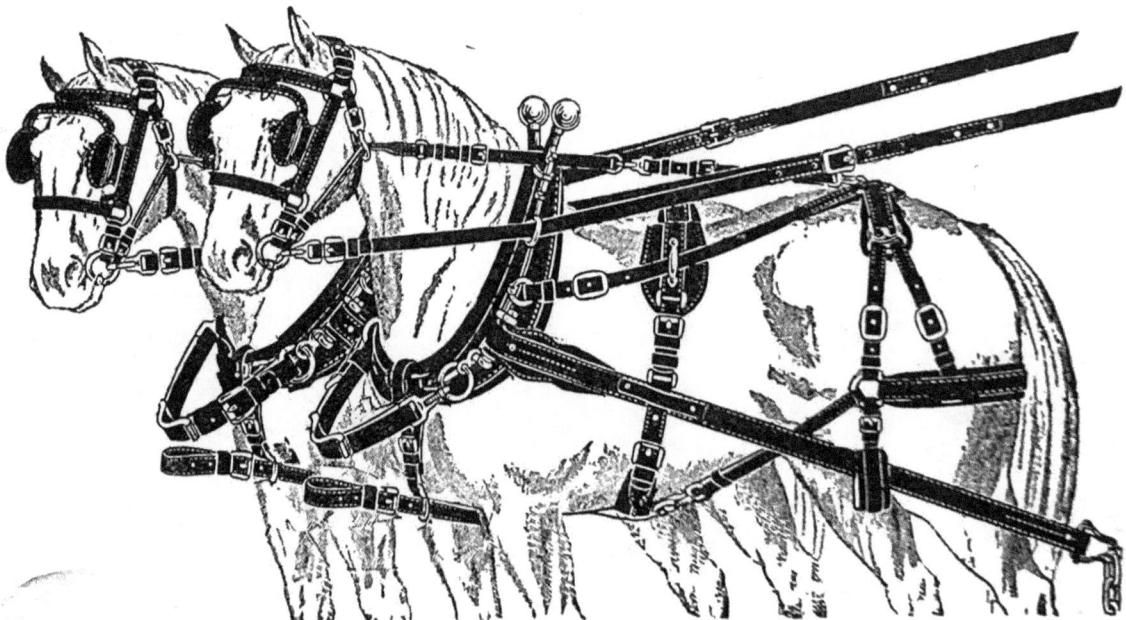

How to Size Horses and Harness.

In 1913 there were millions of horses in harness in North America. The majority of the farm work horses averaged 1400 - 1600 lbs. and were the result of crossing 1800 to 2000 lb. Draft stallions with (or on) light mares (usually around 900-1150 lbs). Back during that golden age of agriculture and horse farming, only breeders of registered animals, commercial enterprises (for example, beer companies) or wealthy enthusiasts had 17-18 hand one ton horses in harness.

In 1917 'large' in work horses and mules was 1600 lbs. When ordering harness, neck size, heart girth, height at withers, and collar or hame size are needed by the makers. For **neck** sizing see pages 26, 44. For **heart girth** run a cloth measuring tape around the animal from withers to just behind the front legs. For **height,** with animal's head held up, measure from the ground to the top of the withers. Divide that length by 4 and you have the **hand's high** measurement. Each 'hand' is four inches in length, so 16 hands equals 64" from the ground to the withers.

THESE HARNESSES FIT LARGE PERCHERON HORSES

LARGE TEAM HARNESS FOR 1600-LB. HORSES

QUALITY GUARANTEED—
Black or Oiled Tan

Bridles—No. 1840F, ⅞-inch cheeks, ¾-inch bit straps, ⅞-inch spotted braced nose band, 1-inch spotted front, round winker stay with drop, layer on crown, gag swivels on throat billet, ⅞-inch flat reins. Per Pair. $10.50

Lines—No. 1E heavy steerhide, 1 inch x 20 feet with 6½-foot cross checks, with No. 150 roller buckle at center and No. 1880 buckles and snaps at front. Per Pair. 8.20

Hames—No. 94, steel, bolt style, brass or nickel ball tops. Per Set. 6.20

Hame Ears—Wear leathers under trace bolt. Per Set. .50

Hame Straps—No. 17, 1 inch x 2½ inches with roller buckles and twin loops. Per Set. 1.30

Spread Straps—¾-inch with 1¾-inch ring. Per Pair. .90

Traces—1½-inch x 6 feet, 3 ply, 2-row stitched, dee in keeper, advance attachments on front end; No. 488 screw dee and 8-link heel chains on rear end, 1¼-inch belly band billets. Per Set. 24.10

Belly Bands—2-inch double and stitched with 1¼-inch buckles and leather loops. Per Pair. 2.20

Pads—4¾-inch x 21 inches double and stitched, spotted, metal bridges, 1¼-inch layer with thill loops in end and 1¼-inch market straps to reverse. Per Pair. 7.30

Turnback and Hips—1-inch crotch turnbacks, 1¼-inch spotted hip layer, with 1-inch hip straps to reverse. Per Pair. $ 9.30

Breeching Body—2½ x 42-inch single strap body, with 1½-inch full-length layer with wear leather around dees. Per Pair. 6.50

Flank Straps—1½-inch with snap one end and Conway buckles on both ends. Per Set. 3.30

Lazy Straps—¾-inch with safe. Per Set. 2.20

Martingales—1½-inch with reverse buckle and slide loop on front end with ring in lower end, loose collar strap slide. Per Pair. 2.10

Nos. 570 and 0572

Collar Straps—¾-inch x 32 inches with roller buckles and twin loops. Per Pair. $ .80

Breast Straps—1½-inch made up 4 feet, 6 inches long with snap and slide. Per Pair. 3.20

Trimmings—Black harness with Japan hardware and brass ball hames. Oiled tan harness with cadmium hardware and nickel ball hames.

Finish—Creased edge.

No. 570 BLACK or 0572 OILED TAN, 1½-inch Standard, creased edge finish, complete less collar. Per Set. $88.00

No. 571 BLACK or 0573 OILED TAN, 1¾-inch Standard, creased edge finish, as above with following changes—1½-inch lines; 1⅛-inch back strap and hips; 2¾-inch breeching, 1½-inch belly bands, 1⅛-inch side straps, 1¾-inch traces, martingales and breast strap, complete less collar. 94.00

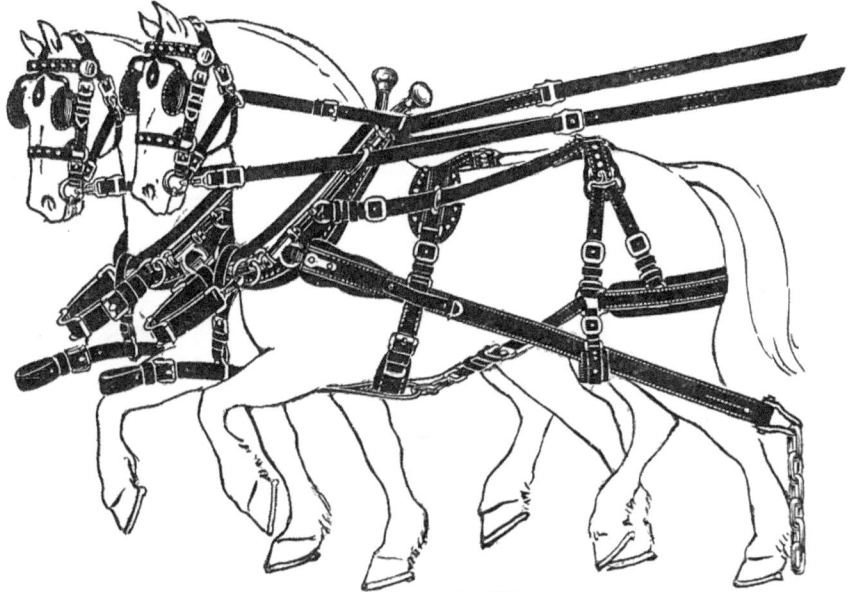

Six Suffolk Punch draft horses working in the field. The team in the center is hitched to the pole. The outside four are tied back in with jockey sticks. For greater detail on such larger hitches we recommend you find a copy of the Work Horse Handbook or Training Workhorses / Training Teamsters by Lynn R. Miller.

TEXAS PLOW OR TEAM BODIES

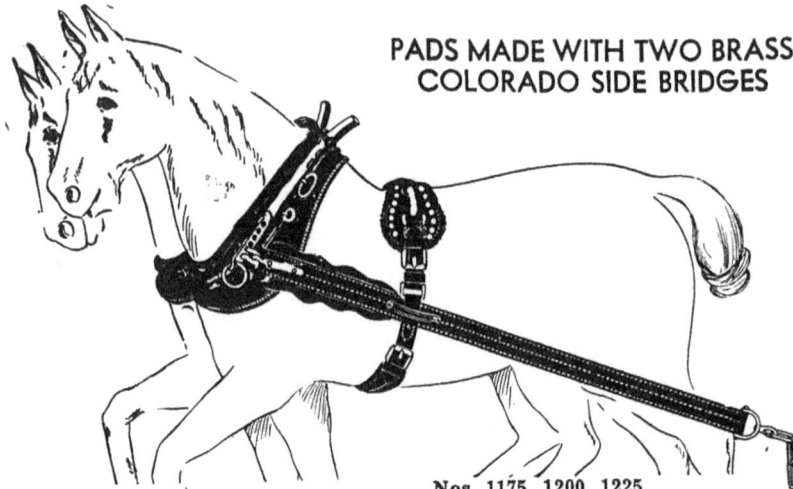

13

PADS MADE WITH TWO BRASS COLORADO SIDE BRIDGES

Nos. 1175, 1200, 1225

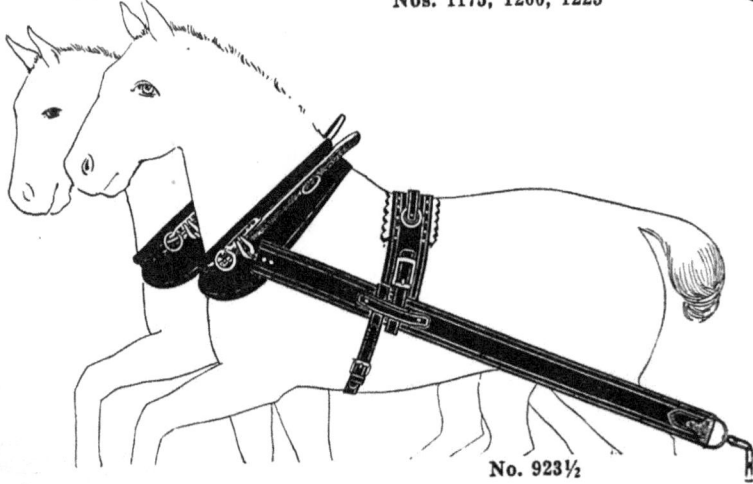

No. 923½

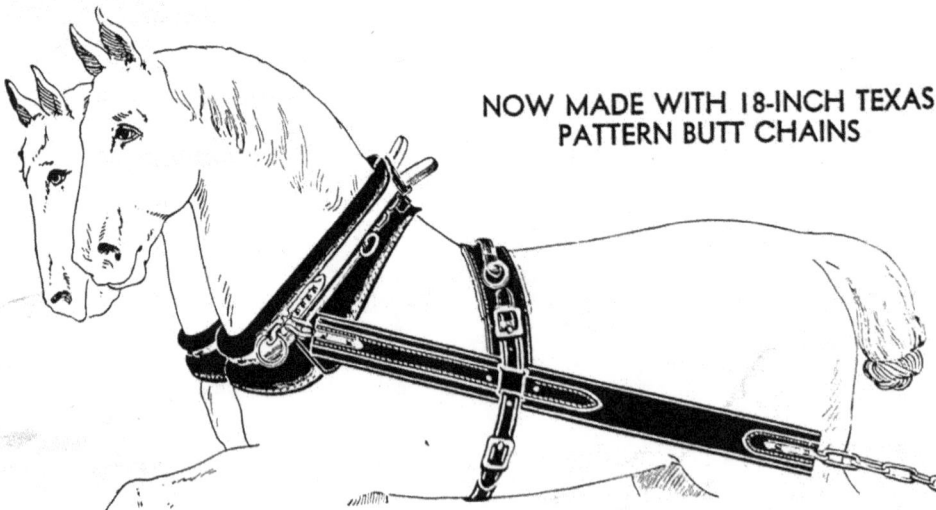

NOW MADE WITH 18-INCH TEXAS PATTERN BUTT CHAINS

No. 1175 TEXAS TEAM OR PLOW BODIES

Traces—1¾-in. x 6-ft., 3-ply, 3 rows stitching, with safes and clips on for hames, 6-link toggles, sewed in, hand riveted. 1¼-in. belly band billets. Set $25.00

Pads—No. 510, 6x20-in., swell flat harness leather, felt lined and spotted, 1½-in. plain layer with dees in end, brass Colorado bridges. 1¼-in. market straps to reverse in. Pair 8.50

Belly Bands—No. 923½. 2¼-in. folds. 1½-in. layer and buckles. Pair 2.50

No. 1200

Traces—2-in. x 6-ft., 3-ply, 3 rows stitching, with safes and clips on for hames, 6-link toggles, sewed in, hand riveted, 1½-in. belly band billets. Set $27.00
Pads—No. 510, per pair 8.50
Belly Bands—No. 923½, per pair .. 2.50

No. 1225

Traces—2¼-in. x 6-ft., 3-ply, 3 rows stitching, with safes and clips on for hames, 6-link toggles sewed in, hand riveted, 1½-in. belly band billets. Set $29.00
Pads—No. 510, per pair 8.50
Belly Bands—No. 923½, pair 2.50

No. 923½ TEXAS PLOW OR TEAM BODIES

Traces—3½-in. x 6-ft., single strap, with clips riveted on. 3-link toggles riveted on, 1½-in. belly band billets. Set $27.00
Back Bands—3½x40-in., 18-in. felt lined center, spotted 1½-in. loose ring in top loop, one 1½-in. bar buckle and billet riveted on each side of band. Pair 4.70
Belly Bands—2¼-in. folds, 1½-in. layer and buckles. Pair 2.50

No. 924

Traces—4-in. x 6-ft., single strap, with clips riveted on, 3-link toggles riveted on, 1½-in. belly band billets. Set $29.00
Back Bands—4x40-in., 18-in. felt lined center, spotted 1½-in. ring in top loop, two 1-in. bar buckles and billets riveted on each side of the band. Pair 6.00
Belly Bands—2¼-in. folds, 1½-in. layer and buckles. Pair 2.50

STANDARD TEAM HARNESS

CUSTOM MADE FANCY SPOTTED WESTERN TEAM HARNESS

CUSTOM MADE BUTT CHAIN HARNESS

STANDARD LUMBER HARNESS

Author backing Polly and Anna with wagon load of loose hay.

The only clear difference between these two harness setups is that the bottom image shows snap and roller on breast straps plus heavy snaps on pole (or martingale) straps while the one to the left shows none. As mentioned previously, this arrangement is for neckyokes designed, in their shape, to hold the pole strap to the ends. (See page 139). And these neckyokes also have larger rings on the ends to allow that the breast straps be threaded through and then taken back to the off side hame ring.

STANDARD HARVESTER OR TEAM HARNESS

STANDARD HARVESTER OR TEAM HARNESS

STANDARD HARVESTER HARNESS

The market tug harness to the right features hip straps that snap into the first trace ring rather than forming a lazy strap to allow the trace to slide.

The harness above has hip straps which are fastened by Conways to the end of the butt chain tugs.

STANDARD FARM CHAIN HARNESS

DUNCAN & SONS, Inc., SEATTLE

CUSTOM MADE LUMBER HARNESS

CUSTOM MADE WESTERN TEAM HARNESS

The Case of Walsh Harness

Made in 10 Standard Styles
For Prices See Pages 40 to 50

Walsh no-buckle Back Pad Harness.

No Rings To Wear

The Walsh Way

The Ordinary Way

12 Big Reasons Why Your Next Harness Should Be a *Walsh*

1. No buckles to tear straps—no buckle holes to weaken them.
2. No rings or dees to wear the straps in two.
3. Walsh Harness can be easily and quickly adjusted to fit any size work horse or mule.
4. Only harness in the world using tested leather.
5. Nothing but the best Irish linen thread used in stitching.
6. All hardware is double zinc-galvanized before being enameled, to prevent rust.

7. Tremendous repeat business proves the merit of Walsh No-Buckle Harness.
8. Lowest harness repair cost in the world.
9. Walsh Harness costs no more than the ordinary harness.
10. Thousands of farmers in every state have tried and proved the merits of Walsh Harness.
11. Strongest and most liberal guarantee ever placed upon a harness.
12. Thirty days' free trial gives you an opportunity to try out my harness.

THE WORLD'S STRONGEST HARNESS

No Buckles To Tear

The Walsh Way

The Ordinary Way

One of the most dramatic and successful wholistic innovations in harness design, introduced in the early twentieth century, was the **Walsh No Buckle Harness**. While the functioning aspects of the harness remained the same, the bulk of the engineering advances went into hardware, hardware function and the interface of all that with the leather components.

This new approach to harness design was so well received initially that the new company spared no expense in aggressive and intelligent advertising to get the word out even further. We present to you a bit of that to illustrate the many features that were being developed. I am certain that the Walsh effort might have spawned competition in these directions but we can only wonder at what they might have been because, as happened so often during the dawn of the industrial age, good ideas were rolled over by the rapid pace of change. With the advent of the internal combustion engine, millions of work horses and mules were unceremoniously carted off to slaughter and the golden age of draft animal power was squeezed off gradually over a long and painful quarter century.

The oft-stated purpose of this book, and the others in the **Workhorse Library**, is to preserve and protect the technological information as it may very well be called upon, enmasse sometime in our difficult to predict future.

Back to our example: With limited or casual familiarity of North American working harness you might come to the erroneous conclusion that, except for minor style variations, all harnesses were and are much the same. While a few quality and material issues are accounting for substantive differences in the modern harness, there were also interesting and important variations back in the early twentieth century which have slipped from view and conversation. As I have noted, the most significant example is the Walsh No Buckle Harness.

The Walsh Way

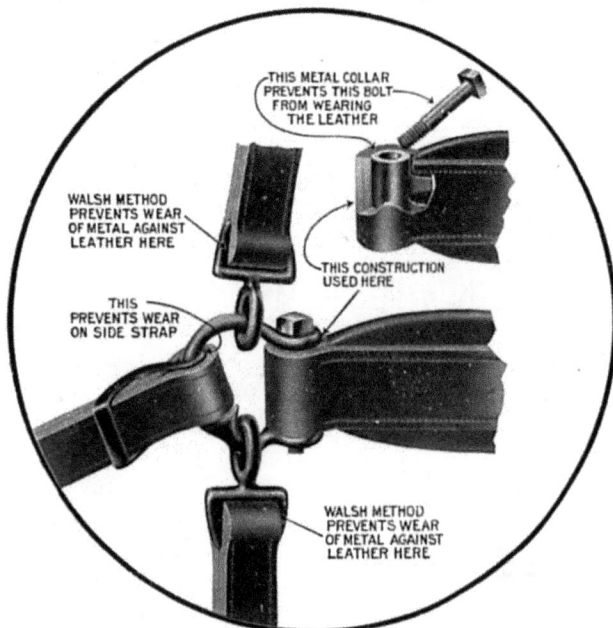

THIS METAL COLLAR PREVENTS THIS BOLT FROM WEARING THE LEATHER

WALSH METHOD PREVENTS WEAR OF METAL AGAINST LEATHER HERE

THIS CONSTRUCTION USED HERE

THIS PREVENTS WEAR ON SIDE STRAP

WALSH METHOD PREVENTS WEAR OF METAL AGAINST LEATHER HERE

In the view below, we see a common problem, the wear portions of the leather attachments have thinned.

Now notice, in the illustration to the left, that various components are made to protect all the interconnecting straps with bushings, bearings, collars and cockeye attachments. Slide buckles, as seen at the top of the next page, were also called into service to allow for speedy adjustments.

And ingenious crossed back straps allowed that the customary brichen spider be replaced by comfortable rump straps. This served several purposes, one of which was to prevent the back of the harness from slipping to one side or the other.

Today we run across very few Walsh harnesses, yet at one time they lay claim to being the world's largest producer utilizing modern day mass sales approaches to get their harness into every barn in North America. Starting out strong in 1914 and then subject to the industrial hiccups of two world wars and the great depression, Walsh No Buckle Harness had a short life. It was set aside, as so many other inventions and designs, as we all got drawn into the pursuit of the next best thing, which in this case was the internal combustion engine.

The beginning premise of the Walsh approach was to question a design which asked folded leather to wear constantly against metal as the straps went about their business.

Walsh Strap Holder

There are no such line catchers on the Walsh Harness.

Walsh Harness adjusted from large to small team in ten minutes

Walsh Hames unfasten from the side—you do not have to shout at a hungry horse to back away from manger to get at the hame strap.

Walsh Traces

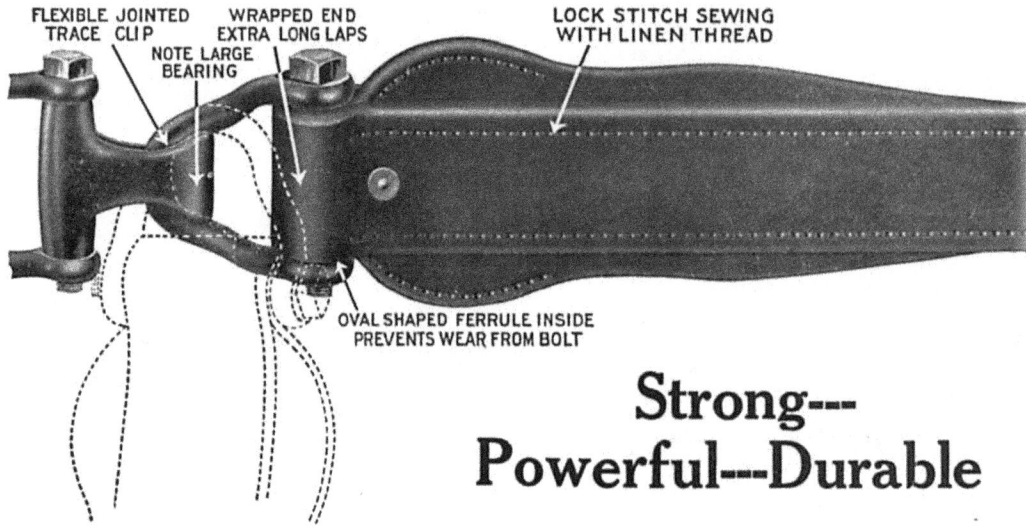

FLEXIBLE JOINTED TRACE CLIP

WRAPPED END EXTRA LONG LAPS

NOTE LARGE BEARING

LOCK STITCH SEWING WITH LINEN THREAD

OVAL SHAPED FERRULE INSIDE PREVENTS WEAR FROM BOLT

Strong--- Powerful---Durable

The Walsh Way

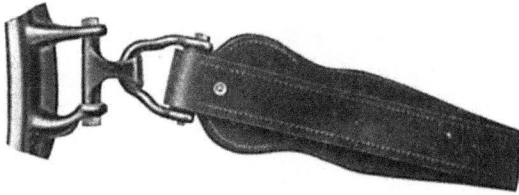

The Old Way

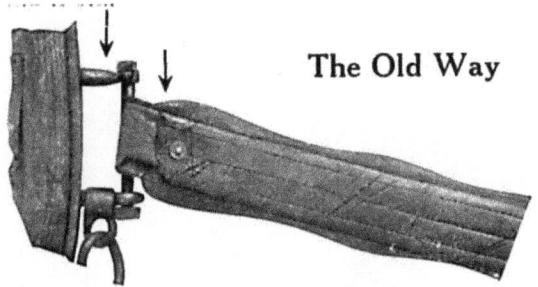

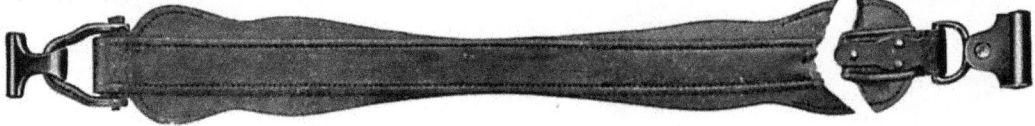

The Walsh Clip.

Ordinary Clip.

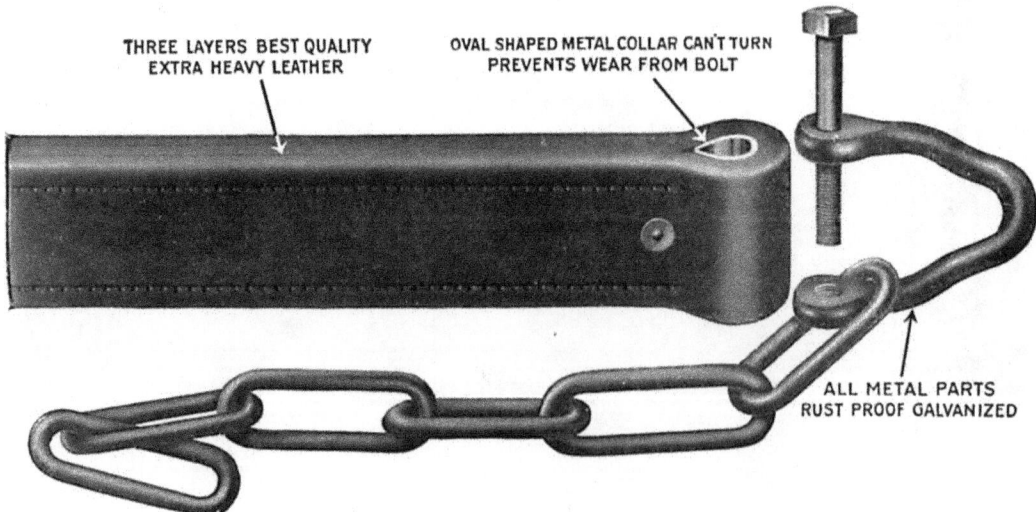

THREE LAYERS BEST QUALITY EXTRA HEAVY LEATHER

OVAL SHAPED METAL COLLAR CAN'T TURN PREVENTS WEAR FROM BOLT

ALL METAL PARTS RUST PROOF GALVANIZED

Raise or lower breeching by adjusting back straps longer or shorter.

Rump Strap enables horse to back with full strength, also raises or lowers breeching when back straps are adjusted.

Lead up straps sewed to fenders, no wear on end of straps.

Nothing to catch lines.

World's strongest breeching. No buckles to weaken straps.

Crossed Back Straps make easy to adjust breeching.

Breeching body folded leather easy on the horse.

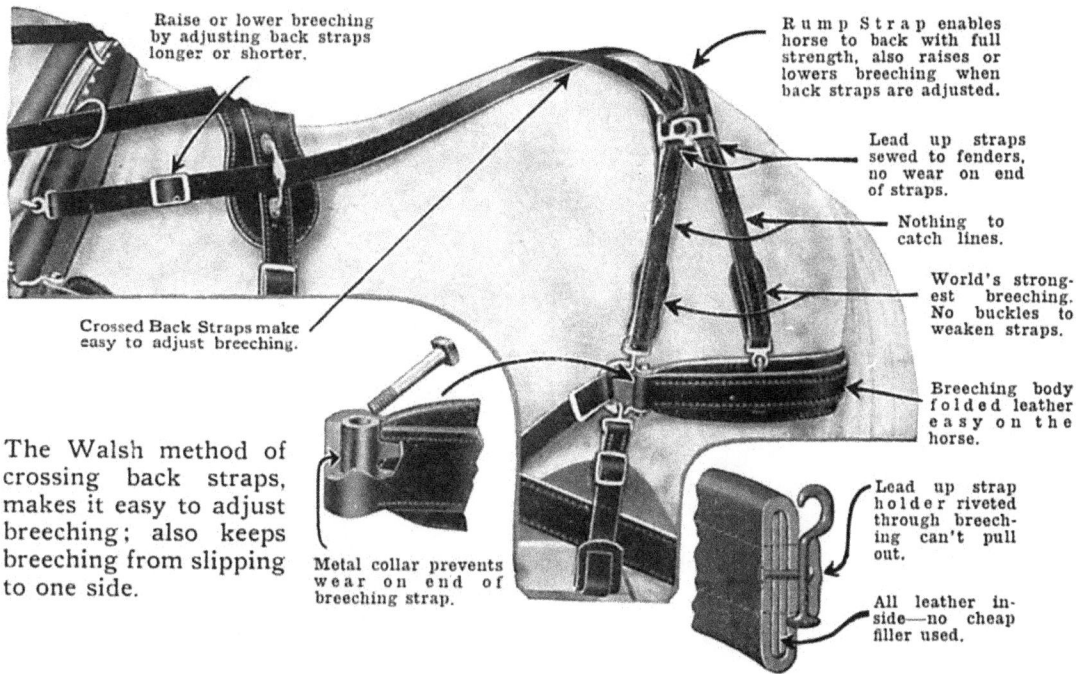

The Walsh method of crossing back straps, makes it easy to adjust breeching; also keeps breeching from slipping to one side.

Metal collar prevents wear on end of breeching strap.

Lead up strap holder riveted through breeching can't pull out.

All leather inside—no cheap filler used.

Easily Adjusted to Fit Perfectly

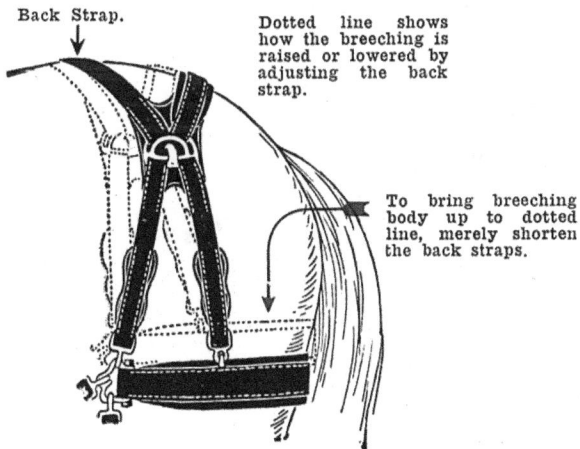

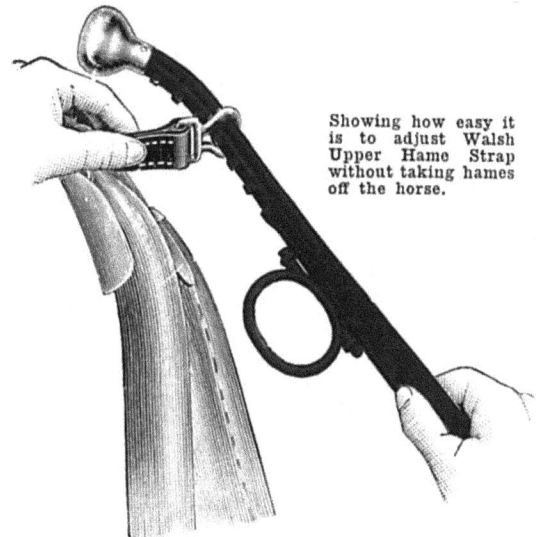

Back Strap.

Dotted line shows how the breeching is raised or lowered by adjusting the back strap.

To bring breeching body up to dotted line, merely shorten the back straps.

Showing how easy it is to adjust Walsh Upper Hame Strap without taking hames off the horse.

The Walsh custom and patented cast hames included a binder-like chain attaching system (see next page) as well as a wider range of adjustability at the top of the hame.

Easy to adjust hames to fit perfectly all sizes of collars.

Fig. B.

Side cut away to show how end of chain is held. Easy to adjust to all size collars. No chance to lose hame chain, as end of chain cannot pass through opening. Each adjustment one inch.

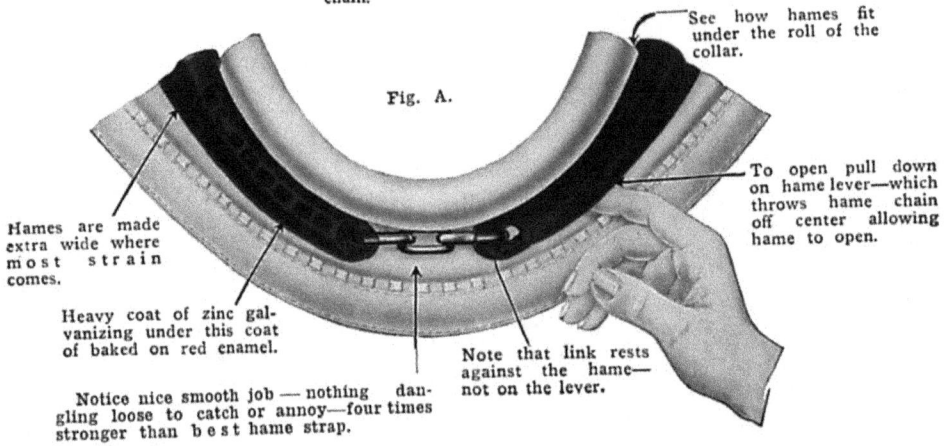

No hame straps to buy.

Extra ½-inch adjustment for real close fit.

Once closed, the hames stay closed until opened by pulling down on lever—Jack knife spring acts only as extra safety.

Lever open ready to receive hame chain.

Walsh Hames The Neatest, Best Looking, Best Fitting, Strongest Hame Ever Made

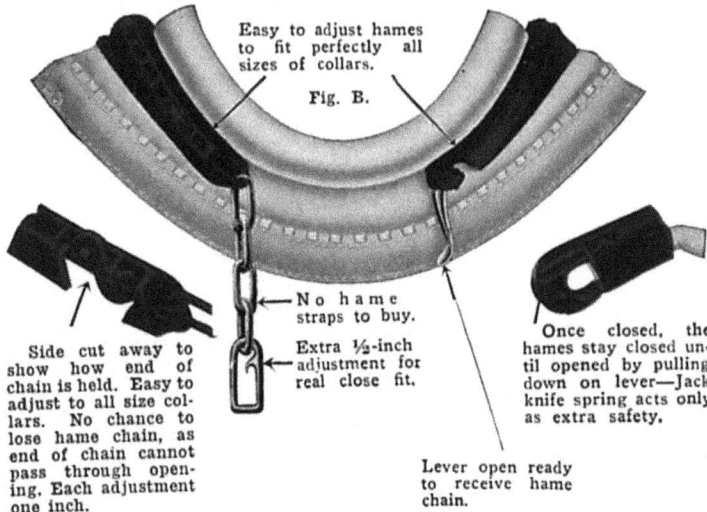

See how hames fit under the roll of the collar.

Fig. A.

To open pull down on hame lever—which throws hame chain off center allowing hame to open.

Hames are made extra wide where most strain comes.

Heavy coat of zinc galvanizing under this coat of baked on red enamel.

Note that link rests against the hame—not on the lever.

Notice nice smooth job — nothing dangling loose to catch or annoy—four times stronger than best hame strap.

Walsh Bridles

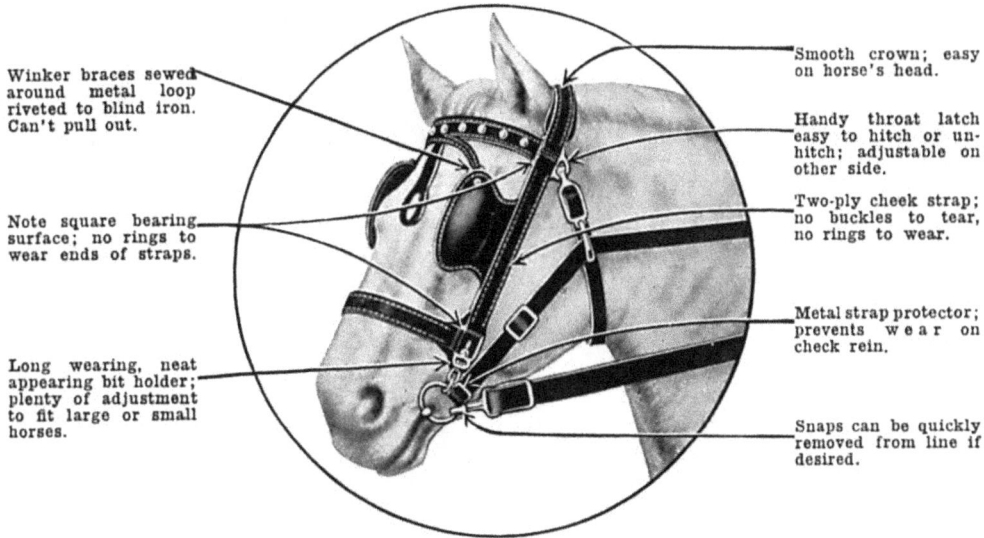

Winker braces sewed around metal loop riveted to blind iron. Can't pull out.

Note square bearing surface; no rings to wear ends of straps.

Long wearing, neat appearing bit holder; plenty of adjustment to fit large or small horses.

Smooth crown; easy on horse's head.

Handy throat latch easy to hitch or unhitch; adjustable on other side.

Two-ply cheek strap; no buckles to tear, no rings to wear.

Metal strap protector; prevents wear on check rein.

Snaps can be quickly removed from line if desired.

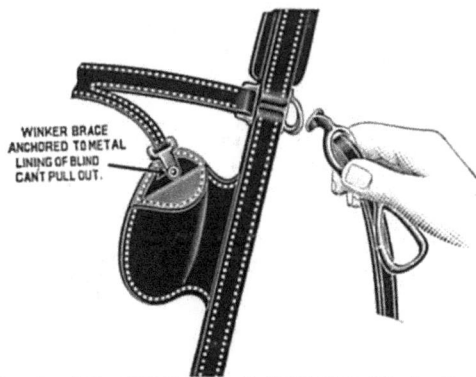

WINKER BRACE ANCHORED TO METAL LINING OF BLIND CAN'T PULL OUT.

Showing how winker brace is fastened to blind. Can't pull out. Also note handy Throat Latch.

How a Broken Strap Led to the Invention of Harness Without Buckles

Full strength double-ply Breast Strap. No buckles to weaken it.

Extra strong Neck-yoke Snap with re-newable spring.

Plenty of adjustment by snapping into either link. Same number links on other side. Links do not wear collar.

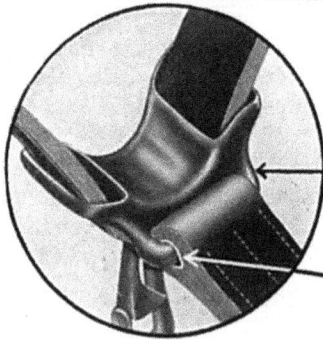

Martingale, thick heavy high-grade leather.

Martingale looped around and sewed directly to breast strap slide. Extra strong.

Oval shaped fer-rule. Can't turn; prevents wear on leather.

Walsh harness featured a unique breast strap system design that was quite handy. Short chains hang on both sides from the tug clevises. The breast strap then can be snapped into any link on either side for a perfect adjustment of length.

The Walsh Harness was so successful during the golden age of agriculture (circa 1914) that it was easy to find photos such as these of entire working lineups dressed in Walsh.

Chapter Eight

Harness Tools

In this modern age there is more than a little suggestion that anyone inclined to work draft animals in harness leans towards the Do-It-Yourself world. While you shouldn't need to have a full compliment of harness making tools in order to actually work your animals, at some point to keep things going, and to have your harness looking the way you want, you might require a few instruments of the trade. Here's but a sampling of what we found in our old hardware catalogs from over a hundred years ago.

My personal penchant is first to know what the tool is for, then to store its image in my brain so that I might be prepared at the next old farmer's garage sale to snatch up that *creaser* or *splitter* or bronze bedecked *riveter* just in case I never see another one.

In the next chapter, where we discuss repairs and care, you might round out your understanding of what should be on your shop bench.

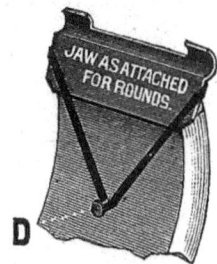

CLEGG.GOESER-CO.CIN.O.

JAW AS ATTACHED FOR ROUNDS.

D

REIN ROUND ATTACHMENT

Fo.

DOERING PATENT STITCHING HORSE.

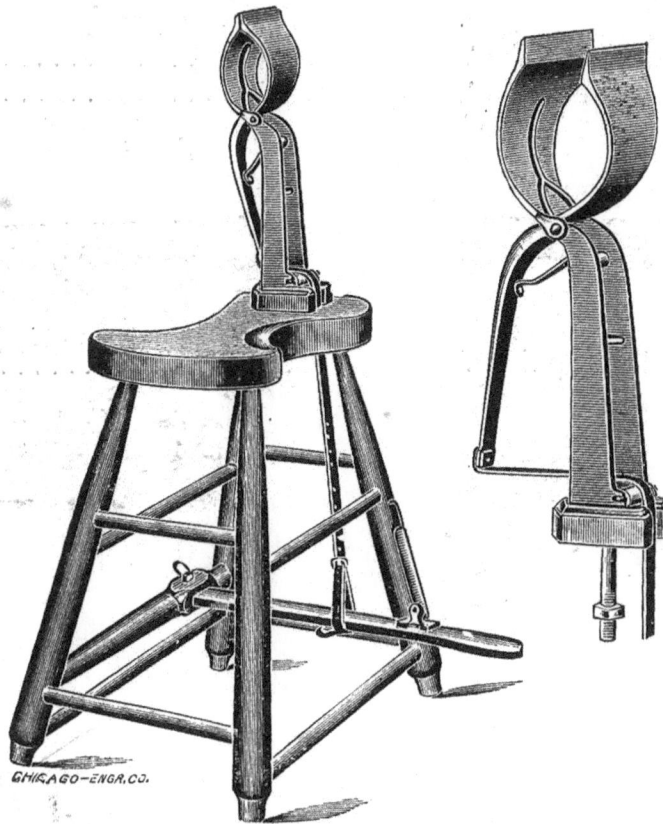

Doering Adjustable Stitching Horse.

Advantages: This jaw holds all work firm. It is quickly fitted to any old seat. It can be set at any angle, right or left. All jaws are 5 inches wide and 17 inches high, but can be lowered to 16 inches or raised higher by placing a block under the base. Any harnessmaker will save the price of this stitching horse by time he saves by holding his work firm.

In days gone by stitching horses were common on most farms. The way they worked was to pinch and hold leather straps so you could work on them, shaving, shaping, creasing, punching holes in, pushing needles through, and even glueing. This basic principle of the standby wooden variety was of a wooden vise to hold leather, such as shown on the previous page. It was direct enough an idea and need that many shade tree carpenters made their own. But then, in keeping with the itch men have always had to come up with mechanized compli-

cations of old standby tools, there were stitching horse models like the one above that featured mass produced intricacies to do the job with bells and whistles while daring anyone to try to replicate the levers and springs on the kitchen table.

Back in the seventies I knew a man who used a post vise on which he had glued wooden pads to soften the pressure on the harness leather. Worked for him. Others have used wooden carpenter vises. You get to decide what you want to use.

RANDALL STITCHING HORSE

HARNESS MAKER'S HANDY ANVIL

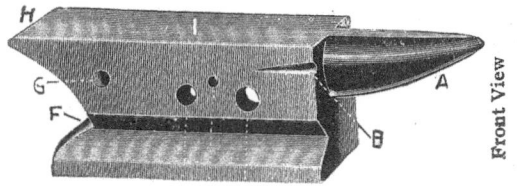

Front View

This is a combination anvil, raise block and bench iron. They are made of highest grade of castings, smoothly finished, top polished, weight 11 pounds, length 11 inches, width 3 inches, height 3 inches. No harness maker's kit complete without one, every one guaranteed to give satisfaction.

Each

Harness Maker's Handy Anvil.............$6.00

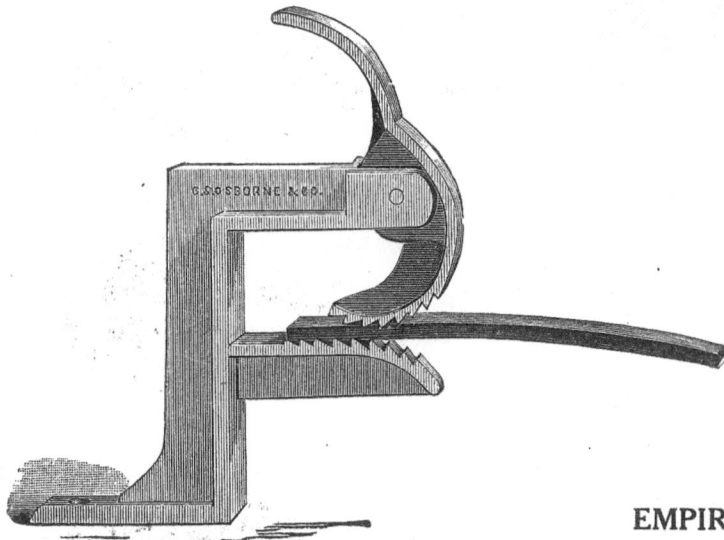

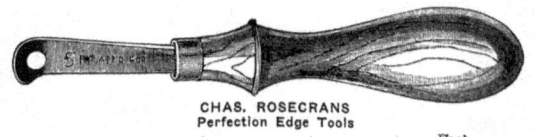

CHAS. ROSECRANS
Perfection Edge Tools

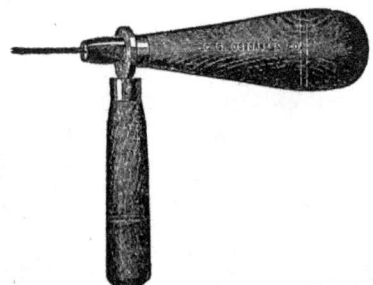

EMPIRE LACE LEATHER CUTTER.

LACE LEATHER CUTTERS
This handy little tool will cut lace, ⅛ to ¾ inch wide. The blade is made of fine tool steel, carefully hardened and tempered.

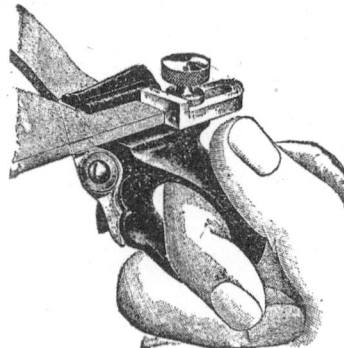

HARNESS RIVETERS

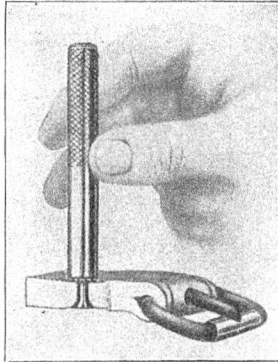

THE "ALPHA" RIVET SET

Will set tubular or split rivets, any length. Simply drive with a hammer and rivet down with the set. Forged from tool steel, knurled head.

Per Dozen_____ $1.20

THE CHICAGO RIVETER

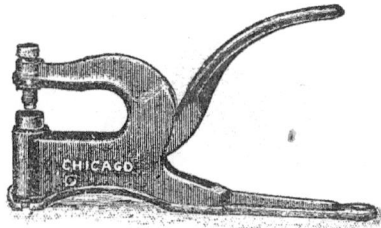

Cast iron, Japanned. Steel plunger. Six inches high. Base 10 inches long. Set screw to adjust to the different thicknesses of leather.

Chicago riveter_____Per Dozen, $ 6.00

THE GEM RIVETER

Will set tubular or split harness rivets perfectly. Automatically adjusts to the various lengths of tubular or split rivets. Cast iron, Japanned. Steel plunger.

Gem riveter_____Per Dozen, $ 9.00

LEATHER GAUGE KNIVES

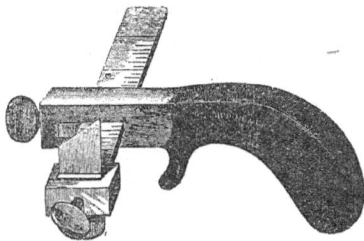

No. 051½

Patent hollow malleable iron handle, 5 inches long, Japanned with polished head. 4-inch polished steel slide. Graduated in ⅛-inch, with set screws. Furnished with one highly tempered steel blade.

No. 051½—Gauge knife_____Per Dozen, $15.00

WASHER CUTTERS

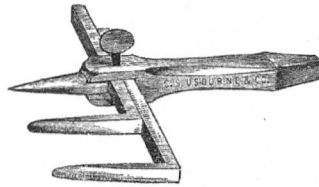

No. 350

Polished steel blades. Used in an ordinary bit brace.

Per Dozen_____ $9.00

BELT OR COLLAR AWLS

No. 164—Handled awl with eye_____Per Dozen, $ 6.00

PATENT PEGGING AWLS

Polished and tempered tool steel.

Pegging awls, assorted_____Per Gross, $ 2.00

HARNESS AWLS—STRAIGHT

Tempered and polished tool steel.

Straight harness awls, assorted_____Per Gross, $ 2.25

BENT SEWING

Tempered and polished tool steel.

Bent sewing awls, assorted_____Per Gross, $ 2.50

LOCK-STITCH SEWING AWL

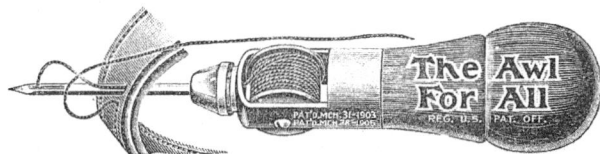

Meyers lock-stitch sewing awl. A practical tool for sewing all kinds of heavy material. Designed particularly for farmers' use. Awl complete with spool of waxed thread, straight and curved needles and wrench.

Per Dozen_____ $12.00

BEST YET RIVETERS

Anvil   Pocket

Adjustable Screw   Spindle   Spring for Raising Pocket

PARTS FOR NOS. 471, 473, 1G and 3G RIVETING MACHINES

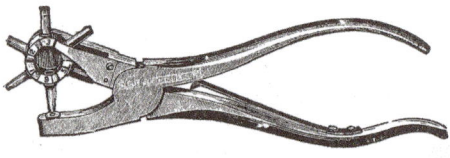

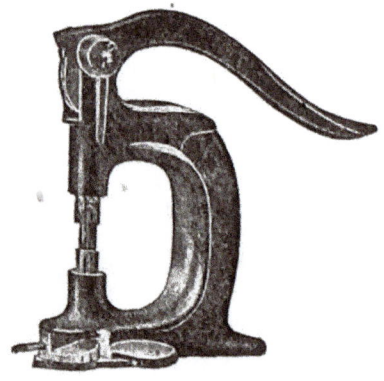

H. F. OSBORNE NEWARK, N.J.

		Each.
No. 622—Spitler's combination splitting and lap skiver, 6-inch......		$13 88
No. 622—Spitler's combination splitting and lap skiver, 8-inch......		17 76

		Each.
No. 171—Chase pattern, 8-inch........................		$12 22
No. 171—Chase pattern, 10-inch........................		14 44
No. 171—Chase pattern, 12-inch........................		17 76
8-inch Chase extra blades for above..................		2 22
10-inch Chase extra blades for above..................		2 55
12-inch Chase extra blades for above..................		2 77

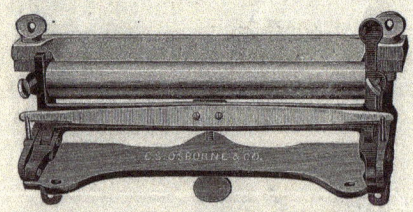

		Each.
Kreb's patent splitting machine......................		$26 66
Extra blades for Kreb's machine......................		5 88

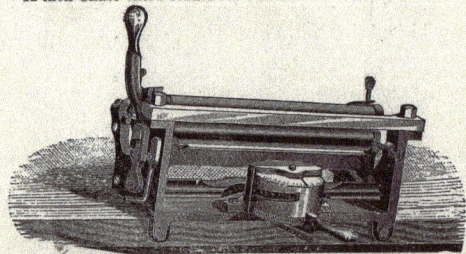

		Each.
No. 86—Iron frame splitting knives, 5-inch...............		$ 8 32
No. 86—Iron frame splitting knives, 8-inch...............		11 10
No. 86—Iron frame splitting knives, 10-inch...............		12 88

		Each.
No. 89—Mallets, hickory or dogwood, 3-inch face.................		$ 1 00

MACHINES FOR SETTING STANDARD SPOTS

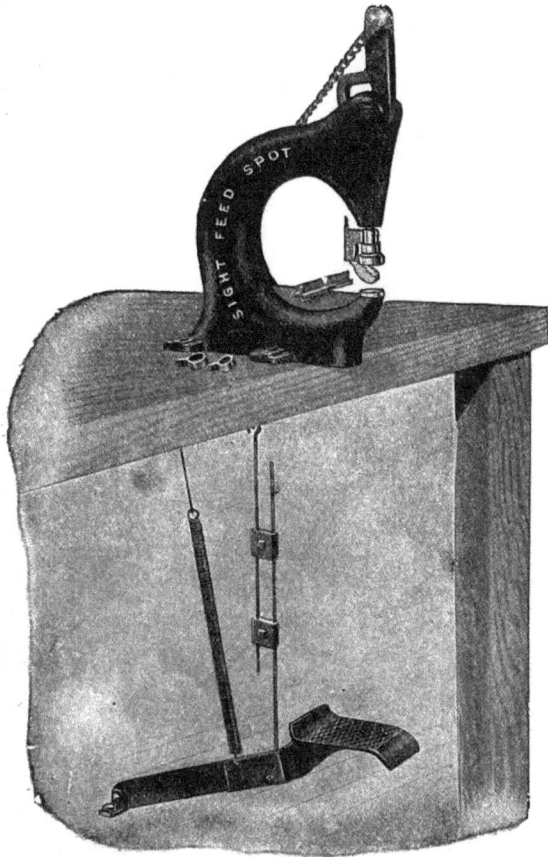

SIGHT FEED SPOT MACHINES, FOR SETTING
STANDARD SPOTS

No. 2. FOOT POWER SPOT MACHINE FOR
SETTING STANDARD SPOTS

Compost-Heated Harness-Shop Water

Ray Drongesen always found a way to convert a poor man's necessity into something elegant, practical and magically appropriate. Here's an example: In 1975 he had a small 8' x 10' wooden harness shop adjacent to his ramshackle barn. In it he had cold running water plumbed to a wide flat sink. He wanted hot water to wash his harness with. He found a discarded hot water heater, one which had no electrical elements that worked. He stripped the pretty outside shield and tangled electrical components off and stood the bare galvanized tank on a heavy duty pallet, plumbing it so cold water would run in and through the tank to the shop faucet. Then he put up three plywood walls around that and setup rails for sliding boards across the front. As he cleaned his work horse stalls, he dumped the wet bedded manure in around the water tank, carefully, poking in old implement handles to stand up. As he first filled around the water tank, he put in front boards to hold the manure in a block. When the manure and bedding got higher he would pull up the implement handles slightly, leaving behind narrow tunnels in the compost block. Every day he would add urine soaked bedding and horse manure. And he would sprinkle the whole thing with water. The pile would keep compressing, airways were maintained by the old broom and pitchfork handles and... the decomposition would generate heat. Before long he could turn on the faucet at his harness repair shop sink and out would come clean 130 degree water! When the temperature of the water would go down Ray would pull the front boards and wheelbarrow out the finished compost to his vegetable garden. Pure poetry.

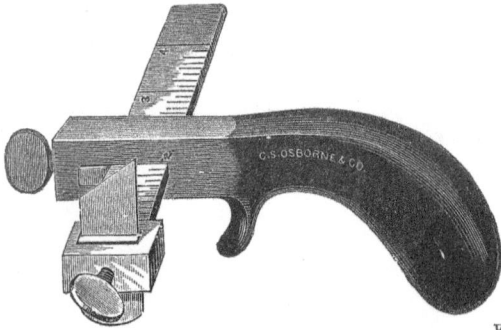

Each

No. 51½—Patent Hollow Handle Draw Gauges $3.50

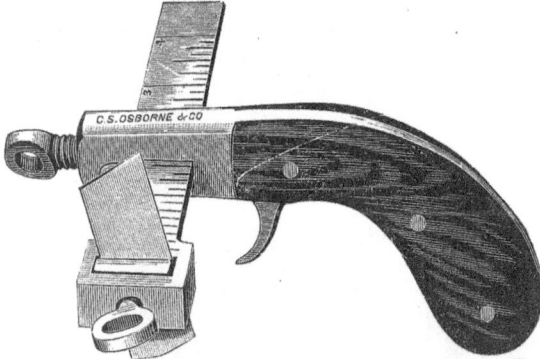

Each

No. 52½—Draw Gauges, improved........$4.40

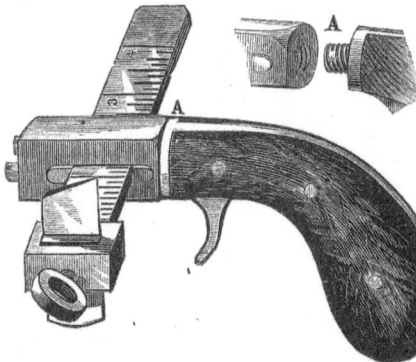

Each

No. 52—Latta's Patent Draw Gauge.......$6.00

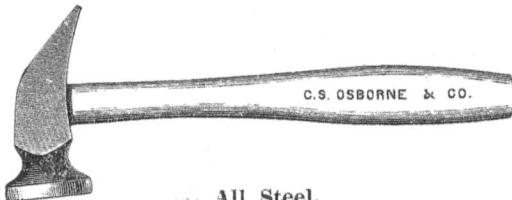

All Steel.

No. 54—Shoe Hammers, white hickory handles.

Nos.	00	0	1	2
Each	$1.90	2.00	2.10	2.20
Nos.		3	4	5
Each		$2.40	2.50	2.60

Each

Gomph Overstitch Wheels and Carriages, Nos. 5 to 10.......................$2.50

Each

No. 56—Saddlers' Hammers, white hickory handles $2.60

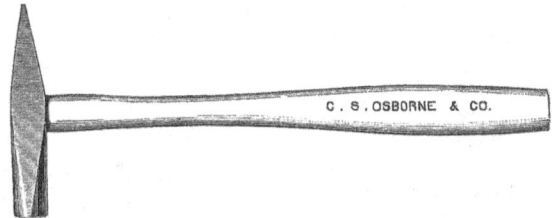

No. 57—Riveting Hammers, steel, white hickory handles.

Nos.	2	3	4	5
Each	$1.50	1.70	1.80	2.20

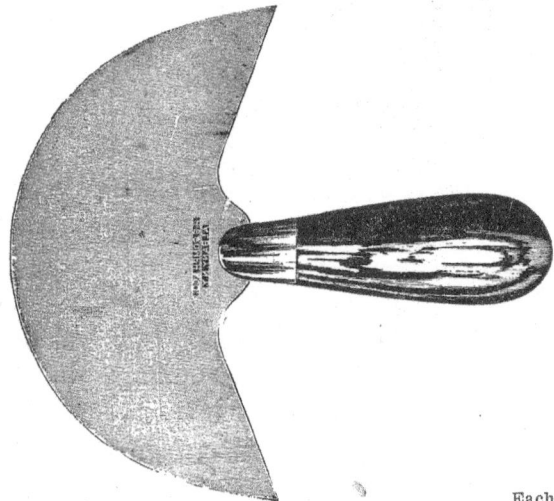

Each

Gomph Round Knives, 4½ inch............$4.00
Gomph Round Knives, 5 inch............ 4.70
Gomph Round Knives, 5½ inch............ 4.70
Gomph Round Knives, 6 inch............ 5.30

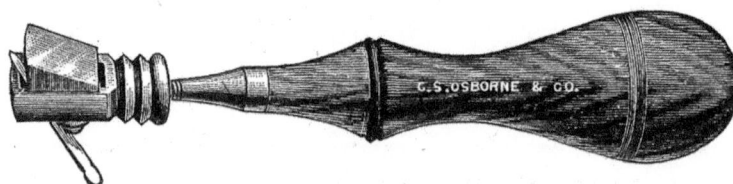

Edge channelers ....

No. 1—Awls, with octagon, ebony handles.. Each $1.30

No. 3—Pad Awls, shouldered, riveted...... Each $1.20

No. 5—Drawing Awls .... Each $1.00

No. 6—Thong Awls .... Each $0.72

No. 8—Collar Awls .... Each $1.00

No. 10—Bouncers, dogwood .... Per Dozen $0.90

No. 26—Saddlers' Steel Compasses, 6 inch.. Each $1.80

No. 30½—Patent Leather Compasses, four points with each............... Each $5.30
Points and Scratches, for above............ Per Dozen $3.30

No. 30—Patent Leather Compasses, best.... Each $2.60

No. 44—Single Washer Cutters........... Each $2.40

No. 45—Double Washer Cutters.......... Each $4.30

No. 51—Improved Draw Gauge, brass....... $6.00
No. 51—Improved Draw Gauges, brass, with guard 8.50
Blades, for draw gauges................ .30

You don't need all of these tools. But there is value in you knowing what they are should you stumble across a few in your travels. Millions and millions of these gems were manufactured and sold throughout the world. Many linger today at the back of shop drawers or storage boxes under benches or back down in dark corners when they fell off the back of shop benches. They are to be valued, and preserved AND used.

Gomph Trimming Knives, 3½ inch, round
    handle, sharp point................**Each**
                     $0.96

No. 79—Cap Knives, round handle........**Each** $0.64

No. 80—Cap Knives, flat riveted handle.....**Each** $1.30

No. 80½—Cap Knives, light, riveted handle.**Each** $1.30

No. 89—Mallets, hickory, 3 inch face.......**Each** $1.50

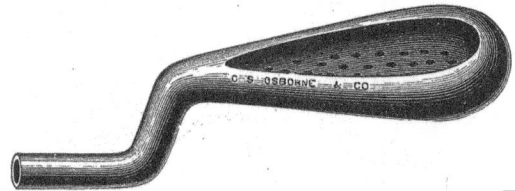

No. 93—Collar Palms ....**Each** $1.50

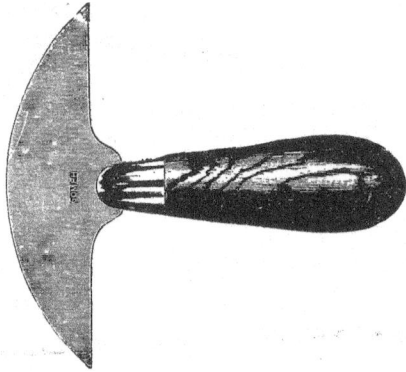

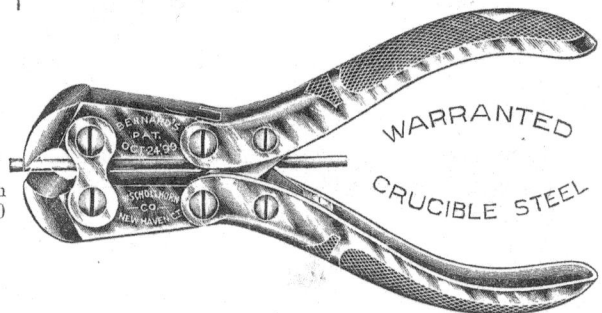

	Each
Gomph Head Knives, 4 inch............	$4.00
Gomph Head Knives, 4½ inch............	4.00
Gomph Head Knives, 5 inch............	4.70

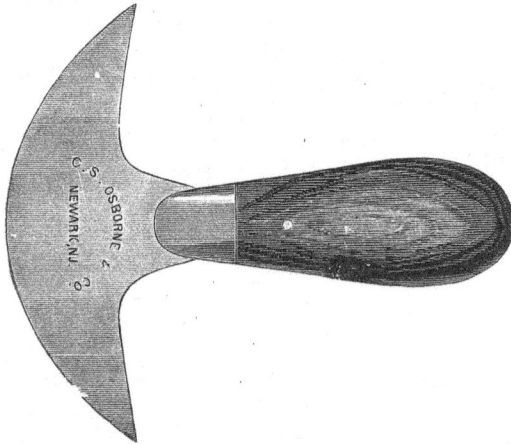

No. 75—Philadelphia Pattern Head Knives,
    oval handles, 4⅝ inch............**Each** $2.20

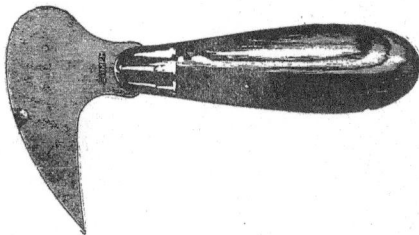

Gomph One-Point Head Knives...........**Each** $3.20

Gomph Trimming Knives, 3½ inch, flat riveted
    handle, sharp point..................**Each** $1.20

No. 1—Bernard's End Cutting Nippers, 7
    inch ....**Each** $4.70

I have never seen a **Gomph one point head knife** *other than in pictures.*
Even as an old man, if I do, and it needs a home, I will adopt it.

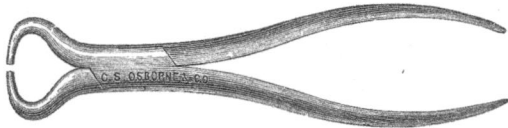

No. 102—Pad Screw Plyers, steel black handles .... Each $3.10

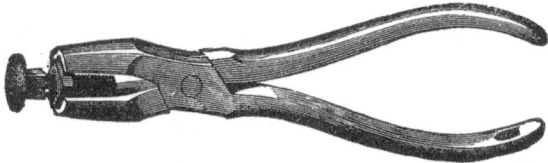

No. 102½—Bolt nut plyers .... Each $3.50

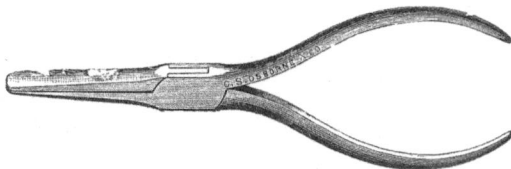

No. 103—Duck Bill Plyers, steel .... Each $2.50

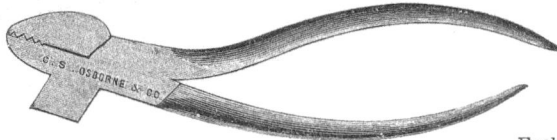

No. 104—Saddlers' Pincers .... Each $4.00

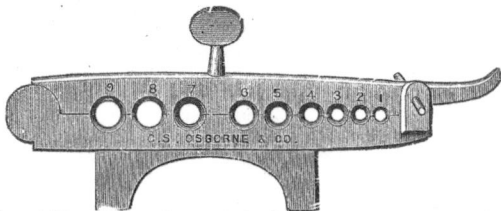

No. 113—Rounders, 9 holes .... Each $10.00

No. 120—Rivet Sets, Nos. 7, 8, 9, polished face only .... Each $1.10

No. 122—Claw Tools, best steel .... Each $1.00

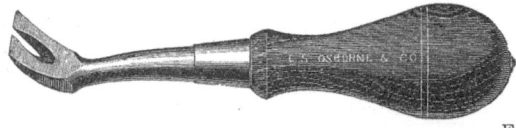

No. 123—Claw Tools, extra heavy, riveted ... Each $1.80

No. 124½—French Claw Tools, riveted ...... Each $1.30

Gomph Common Edgers, Nos. 0 to 5 .... Each $1.44

Gomph Round Edge French Edgers, Nos. 1 to 4 .... Each $2.70

Gomph French Edgers, Nos. 0 to 3 .... Each $2.40

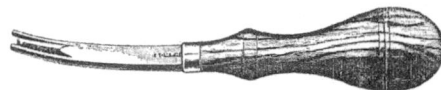

Gomph Round Edgers, Nos. 0 to 5 .... Each $1.44

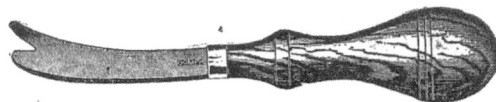

Gomph Rein Trimmer .... Each $1.60

No. 127½—Perfection Edge Tools, 2 to 5 .... Each $0.76

*As so much of the very old harness and equipment featured square head nuts, **bolt nut plyers** would be most handy on the bench. Good luck ever finding one. I keep thinking I might make a pair out of a hefty set of side cutters. Heat up the ends and tap against a square hardy perhaps?*

C. S. OSBORNE & CO. TOOLS

Each
No. 1   Awls, with octagon ebony handles........$1.40

Each
No. 3   Shouldered pad awls, riveted handles......$1.30

Each
No. 6   Thong awls, riveted handle..............$1.10

Each
No. 8S Small collar awls, riveted handles, 8 inch $1.40
No. 8   Large collar awls, riveted handles, 9¾.....
        inch 1.40

Each
No. 22 Single edge creasers, octagon, sizes 1, 2,......$1.90
        3, 4, 5

Each
No. 22½ Double edge creasers, octagon, sizes 2,.....$1.90
        3, 4, 5, 6

Each
No. 24½ Layer creasers, single, sizes 1, 2, 3, 4.....$1.90
        and 5

Each
No. 27 Saddler's compasses, with wing and set
        screw $3.50

Each
No. 28 Common compasses, 6 inch.............$0.80

Each
No. 29½ Spring patent leather compasses $1.80

Each
No. 30 Patent leather compasses (best) sizes 1, 2,
        3 and 4 ....$4.20

Each
No. 30½ Patent leather compasses, four points
        with each...........................$5.50
No. 30½ Points and scratches for above..........   .56
No. 30½ Wing screws 20
No. 30½ Point screws 20

Each
No. 31 Channelers $3.20
No. 31 Screws for above 30

Each
No. 32 Hall's patent channelers $4.50

Each
No. 208 Edge channelers $6.00

PRICES SUBJECT TO CHANGE WITHOUT NOTICE

Collar needles and sack sewing needles could be found in any old pre-WWII general store all across North America. I've seen a few in sliding-drawer cardboard boxes such as matches come in, only these little boxes are larger. I want to sit my grandchildren down and somehow impress on them that a small stash of things, such as these magnificent big needles, will have long life value far in excess of computer thumb drives or wireless ear buds. Needles such as these will be valuable thousands of years from now.

NICHOL'S COLLAR NEEDLES
Usual Bend

		Per Dozen
No. 66	3 inch	$2.30
No. 67	4 inch	3.10
No. 68	5 inch	4.20
No. 69	6 inch	4.90
No. 69A	7 inch	6.00

NICHOL'S COLLAR NEEDLES
Half Moon Bend

		Per Dozen
No. 133	3 inch	$2.60
No. 134	4 inch	3.40
No. 135	5 inch	4.50
No. 136	6 inch	5.30

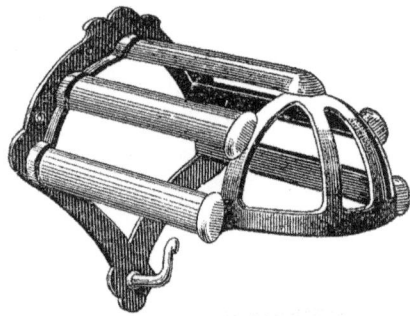

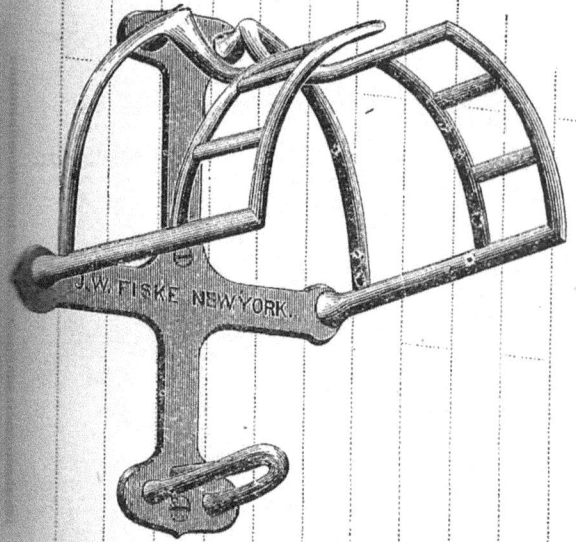

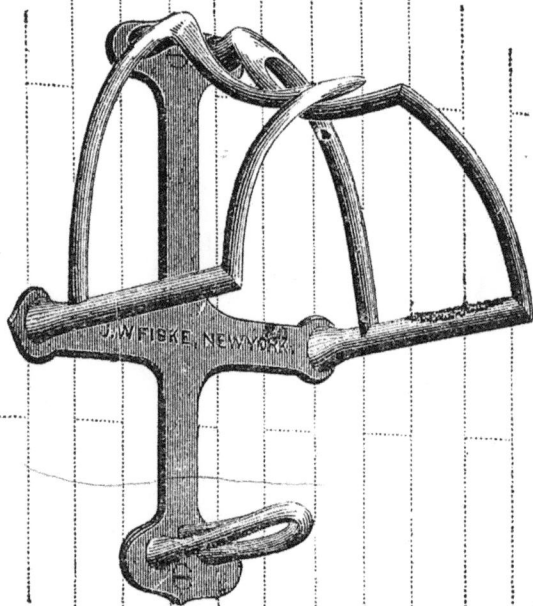

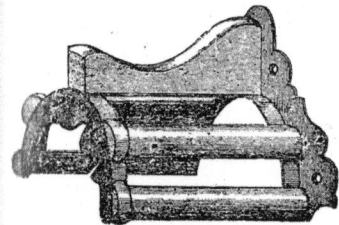

FIG. 356 K.

Collar Bracket.

Brass or Bronze

HARNESS HOOKS

J.W. FISKE, N.Y.

J. W. FISKE, N.Y.

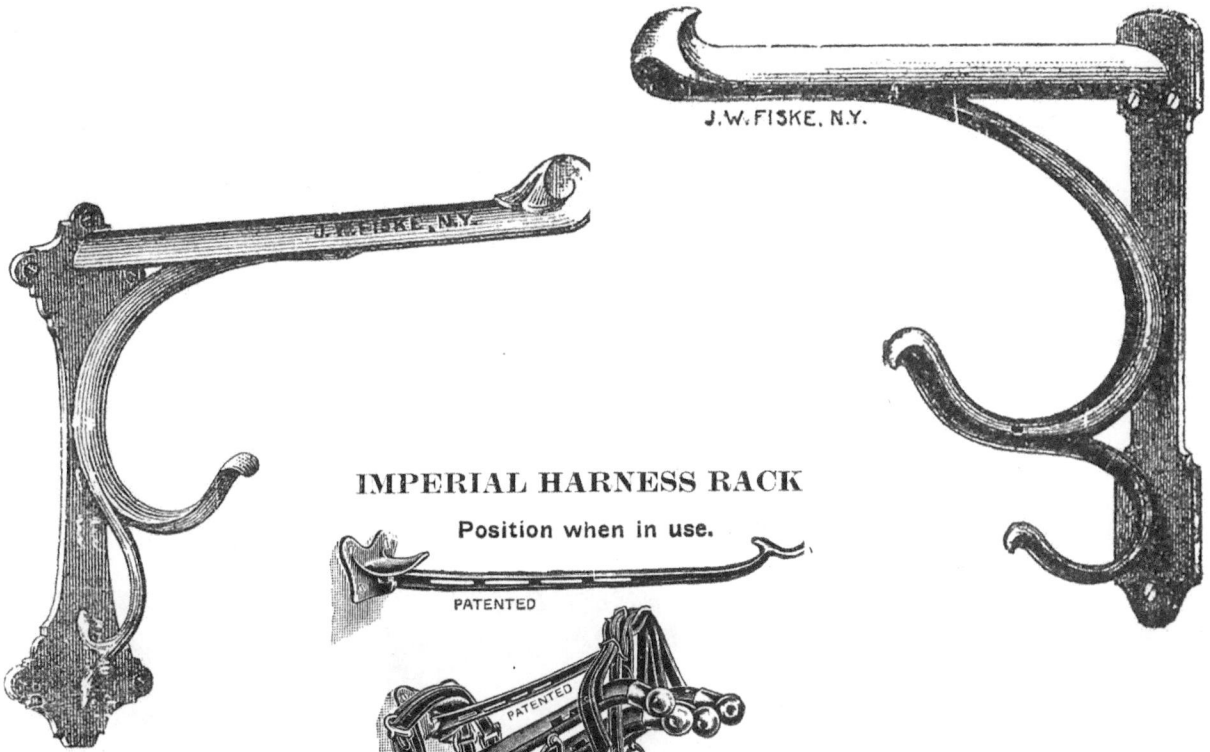

IMPERIAL HARNESS RACK

Position when in use.

PATENTED

PATENTED

Not in use. Out of the way.

PATENTED

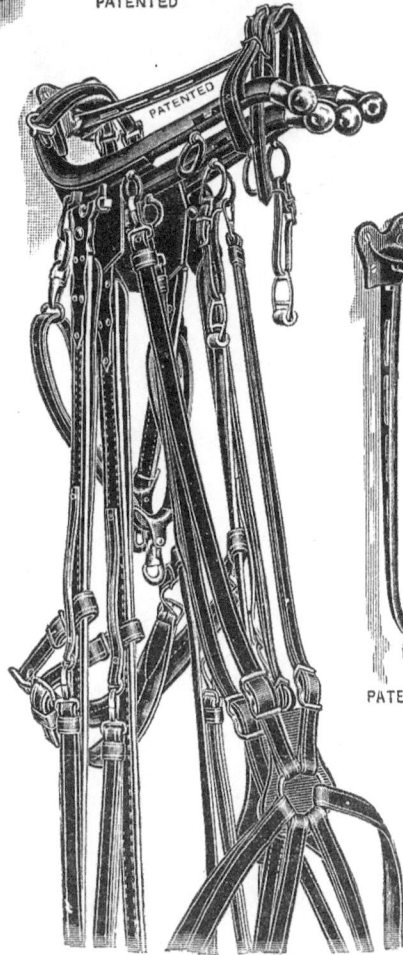

Team Harness Rack

I use the two cast iron harness hangers you see pictured above. The one on the top right is generally available, mail order, and comes in slight variations. The one on the top left, and immediate right, is something I found many years ago at auction. Wish I could find a small foundry that would cast these. I like the way they keep sharp bends from forming with lines and straps that have hung for a while.

I use **Neatsfoot oil compound** to keep my leather as pliable as heavy weight leather can be. I have known people who took dried and cracked leather and dunked it in vats of used motor oil before bringing it to auctions. It was a mess and turned that leather into a weak structure that would rip easily.

Look to the chapters 10 & 11 for some thoughts on maintenance and salvation (of harness).

Spotted parade harness draped over livestock panels for display at one of our former auctions.

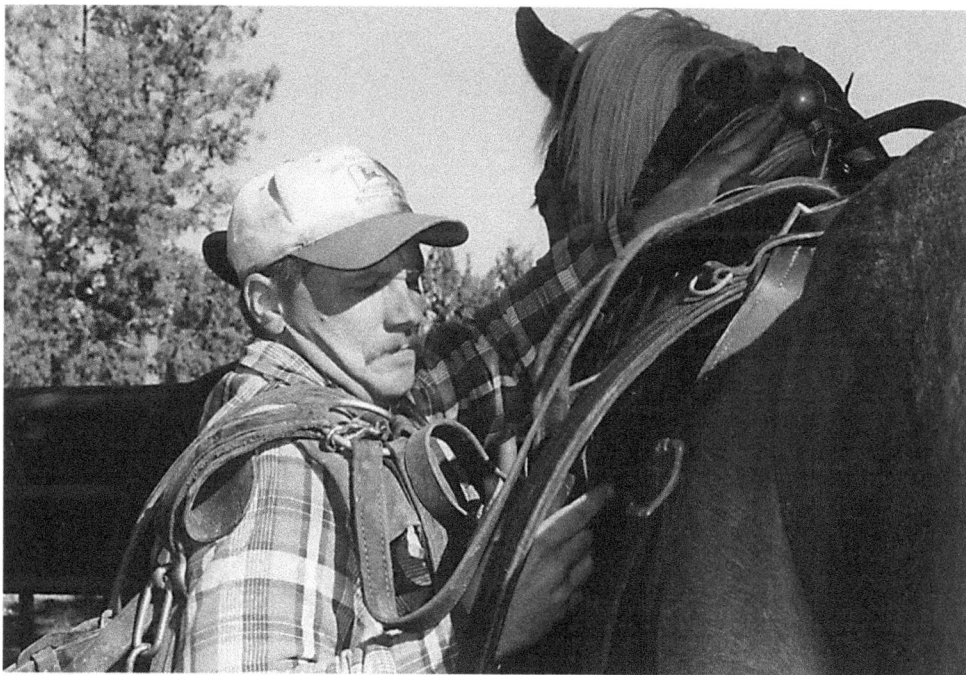

A very young Ed Joseph harnessing at a workshop years ago.

Chapter Nine

Harnessing

The photos in this sequence were taken in the 1970's and they feature the legendary crossbred gelding King, and the immortal small farm horseman Ray Drongesen of Harrisburg, Oregon. The accentuation and diagramming-arrowed captions are by the author.

This volume was not meant to be a 'how-to' but a 'what it is' treatise. But I recognize that there is an opportunity, within the context of 'this is a harness,' to try to make sense of how it is put on. The reader is encouraged vigorously to gather more information on 'how-to' with work horses.

First caution: If you are new to working horses or mules in harness, and/or your animals are new to the process, DO NOT attempt to harness them without a knowledgeable person in attendance to help you.

Second caution: At the risk of it seeming impossibly contradictory, you need to learn to harness a horse with a horse that is familiar with the process and using a proper harness and collar that have already been identified as the right size. A first time harnessing event is NOT the right time to teach yourself about whether or not this or that harness will work. Subjecting a green horse to a first time harnessing might result in frightening experiences for the animal that may be hard to erase. A knowledgeable teamster cannot help but reassure a novice horse, and an experienced horse will calm the novice teamster.

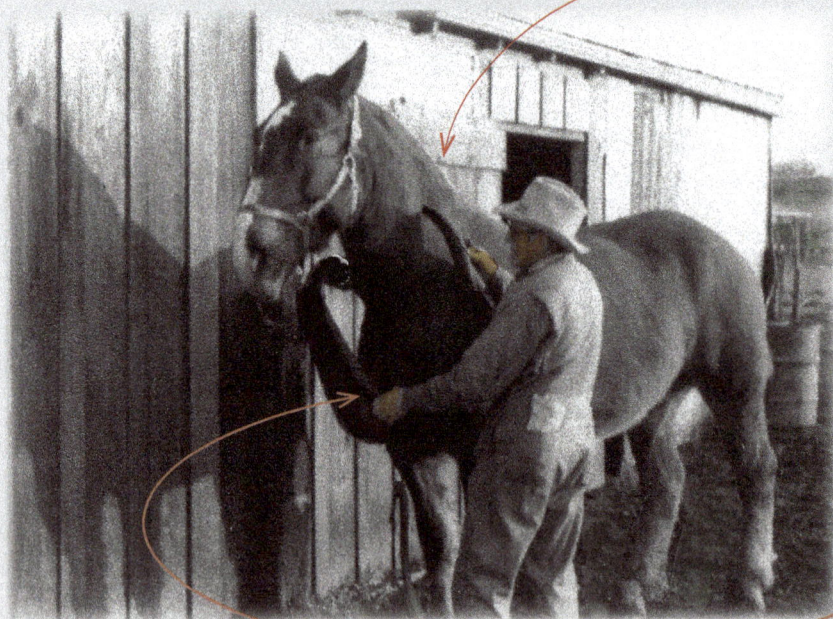

1. PUTTING COLLAR ON. Having identified the right collar for the equine, open it at the top (strap or clamp) and, supporting both sides as Ray does, spread it enough to pass it up and around the horse or mule's neck ahead of the shoulders, where the neck is narrower...

2. ...and bring the top ends together, clamping the collar or buckling the top collar strap. If you do not support both sides of the open collar while putting it on, it can develop a creased hinge at the throat that will eventually crack and break the support which holds the integrity of the collar shape.

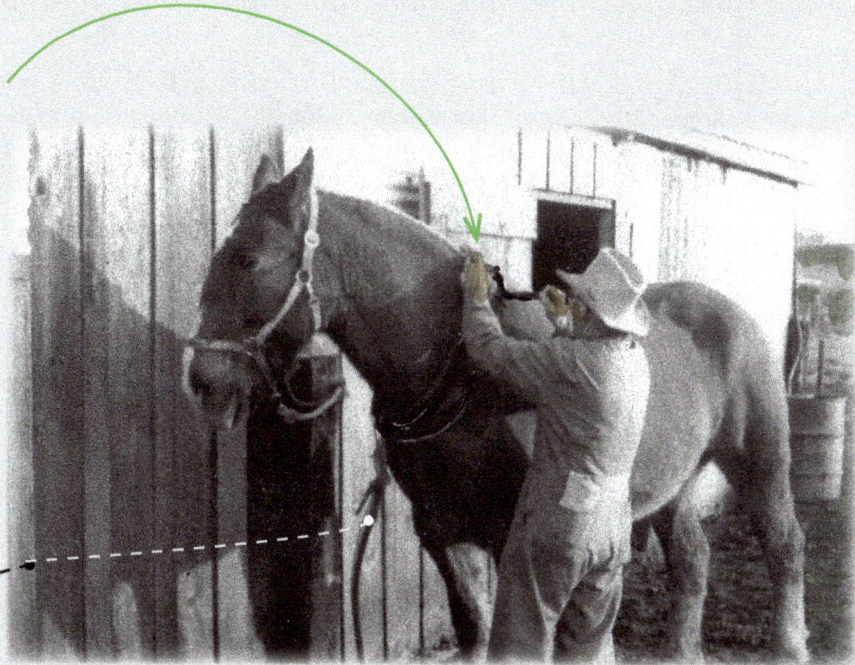

Ray hangs his pole strap, semipermanently, from the throat of the collar with a light, thin, buckled strap. This is but one possible variant in how harnesses go together.

See pages 26, 44 of this volume for fitting collars to horses.

3. Seat the collar back on the shoulders and check it for length.

4. **PUTTING HARNESS ON.** Whether the main body of the individual harness is hanging from a single hook or peg, or hanging from two pegs stretched across a wall – you want to begin by passing your right arm under the brichen and all the way forward, passing also under the back pad or saddle, finally reaching the right side hame at just below the center. The left hand will then take hold of the left hame at the center. Now, lift the harness free of the pegs. Make sure that you can spread the hames apart and no straps or lines are crossed at the center. You want the bottom center path of the harness, from front to back, to be able to open up as you lift the hames over the withers and place them on each side of the waiting collar.

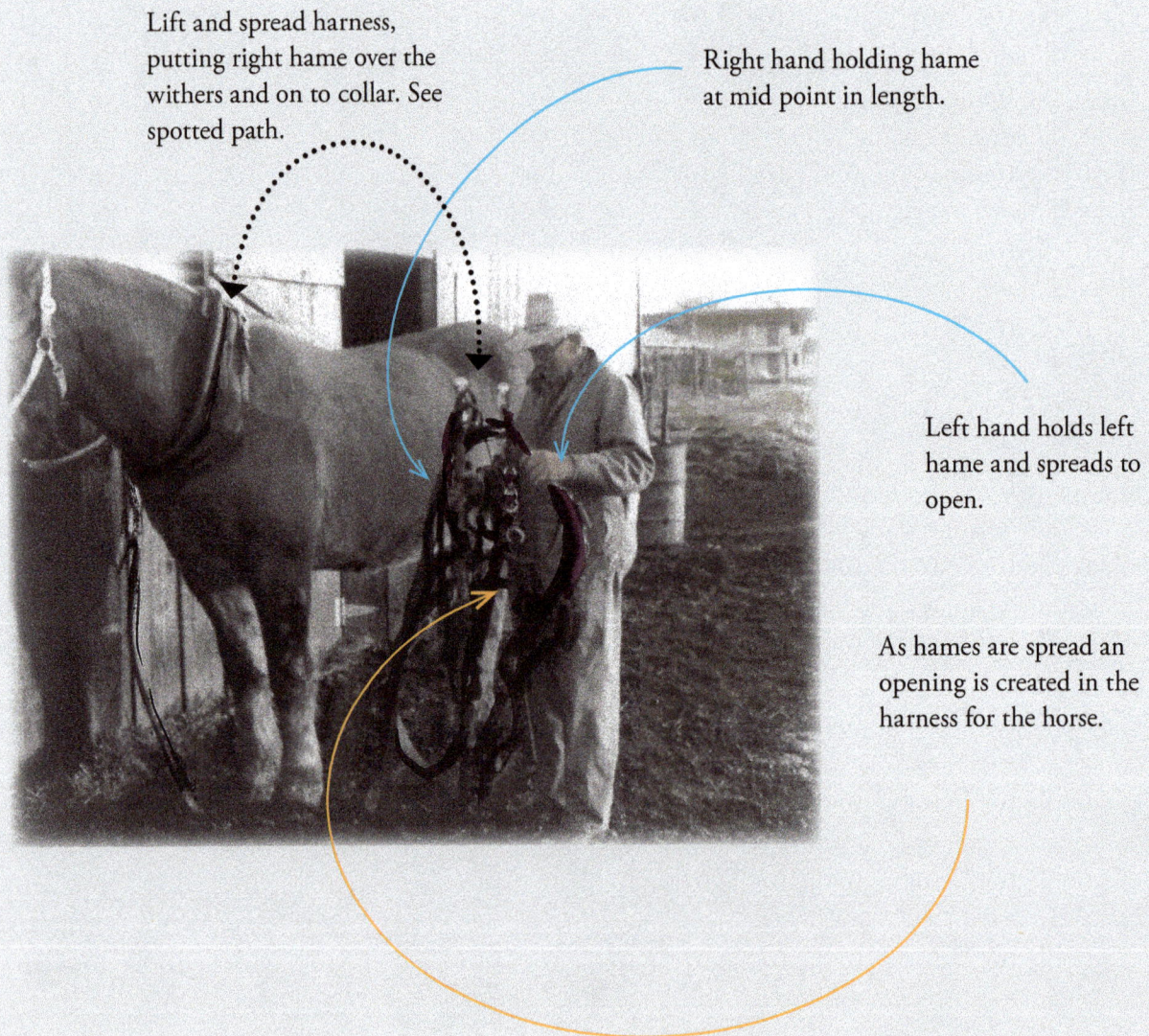

Lift and spread harness, putting right hame over the withers and on to collar. See spotted path.

Right hand holding hame at mid point in length.

Left hand holds left hame and spreads to open.

As hames are spread an opening is created in the harness for the horse.

5. A strong and knowledgeable teamster may prefer to flip the right side of the harness up and over the horse's back, careful to keep hold of both hames. At that point, careful to balance the bulk of the harness, the teamster will position the hames in the groove of the collar.

Notice that Ray has positioned the brichen, spider and back straps ahead of the hips, making sure that weight is evenly distributed, one side to the other.

Some individuals will be too short and/or not strong enough and will prefer to pass the right hame up the horse's side pushing it over the collar, then pushing the remainder of the harness up the side, careful to keep the bulk of the straps along the middle back. This works.

To repeat: pulling the weight of the harness backwards, over the hips and tail before the harness is secured at the collar, should be avoided. It may make the harness prone to, before and during the fastening of the hames, slipping back and falling down around the back legs. Quick frightened movements on the horse or mule's part might cause injury.

6. **SECURING HARNESS.** Again, having the bulk of the harness evenly divided across the back of the horse and ahead of the hips, place each hame in the collar groove, being careful to even them up. If the harness features hame binders (see page 80), fasten it up. If the harness features hame straps, thread that bottom strap through the buckle...

7. ...and pull the strap end HARD to tighten. This is important because IF the hames work themselves loose at that bottom strap, the hames could come loose on a pull, move forward, and choke the horse as it moves ahead.

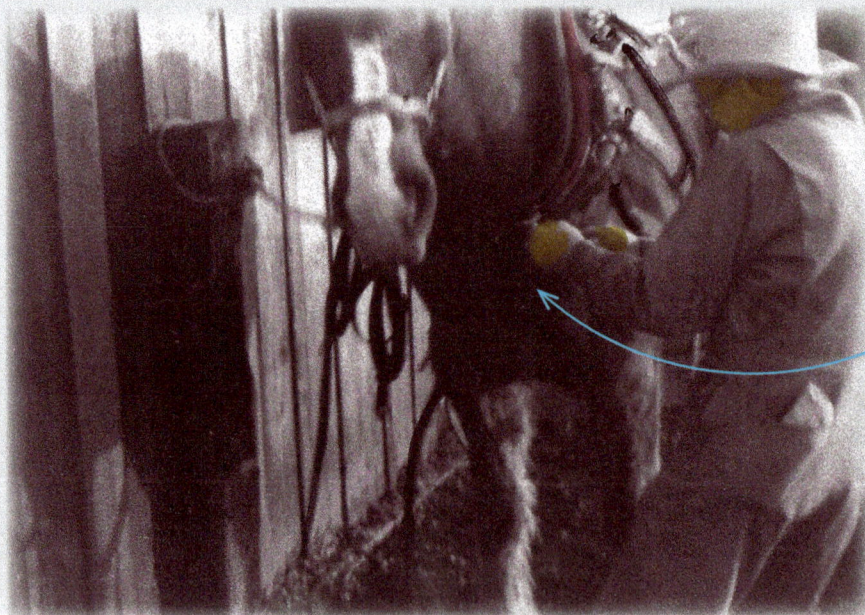

8. Hames secure, now pull the bulk of the harness back along the animal's back, getting the tail out, free of and on top of the brichen. At **A** Ray's left hand is resting on the spider. At **B**, out of sight, Ray's right hand has pulled the brichen down over the hips.

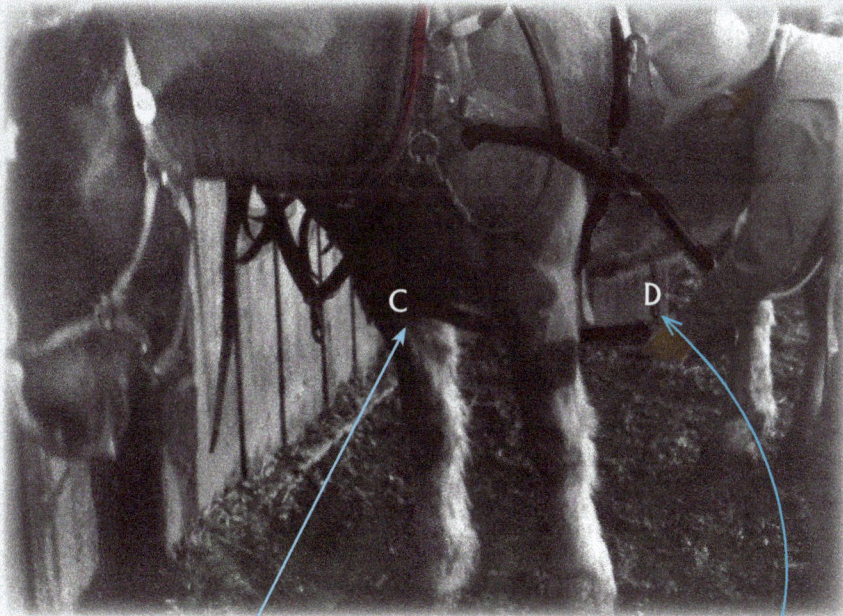

9. Ray pulls the pole strap (**C**) back between King's front legs and snaps the two quarter straps to it (at **D**).

10. Then he reaches for the belly band and brings it towards him to buckle in place. There should be at least a hand's thickness in between the belly band and the horse. The pole strap should move freely in this space.

11. Notice that traces are hanging by chain links from the trace carrier at the tops of the hip drops, from each side of the brichen spider. At this point in harnessing, this is what a handsome harnessed horse should look like. (Ray used to chuckle and say, "you're the only one sees him as handsome.")

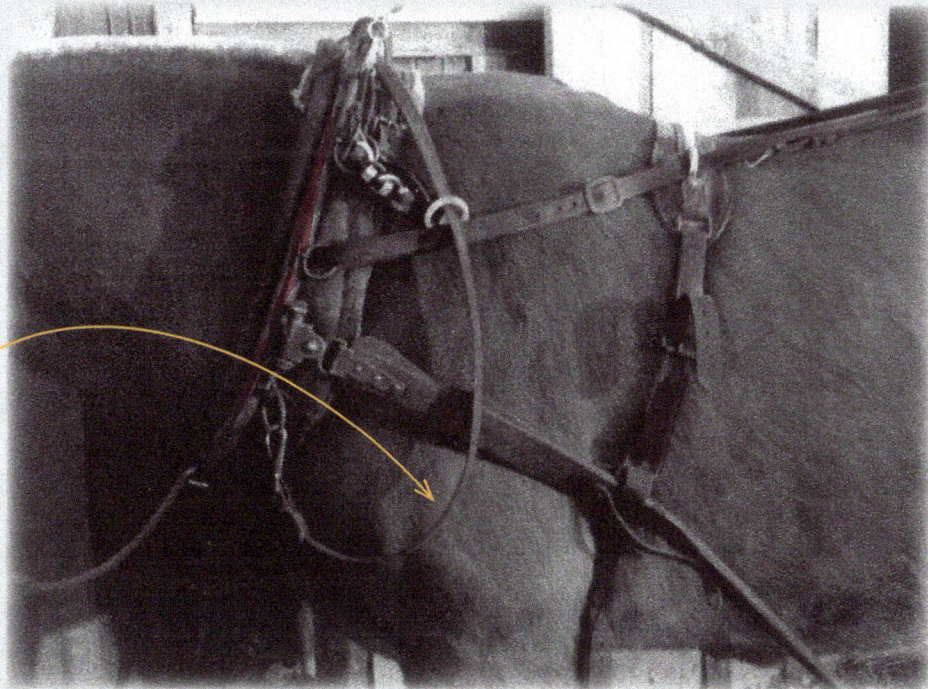

12. This loop is the inside cross check of King's line. Ray has run it through the spreader ring and temporarily snapped it to the breast strap ring of the harness where it's easy to get to when he sets up the team lines on King and Ruby.

13. **BRIDLING.** This is Ray's system. Take the top of the bridle in your right hand and lift it gently up the face of the horse. Place your left hand inside the noseband to keep the bridle from dragging up the horse's nose. Lift up and over the ears.

At the same time, you use your left thumb and finger at the corners of the mouth to get the horse to relax and open so that the bit can slide in – this while, with the right hand, you pull the bridle back over the ears and, with one or two fingers, you gently pull an ear forward of the bridle pole strap.

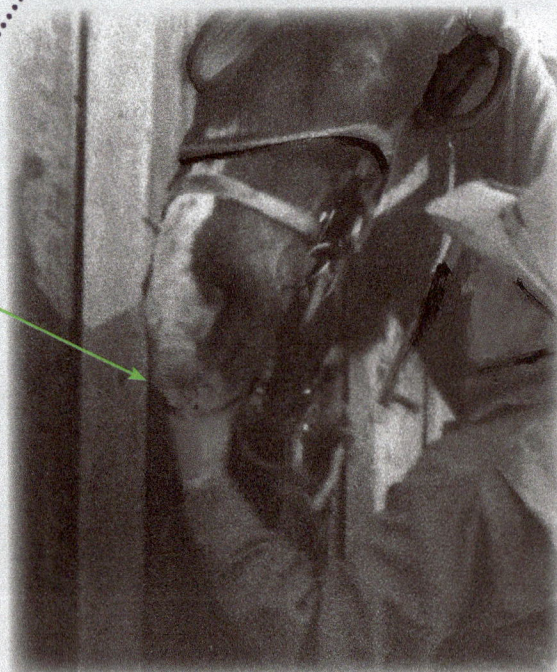

Bit in the mouth, now Ray moves to get the second ear free of the pole strap. Last, he will buckle the throat latch.

The whole bridling business is a bit of a dance that requires coordination, patience, and a sense of humor. A smart horse like King can sense if you're hesitant, impatient or grouchy. Bridling wants to be a matter-of-fact fluid transaction.

Side note: Ears should not be crowded by forelock or pinched by a too small bridle. It would just add needless aggravation.

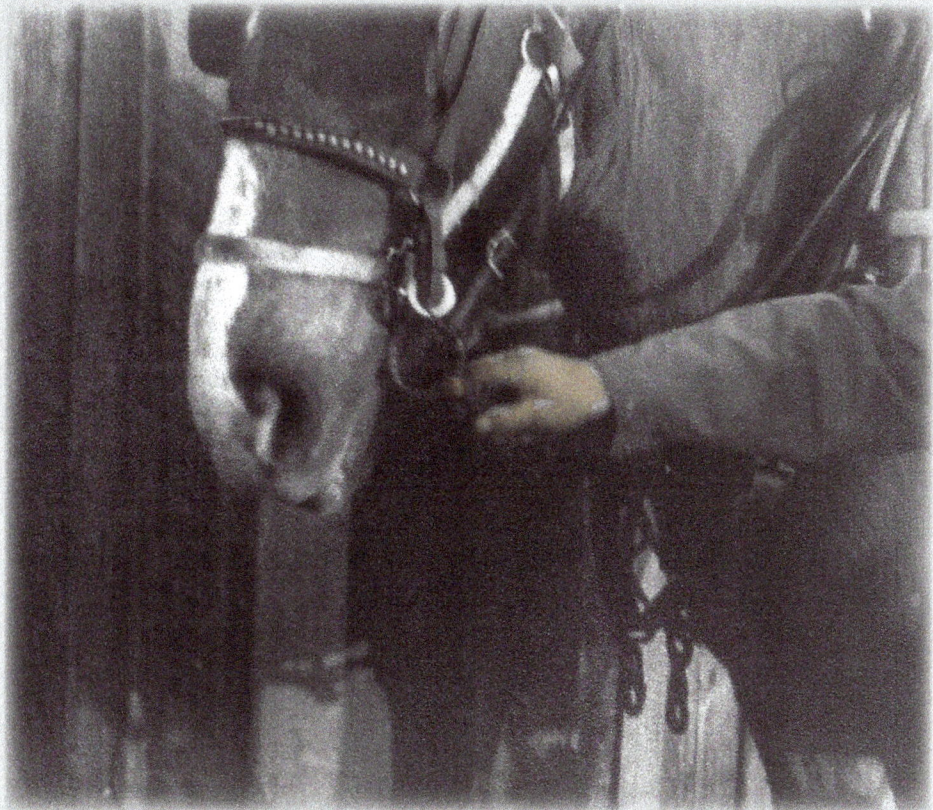

King is bridled. Notice that the bit very slightly curls the lip. A comfortable horse or mule is far more likely to be a willing and companionable workmate. But a bit that is loose in the mouth is a hazard as the animal might get its tongue over it, or be more easily able to take the bit in its mouth and negate the teamsters control.

14. With King's harness the check rein passes through **gag swivels** and runs between hame tops to snap into a strap that runs back from the brichen spider. Some harnesses would have the strap run clear back to a crupper that encircles the tail. Other harnesses might have a check rein suitable to put over one hame top. And the older systems often had the overcheck or check rein hook at the back pad of a button or hook (see page 117).

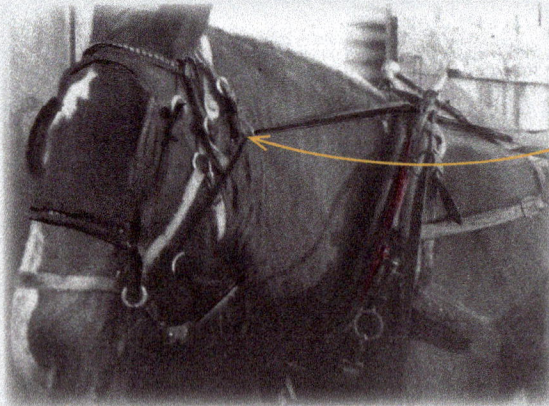

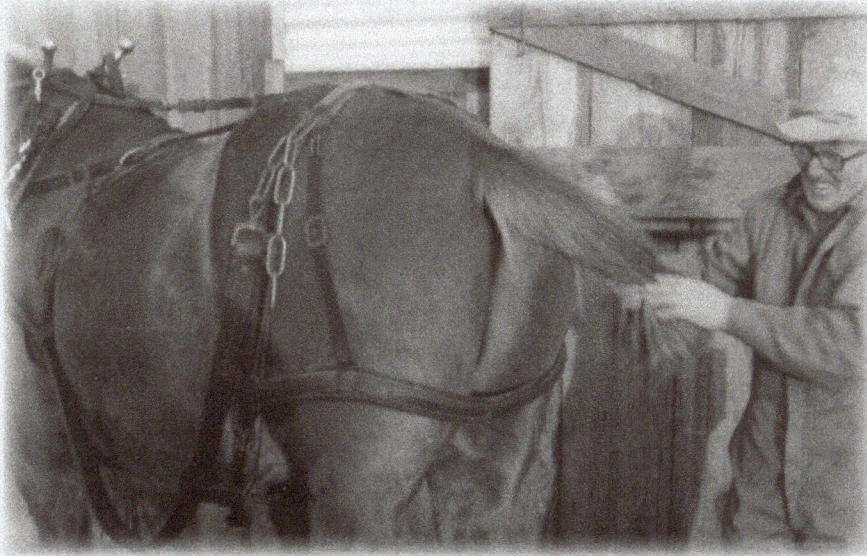

Ray's smile says it all, as does King's relaxed posture.

FIRE!

The late Forrest Davis of the San Francisco Ranch in Montana told me about an apparatus he saw in Chicago that was used in fire halls to harness horses in a hurry. Ropes and pulleys were arranged in the stalls up over each standing horse. The harness was held at strategic points and suspended, spread apart like a butterflied roasting chicken. When the fireman released the load rope, the harness dropped onto the waiting horse and was then unsnapped from the carrier and quickly buckled on to the horse.

When it came time to take the harness off, the lifting rig was again lowered and snapped to the harness, then the bottom hame strap, breast strap, belly band and quarter straps were all undone. If it was a brichen harness, the tail was set under and the load rope pulled until the harness was completely free of the animal, hanging above it in wait for the next fire call.

Records indicate horse drawn firewagon response times to big city fires was truly amazing.

Chapter Ten

Harness
Care and Repair

The material in this chapter includes quotations from an excellent article which appeared in a past issue of our publication, Small Farmer's Journal. This chapter is about and for the home shop repair and maintenance of leather harness. Synthetic materials such as Biothane will require research elsewhere, as not all of what you read here will apply.

The efficiency of draft animal work depends a great deal on the strength and serviceability of the harness. A broken trace or hame during the busy season may cause an expensive loss of time, besides it's a nuisance and inconvenient. Lumpy, cracked collars of the wrong size, as well as other unfortunate parts, can put a horse or mule out of service with sore neck and shoulders. A rotted and weakened line or hame strap could result in a serious accident and injury to both horse and teamster. Also, because clean well-kept harness adds a great deal to the attractiveness of a team, a farmer can take pride in keeping his outfit in first class condition.

Good leather harness will give many years of service if properly taken care of. If no steps are taken to keep it in good condition, leather will soon dry out and get hard and brittle, thus breaking easily.

"Harness should be washed occasionally and a good quality of harness oil applied in order to keep it soft and pliable. Manure and sweat, if allowed to accumulate, can

cause the leather to dry out and crack. Repairs should be made before breaks occur, all harness should be thoroughly overhauled, cleaned and repaired before the busy season starts. The farmer should take advantage of rainy days during the working season to inspect his harness, making necessary repairs and taking all possible precautions to prevent further breakdowns in the field."

Equipment

You will need some sort of clamp to hold the leather for sewing. You can make one, akin to the wooden paddle vises used for wood work, or get yourself a stitching horse as pictured on page 252, or you can improvise a vise with rubber or wooden jaw pads attached.

Sewing is done with No. 10 linen shoe thread. Wax the thread by hand and use ordinary harness needles that you pass through holes made with an awl in the leather. A store-bought sewing awl, using prepared thread, is sometimes used. It can be more convenient to use the awl, but the stitching might not be as strong as when done entirely by hand. A waxing pad is made by melting some of the wax and placing it on a piece of soft leather or canvas which is held in the hand, like an open sandwich, and rubbed along the thread giving it an even coating of wax.

You shouldhave a small riveter. Different types of rivets are used. Riveting is discussed in greater detail later.

If you plan on doing repairs yourself, here's a list of tools and supplies sufficient for most repair jobs:

1 ball of No. 10 linen shoe thread
1 ball of shoemaker's black wax
1 ball of beeswax
2 awl handles
3 two-inch awls
1 set of assorted harness needles
1 pair of dividers
1 four-tube revolving harness punch
1 riveter and rivets
1 knife
Necessary repair parts such as: leather, straps, snaps, buckles, rings, hame staples, conway loops, concord clips, etc.

Over the years, whenever an old harness with rotting leather is found, I have disassembled it and put the

Riveter and punch

Sewing awl and thread

resuable hardware in containers to save for future repairs. If you do this, be careful that rust is not hiding cracks in the metal.

*"**Preparing Thread for Sewing** – An important step in harness repair is the preparation of the sewing thread or 'waxed end,' as it is commonly called. A waxed end consists of three or more linen threads waxed tightly together to form one strong, uniform, durable thread with*

a sewing needle attached to each end. The following steps should be followed carefully.

1. *Measure off a suitable length of thread for sewing.* The length needed for a particular piece of work can be estimated only after practice. A thread five to six feet in length is about as long as can be handled conveniently. Do not cut or break the thread.

2. *Tear the thread* – The thread should not be cut or broken off squarely, but it should be untwisted and the fibers torn off in such a way as to make a long tapering end. To do this, lay the thread across the right thigh, grasping the thread tightly with the thumb and fingers of the left hand. With the palm of the right hand roll the thread against the thigh in such a way that the fibers become loose and untwisted at that point. When completely untwisted, the thread may be pulled apart, making ends that are long and tapering, suitable for waxing together. Repeat this procedure until three or four threads of the same length with tapered ends have been torn off. Three threads are usually sufficient for light work. For heavy pieces, such as traces, five or six threads should be used in making a waxed end.

3. *Place the threads together* – The torn threads should be of the same length but should be placed together unevenly, that is, one end of the first thread should extend about an inch or so beyond the end of the second, the third should extend beyond the second, etc., so that when twisted together the combined ends will be long and tapering to a fine point which can be threaded into the needle readily.

4. *Wax the ends* – Place the middle of the assembled threads over a hook, holding both ends together with the left hand. Holding the waxed pad in the right hand, draw each of the tapering ends through the pad until each is slightly coated and sealed

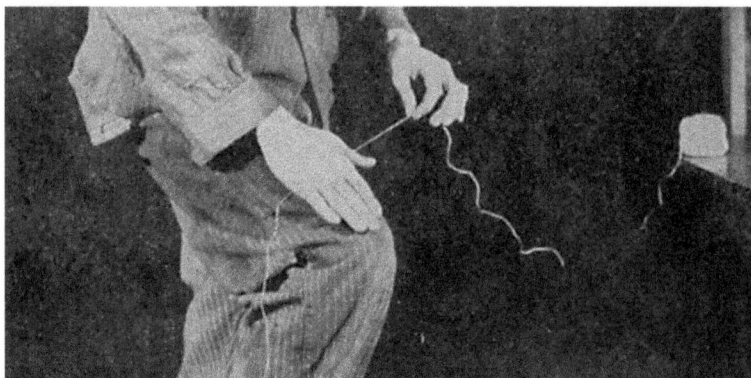

Untwisting the thread preparatory to tearing.

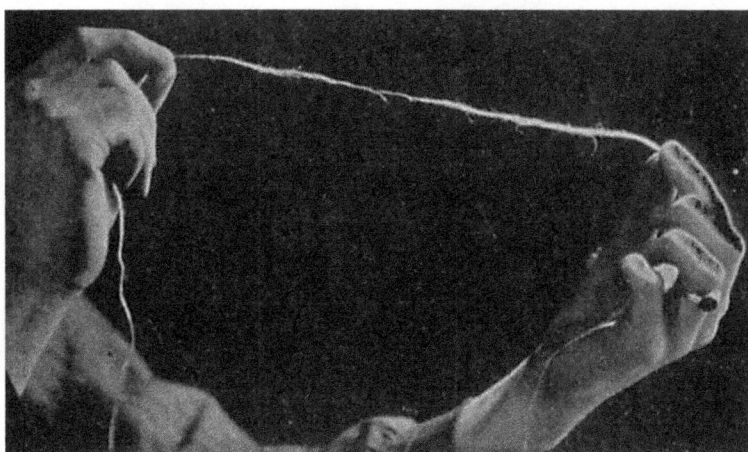

Tearing the thread.

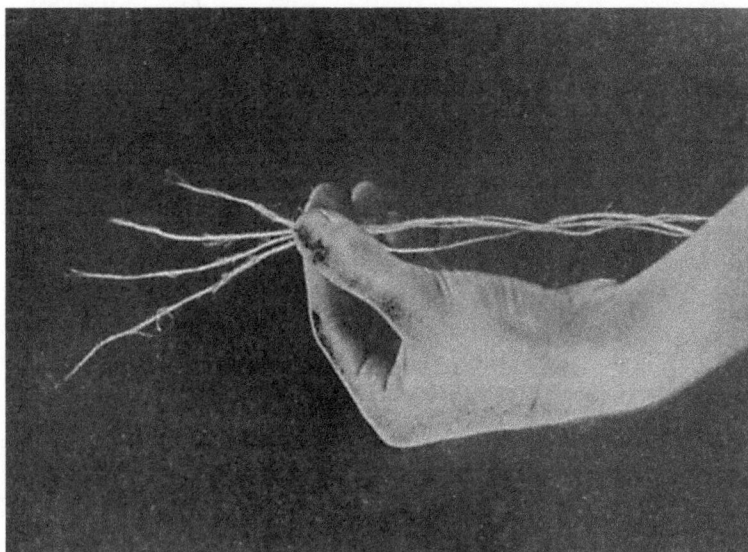

Assembling linen threads. Each succeeding thread should project slightly beyond the one before.

together with the wax.

5. *Twist and wax the threads* – Still holding the ends with the left hand, twist each end by rolling on the thigh with the palm of the right hand to twist the assembled threads together uniformly. Move the thread back and forth on the hook to equalize the twist. Keep the thread drawn fairly tight to prevent snarling. Do not release the ends or the twist will be lost. Apply the wax by pinching the thread between the folds of the wax pad and rubbing the pad briskly up and down the thread. The friction melts the wax and causes it to be distributed evenly into the thread. Work all parts of the thread to get it smooth and uniform. The thread when finished should be round and hard and black. After sufficient wax has been applied the thread may be smoothed by drawing it between the thumb and forefinger of the right hand. To give the thread a hard finish making it slip through the leather more readily, rub with beeswax. Do not get beeswax on the ends of the thread or the needles cannot be fastened on securely.

6. *Thread the needles* — If the ends of the original threads have been properly torn and placed and waxed, the waxed ends should be long and finely tapered so that they may readily be threaded through the eye of the sewing needle. Draw the fine end of the thread through the eye for at least two inches. Double the end back along the thread and hold it with the left hand. With the thumb and forefinger of the right hand twist the needle so that the short end of the thread wraps itself closely around the thread. The end should twist down into the wax so as to make a smooth round wrap that will hold securely and is no larger than the main part of the thread. Attach a needle to the other end of the thread in the same manner; the thread is now ready to be used for stitching."

Making a Splice – The best method for splicing a strap is by stitching (or sewing) two tapered ends, in sufficent and appropriate overlap, with a waxed thread. A stitched splice is

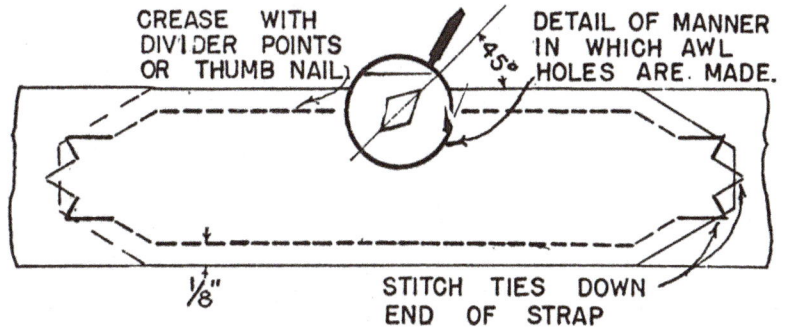

CREASE WITH DIVIDER POINTS OR THUMB NAIL)

DETAIL OF MANNER IN WHICH AWL HOLES ARE MADE.

1/8"

STITCH TIES DOWN END OF STRAP

MAKING A STITCHED SPLICE

This illustrates an excellent method of stitching down and across the ends of the splice. It's important that the thin edge be securely fastened to keep it from curling up and catching in fly nets, rings, etc. The comoon long cross stitch frequently used across the end leaves a large amount of thread exposed that could become frayed and broken.

Turn the strap over, end for end, and stitch down the other edge and across the other end. Reverse again to bring the stitching back to the starting point.

stronger than one made with rivets and it can be smoother if care is taken. But a sewn splice does take longer.

A. Get the ends ready to be spliced – The ends to be sewn should be lapped 2 to 4 inches. For most of this distance they should be 'skived' or shaved off to a long tapered, beveled edge so that when lapped together the sewn ends will fit down together smoothly. Shaving may be done with a sharp knife or a small block plane, and should be done on the rough or flesh side of the strap. Square the strap ends and round off the sharp corners.

B. Mark guide lines for the stitching – Place the tapered ends together in the position to be spliced, lapping carefully. Make sure that the smooth comes against the hair side of each strap. As a guide to keep the stitches straight,

Starting a stitched splice.

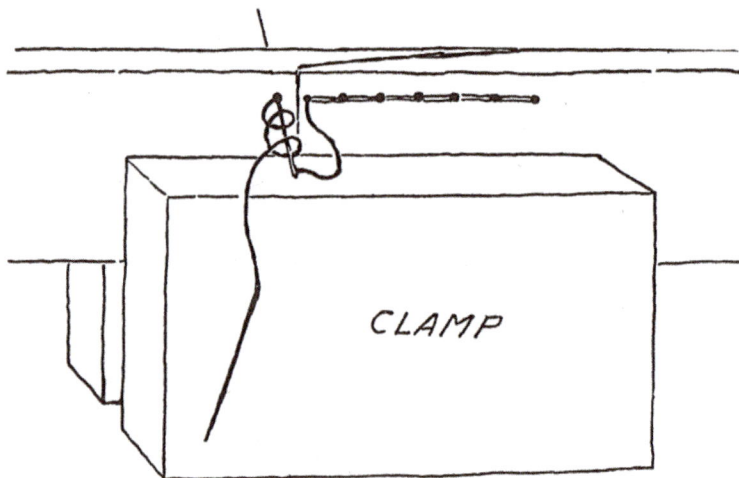

CLAMP

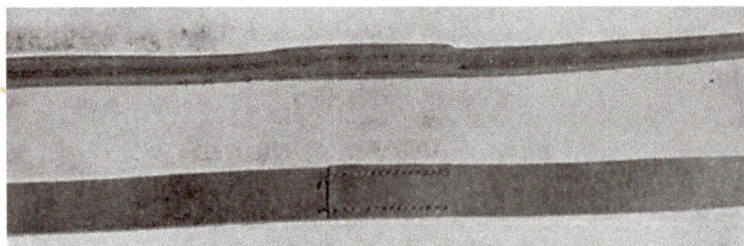

A finished stitched splice.

mark off a crease, a long indentation, in a straight line along each edge of the top strap about ⅛" from each edge. A pair of dividers, or compass, is very useful for this. Set the points about ⅛" apart and guiding one point against the edge of the strap, draw them along the length of the splice, applying pressure so that the other point makes a crease. Press down, do not cut the leather. By 'walking' the dividers along in this crease, marks can be made for the awl holes which will result in evenly spaced stitches. The splice may be held in place temporarily with small tacks while marking and sewing.

C. Set in a stitching clamp or vise with the top strap to the right and nearest you. The marked side should be to the right. Clamp the union of straps firmly with the upper creased guide line just above the edge of the vise. (See top left photo).

D. Stitch the splice – You are going to use two needles simultaneously. The sewing should begin at the middle of the splice, following the creased guide line toward the end of the top strap, threading the needles through holes made with the awl. Keep the awl handle at right angles to the face of the strap when pushing the awl through the leather. The long side of the awl blade should be slanted away from you, cutting across the guide crease at about 45 degree angles. (Top drawing previous page).

Make the first hole near the middle of the top edge of the splice. Push one needle through and pull about half of the thread through, leaving the same length with the other needle on the opposite side. Make the next hole. Push the right hand needle through the same hole in the opposite direction and draw that end of the thread

through. Pull both threads up tight.

Continue in the same way, making one stitch at a time, passing both needles through each hole in opposite directions, pulling each stitch tight until the end of the splice is reached.

E. Tie off the ends of the thread – When you are done with the sewing, to prevent loosening, the ends of the thread should be tied tight, as follows. Push the left hand needle and thread through the last hole, and pull as usual; then push the right hand needle halfway through the hole and while it's in this position wind the left hand thread once or twice around the needle. Pull both threads up tight. The winding sinks down into the hole, forming an overhanded knot which prevents the ends from loosening. For additional security another knot may be tied in similarly, through a second hole just below the last one.

F. Smooth the splice – Cut off the sewing thread close to the strap. Trim off the edges of the straps if necessary to give a tidy appearance. With the round end of the awl handle, or similar smooth instrument, rub the stitching until it lies down flat and smooth. If a fairly deep crease has been made as a guiding line, the threads will lie in it practically flush with the face of the strap. Tapping the splice lightly with a hammer on a flat surface will flatten and smooth it. Or a finishing wheel or marking wheel is sometimes run over the stitches to press the threads down smoothly.

Attaching Snaps or Buckles – To attach a snap or buckle to a new strap, or to repair a strap from which a buckle or snap has been torn out, any one of four methods might be used.

1. By a riveted loop – Trim off the corners of the strap, and taper or bevel the end for about an inch. If a buckle is to be attached, a slot must be cut for the buckle tongue. This is done as follows: Punch two holes with a punch about an inch apart, the first one about 2½" from the end of the strap.

Cut out between the holes, making a slot an inch long large enough for the buckle tongue to move freely. Place the buckle in position and rivet with two rivets, one close to the buckle, the other near the end of the strap. A slide loop may be attached just back of the buckle.

2. Via a stitched loop – The end of the strap is prepared in the same manner as described above and fastened by stitching in the usual manner.

3. Or with a Conway loop – Square the end of the strap. On the center line punch two holes large enough for the tongue of a Conway loop; the first hole from ½" to ⅞" from the end of the strap; the second from 3½" to 5" from the first, depending upon the size of the loop needed. If a buckle is to be attached, cut a slot as described above to fit the buckle tongue. Place the strap through the branches of the Conway loop, around the snap or buckle, and bring the end back underneath itself into the first branch of the Conway loop. Insert the tongue of the Conway loop first into the hole at the end of the strap and then into the other hole. Draw it up tight.

4. And then there is a buckle repair clip – This clip may be used when a strap is too short to spare the extra length necessary for making the other types of loop attachments. Square the strap end. Put the buckle in position in the repair clip and fit the clip over the end of the strap. Mark the places for the holes. Remove the clip and punch the holes. Replace the parts and rivet the clip securely to the strap.

Replacing Hame Staples and Clips – Due to heavy wear the hame staple and clip by which the trace is fastened to the hame often wear out and need to be replaced. (See the worn hame clip rivet in the photo on page 289.) You should check these parts regularly, and, if they are wearing thin, replace them before they break. If they do break under a heavy pull it could result in a serious accident or costly delay.

Hame Staples – With a file, nippers, hacksaw or cold chisel cut off the riveted ends of the old staple and drive it out of the hame. If the riveted ends are sunken in the hame, the staple should be cut in two at the worn place; then by fastening one-half of the staple in a vise and twisting on the hame, the half staple will break off below the shoulder and the remaining end can usually be driven out with a punch. The new staple is then pushed tightly in place, and the ends are cut off to fit the hame. Place a washer over each end of the staple and rivet it solidly.

Hame Clips - Remove the old rivets and the old clip from the trace. If the old rivet holes are badly worn, it may be necessary to cut off a piece of the trace. (Make sure that both traces are of the same length when replaced or repaired.) Place the new clip on the trace, mark and punch new rivet holes if necessary. Run the clip through the hame staple and rivet securely to the trace.

Bottom hame loops and clips should be replaced if badly worn.

Repairing Traces - Check your traces frequently and keep them in good repair. A broken trace puts a horse out of service and it frequently requires considerable time to repair properly.

The prong of each clip, on the side of the trace next to the horse may be inserted in the trace before riveting, especially if the trace is thick.

Restitching – Any broken stitching should be restitched without delay.

When once started to break, the old threads wear out rapidly, the plies of leather separate, and the trace becomes greatly weakened.

Cockeyes which have torn out of their loops may be reattached by means of a Concord clip riveted to the end of the trace.

Splicing – A broken trace may be spliced by any one of several methods.

1. By a stitched or riveted splice, as previously described. The ends should be squared and tapered for five or six inches. The old stitching should be taken out so that when the ends are placed together, the plies on one end can be pushed between and lapped on the layers from the other end. This shortens the trace quite a bit.

2. With two Concord clips riveted securely to the squared ends of the trace and joined together

with a trace square.

3. With two hame clips riveted securely to the squared ends of the trace and joined by a link.

4. With a trace splicer. The ends to be spliced should be squared and fitted together. Place the splicer, which is a flat strip of metal with holes for rivets, in position, on top of the trace, mark and punch the holes. Insert the end splicer into the center of the trace ends and rivet securely.

Cleaning and Overhauling Harness – Overhauling, cleaning and oiling harness is essential toitscare and maintenance. Not only will the appearance of the team and harness be greatly improved but the length of life and strength of the harness will be increased as well. Here's how:

1. Take the harness apart. Remove all buckles, snaps, and other parts or fittings that can be taken off without cutting rivets or stitches.

2. Make all needed repairs, such as restitch-

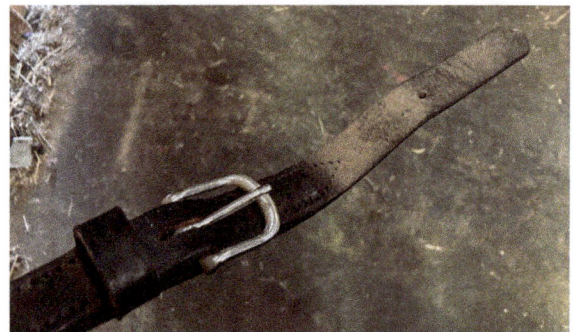

ing, splicing, replacing broken parts, etc.

3. Fill a tub about three-fourths full of warm water. Add an appropriate leather soap.

4. Place the harness in the tub until it is fairly well soaked and the crusted dirt softened.

5. Take out one piece at a time on a drain board and scrub thoroughly with a stiff brush until clean. A dull knife may be used to scrape off the caked material.

6. Lay in clean place in the shade until dry or ready to apply oil.

Oiling the Harness – Look for the best grade of Neatsfoot Oil Compound that you can find, it will make all the difference. Apply the oil to the harness while it is still wet (from washing) – the oil will penetrate the leather as the water evaporates. Use a sponge or rag, rubbing the oil into the leather. Allow the oil to dry overnight. If necessary to soften the leather, put on another coat. I find that to warm the oil, even slightly, helps it to penetrate more evenly. Wipe off surplus oil with burlap or absorbent rags. Assemble the harness when dry.

Side note – dunking: In years gone by, harness shops performed a service for a fee by taking in your harness, hanging it into and on an oversized treble-hook-like apparatus with a block and tackle. Then they would lower the whole assembly into a large vat of warm oil for an hour or so. After that it would be pulled up over the tank and allowed to drip for a day. Though I don't know how to measure my conclusion, I think a better result is accomplished by disassembling buckles and conways and rubbing oil into all the leather, especially where it folds around hardware.

Each part of the harness is scrubbed after it has soaked in soapy warm water. A homemade scrubbing board is shown.

Such as with this conway buckle, it is important to take assembly apart and oil it wherever the leather is 'trapped' against metal.

This handsewn repair, while strong, would have been far better if creases had been made in the leather for the stitch path and then the stitching rolled and pressed into the groove. This way it would be less likely that the exposed thread would catch on things and fray.

Chapter Eleven

Used and Abused
Harness

An experienced teamster can take a quick look at this harness and expect it is good.

How do you tell if a used harness, a discovered and perhaps unrepresented harness, is safe? And what do I mean by *unrepresented*? If you are lucky enough to purchase a secondhand harness from the teamster who has been using it, that person's ability and reputation are excellent advantages to your determination. That stuff is 'representation.' If instead you happen upon a pile of harness in the corner of a shop or barn and there is no one around to answer your stated or silent questions, I hope you know what you are looking at and how to make quick checks on its condition.

Often old harness, such as above, will be found in piles, deep in dark corners of the shop or barn.

Trust your instincts. Old dry brittle harness will look old dry and brittle, such as this above. Oiling may not save this material.

But your instincts may be wrong. The harness on the right, while dry and dirty, is actually not brittle and, if you look closely, shows little or no wear.

See the uniform thickness and pliability of the exposed bit straps? They look good to my eye but be safe and test them for strength..

The noseband stitching looks relatively young and still intact.

Whereas the snaps, and the riveted end clip on the hame, worry me some
I would replace snaps and repair hame clip.

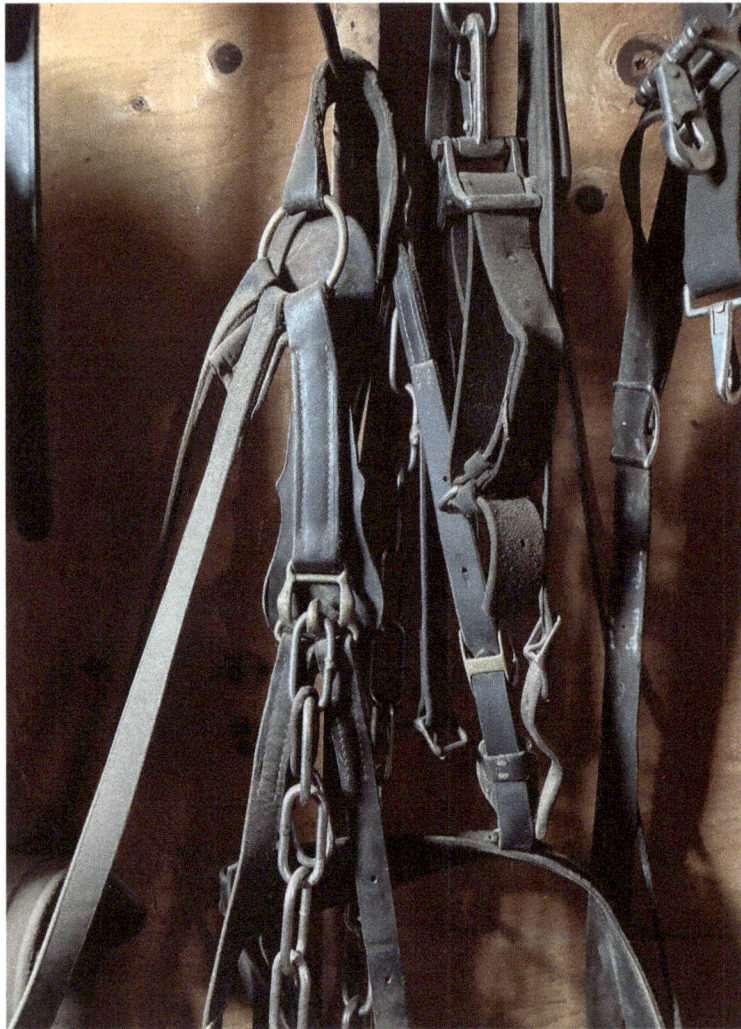

(Left) Obviously a well-oiled harness in regular use, but you still need to check it carefully if you want to protect your animals and yourself.

This closeup (below) of the old lines shows us a sewn and riveted splice, and no obvious detrimental wear at the buckle. Wash this leather, oil it and you may have some excellent gear.

One quick way to test old leather is **the twist test.** Grab both ends of a suspect strap and twist them in opposite directions as hard as you can while pulling. Brittle and raggy leather will surprise you how easily it tears. You don't ever want to trust weak and raggy leather to protect you and your beloved animals.

Critical parts for safety

While you want all of your harness to be strong and correct, there are certain keys pieces of the harness and harness assembly which, if they come apart or fall apart, WILL almost certainly cause a wreck. Using old, unscrutinized and unchecked leather harness is flirting with such outcomes.

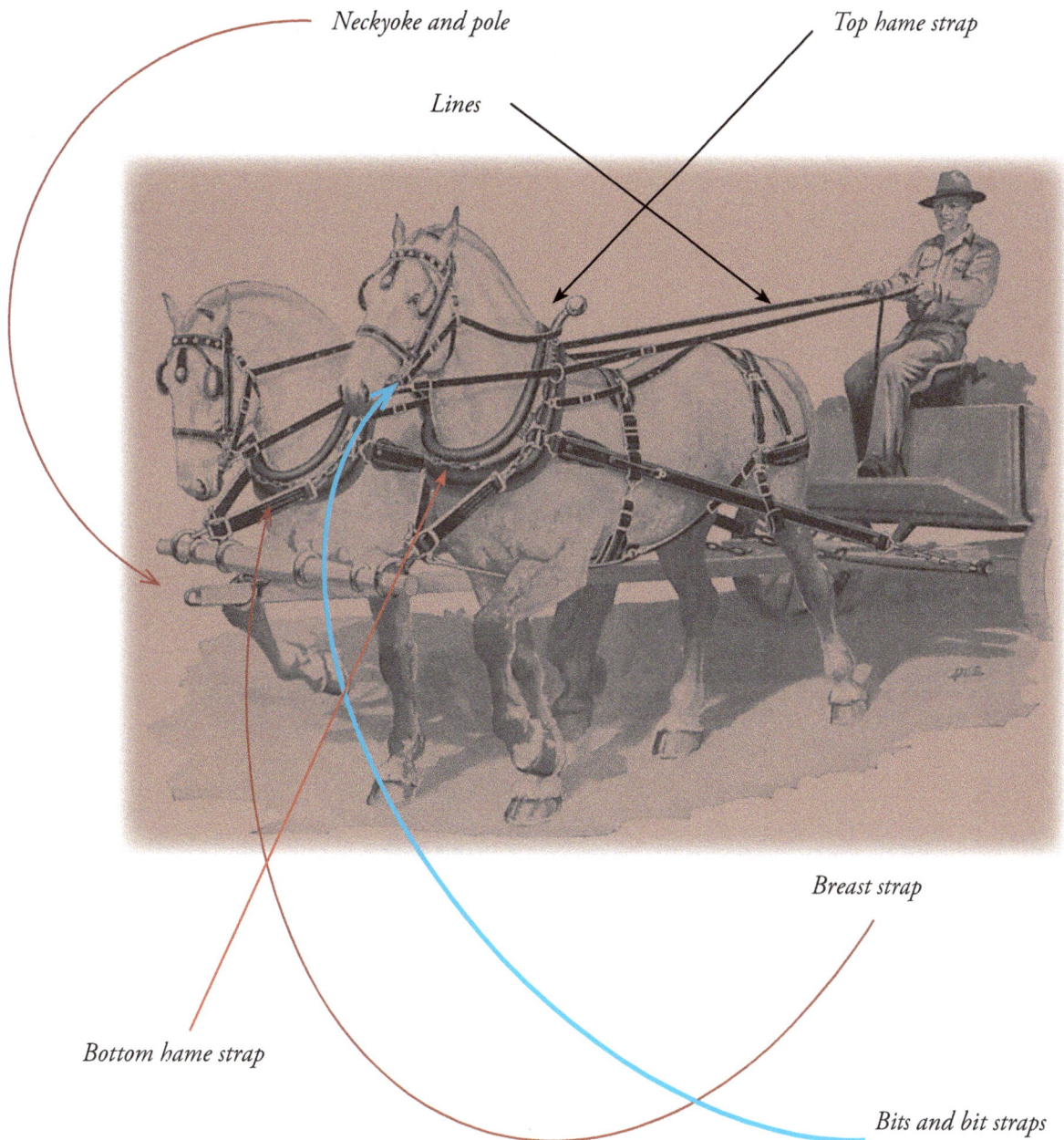

Neckyoke and pole

Top hame strap

Lines

Breast strap

Bottom hame strap

Bits and bit straps

Being smart and checking any old harness COMPLETELY is the best insurance.

MISCELLANEOUS

Depends on the individual of course, but there can be a whole lot of paraphenalia pulled in and scattered around the world of harnessed work horses and mules. There are add-ons like flynets and ear covers, and there are decorations and dress-ups. I have a pair of lovely brass spreaders and two gold tassels waiting should I ever take my horses to town again. Not all of us want to dress up for town or home. Some of us might lean towards spare and uncluttered. While for others doodads, fluff, sparkles, elective punctuation marks, adorned and adorning remedies, and what I like to call *uniquifiers* - those big and little things that say we are different - they can be, for some, essential to the experience. We pay a tiny bit of visual homage to these items because where else would you find them but in a book of this sort.

(Above) Flynets hanging in the tack room...

and on a horse (below left).

BREAST.

NO. 67. BRIGHT WITH MALLEABLE SLIDES.

BREAST.

NO. 69. STAR BRIGHT DOUBLE LINK.

HAME HOUSING BALLS.

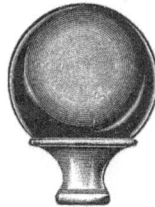

(Above) Some might choose to use chrome, nickel or brass breast chains with a little extra twist.

All sorts of adornments can be affixed to the top of hame housings.

As seen here, and elsewhere in this volume, an easy place to enliven harness is whatever you might want to put on top of your hames. And some will go to the extra extreme of disassembling their hames to change out the terrets to something fancier.

CLINTON CHECK HOOKS.

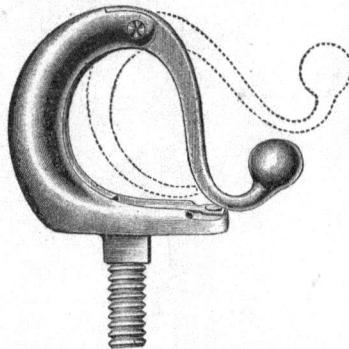

Express Hame Ball.

McKINNEY CHECK HOOKS.

IRON HAME TERRETS.

Band Pattern.

Wire Ball.

Summit Pattern.

The late great Paul Birdsall with his handknit ear covers on his good Suffolk. These were used to do double duty – for distinctive good looks and to repel flies. (Speaking of flies, in the background is the author.) Photo by Kristi Gilman-Miller

Below: My spreaders and tassels.

Index

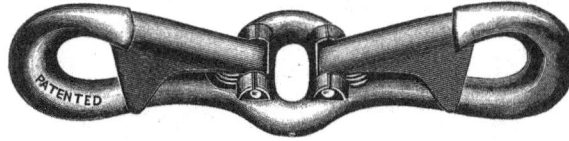

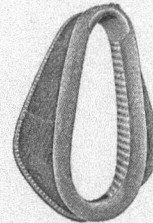

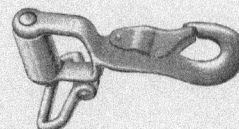

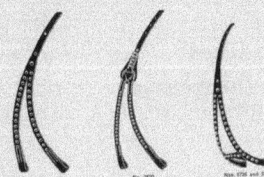

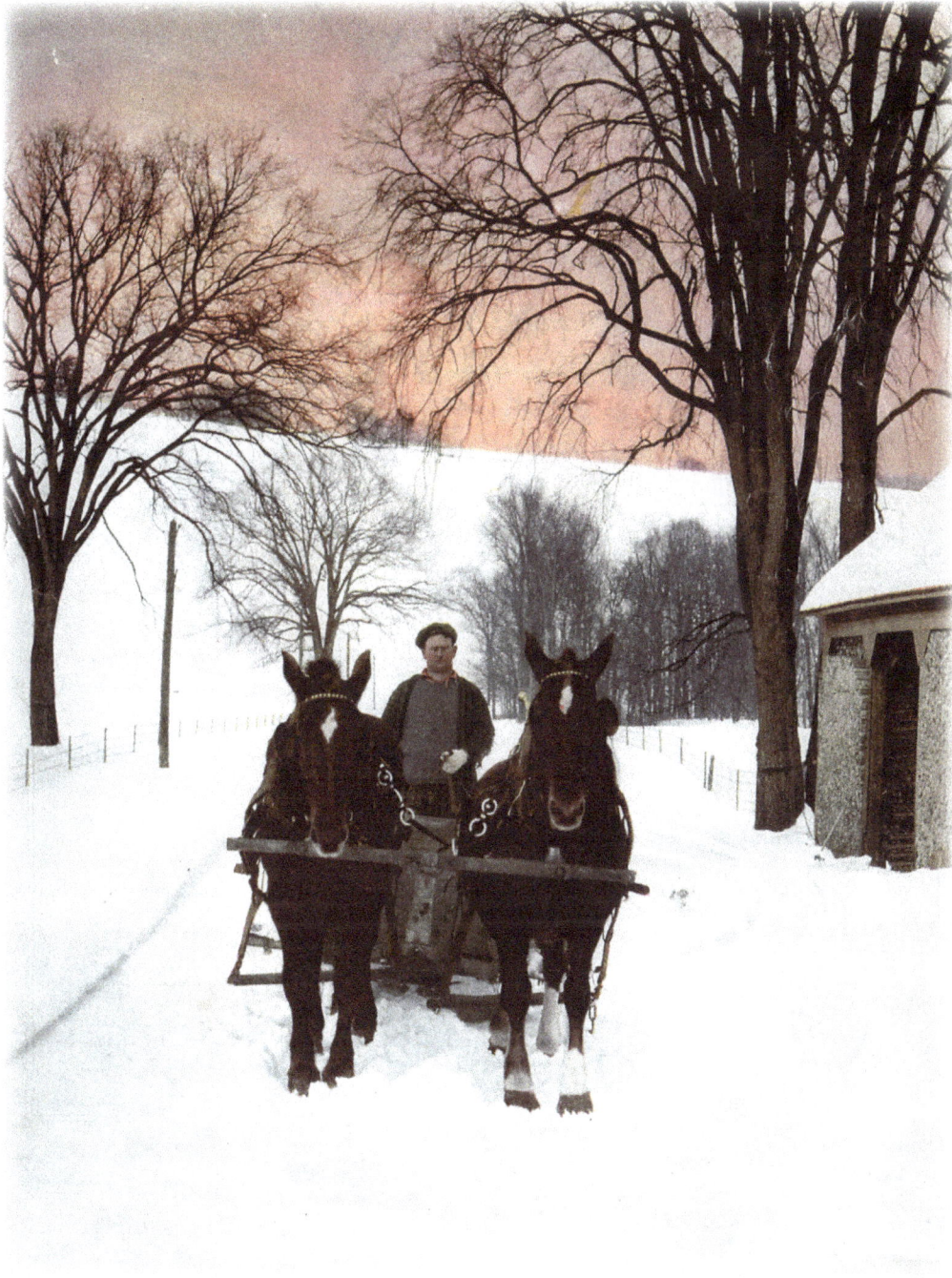

With working horses and mules context is just about everything. Let them put you to work in ways that are guaranteed to make you a better human.

There you have it, a caring capable teamster
and his well harnessed horses.

Conclusion

With this volume I arrive at 42 plus years of writing books and articles on the subject of working horses and mules. This is the ninth book, with eight still in print. In the beginning, with my father's encouragement, we thought I might do two or three volumes in five years and be done with it. But now, I know I will never be done with the subject. There will come a time when it is done with me. No one can know when that might be. So we work on. Right now I think on the subjects I have begun to address in the next volumes - grain binders and reapers, threshing machines, manure spreaders, seed drills and planters, and more - and I feel a wee bit daunted. Old men can't afford to feel daunted, we have to steel ourselves and keep moving ahead, long as we are able.

Thank you for visiting this book. I sincerely hope you have found things here that you can use, for it is in the use that we extend the working life of the technology. The information contained within needs to be kept alive. There are lessons and successes showcased that form a long line of head starts for any future culture finding itself, as I suspect they will, in need of a working union with draft animals.

As so many of my friends, partner teamsters and mentors have said to me on many a given morning, "Well, we've got work to do." To which I say, "Amen to that."

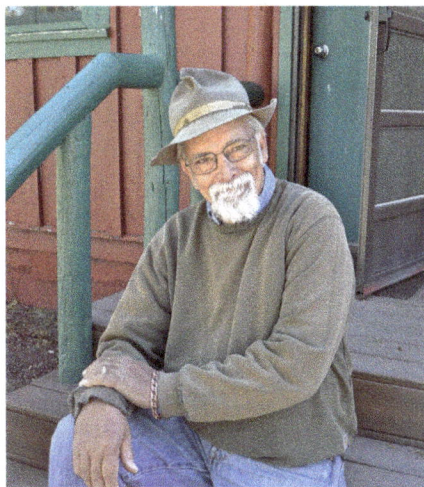

Photo by Kristi Gilman-Miller

Acknowlegements

Wow was this a lot of work. And it would never have happened, it never would have made it between these covers but for the patience, help, cajoling and encouragement of many people. You've heard this speech before, I can't begin to name everyone I should. Thank you all. But some tall thank you's get etched here to my magnificent wife and partner, Kristi Ann Gilman-Miller, who for 40 plus years has stood by me through a long and various life adventure weaving in and through work horses, working horses and all the paraphenalia. And a big thank you to Kema Clark, and before her Shannon Berteau, for their patience with the ping pong game that is proofreading anything I scribble. There was a lot of side-eye and eye-rollin' on both sides of the equation but I stand on record here and now saying that this book would have been a sloppy unreadable mess without their help and especially Kema's passionate long hours proofreading. If there are errors here (and there are) it is because I snuck them in without her looking. Plus a smile, handshake and big thank you to Eric Grutzmacher my workmate in layout and graphic computer wrestling. In the fifth dimension universe of software massage, navigation can be everything. Thank you Eric. You know I would have thrown in the towel but for your help. Now, if you will pardon me, it's on to the next project. LRM

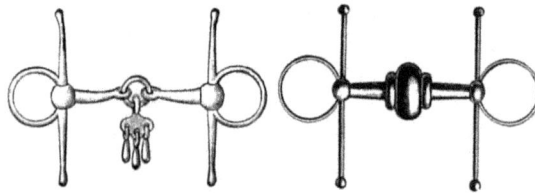

Photo credits
Brisk, Marvin 121A
Castle, William   20B, 65, 68A, 105B, 121B, 213A-B, 218
Foxley, Ryan 221, 222
Gilman-Miller, Kristi 18, 20A, 56, 71, 77, 78, 94B, 101, 115B, 120A, 127, 130, 133A, 141A, 164, 165, 166B, 168A, 183, 193, 200, 239, 267, 294A, 299
Grutzmacher, Eric 266A
Hagen, Steve 160, 225
Hunter, Jerry 147, 155B, 168B, 169A, 182, 212, 219A-B
Livingston, Ida 81, 157
Mascardo, Albano 75
Miller, Lynn R.   11, 27, 28, 44D-E, 48, 49, 50, 51, 52, 53, 67, 69, 70, 79, 80, 93, 102, 103, 104, 111A, 120B, 145, 148, 155A, 170, 197, 266B, 268-279, 280, 281, 288, 289, 290, 294B, 298
Russell, Carl 44A
Russell, Renee 68B, 133B
Sheetz, Doug 14, 172A
Sheldon, Fuller 131
Whitehorse Machine 111B, 155C, 172B, 214

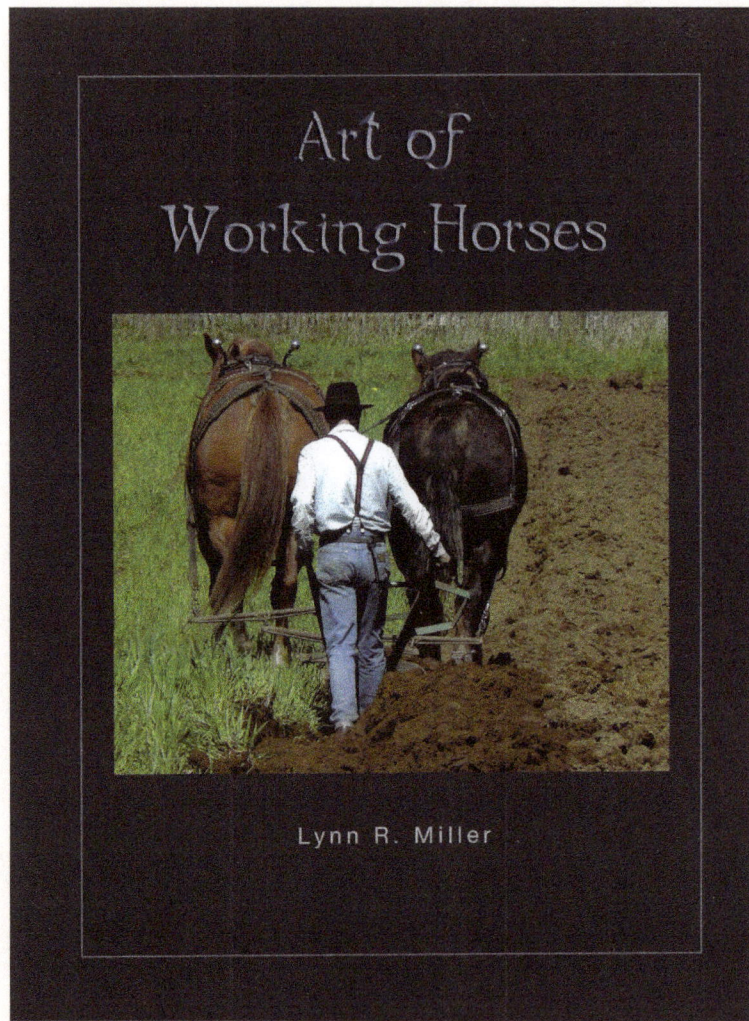

From the Work Horse Library by L.R. Miller

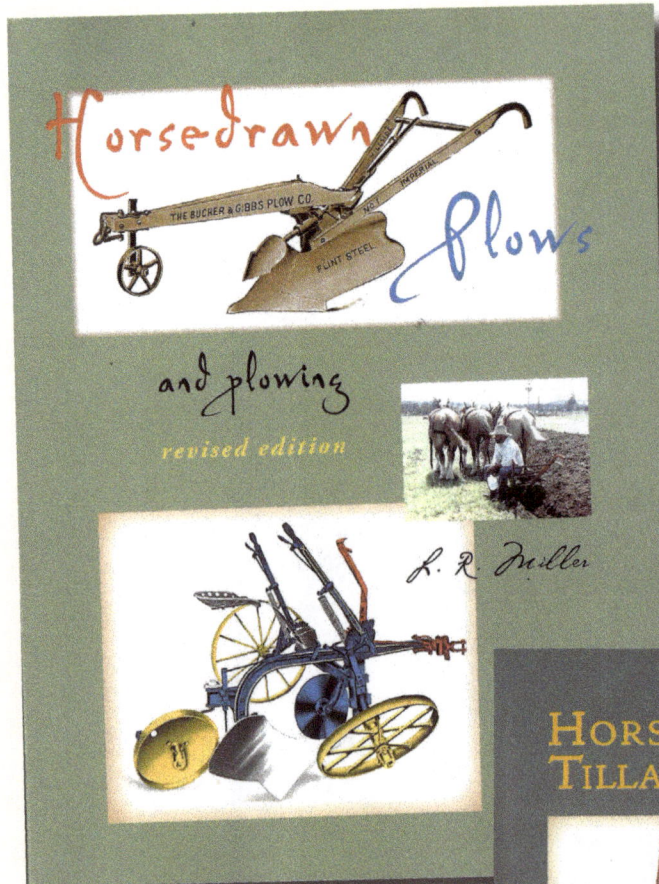

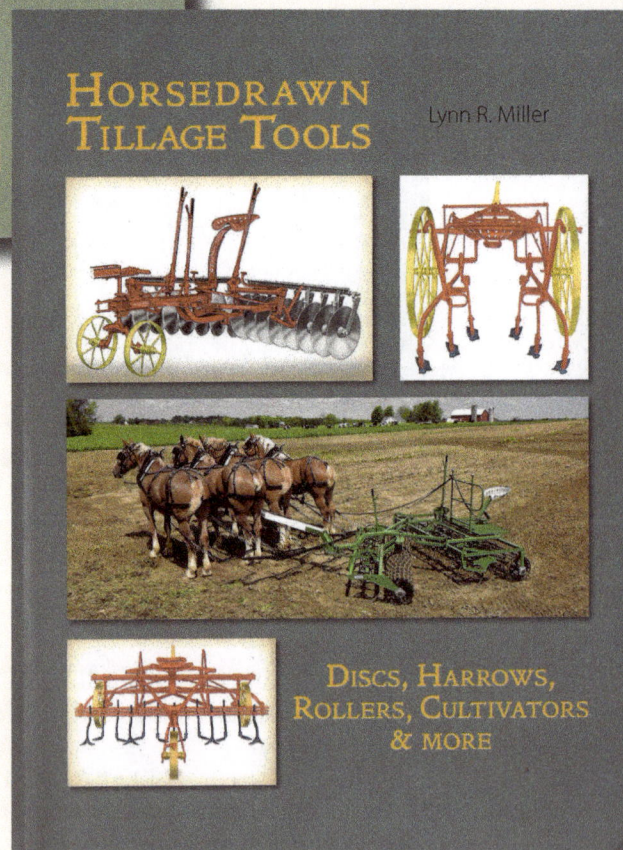

From the Work Horse Library by L.R. Miller

These highly acclaimed volumes are **the** most comprehensive and useful books on the subject. Over 1,000 illustrations.

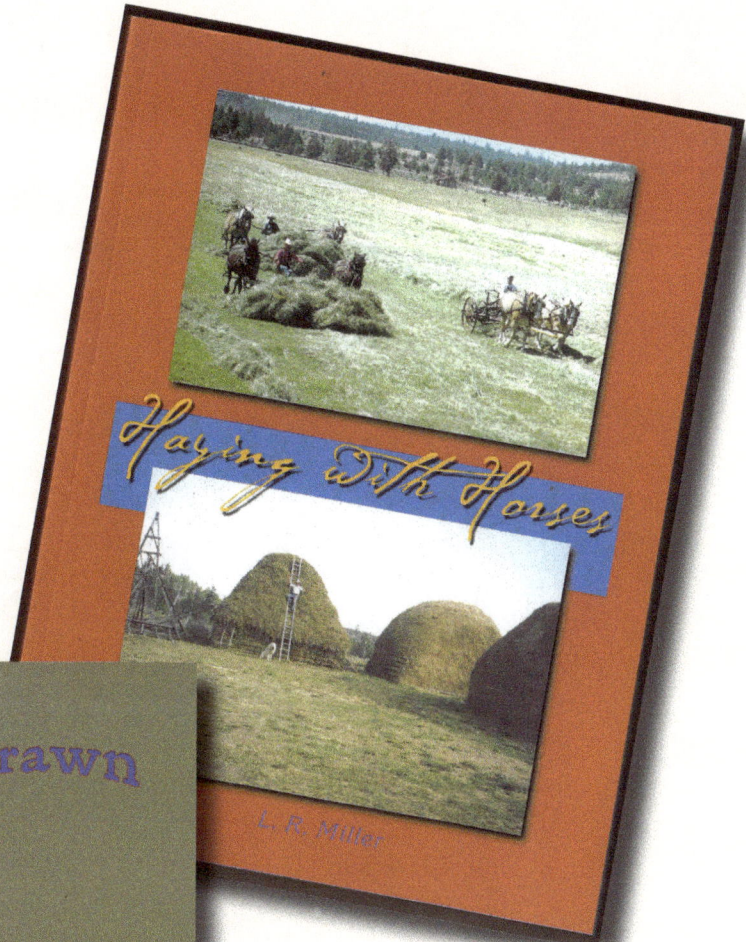

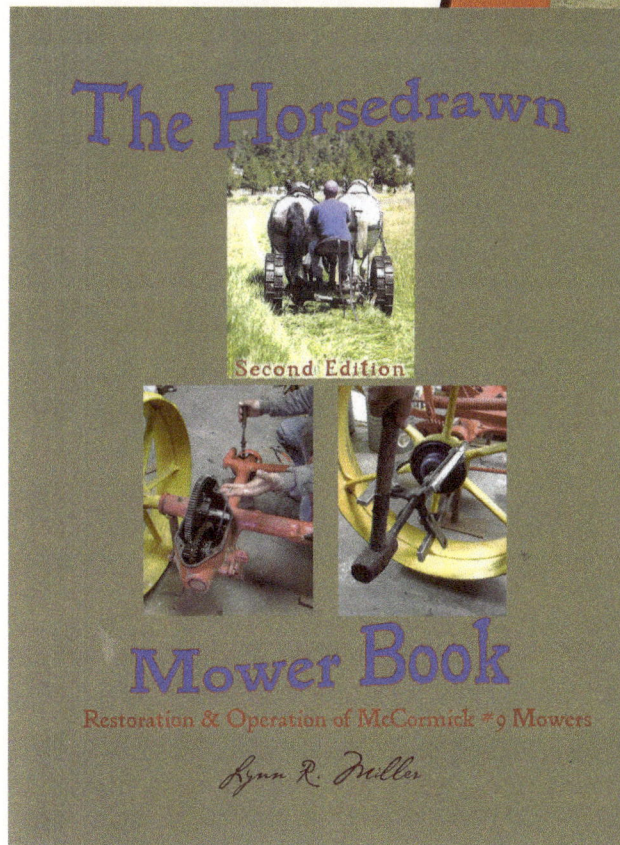

Each 366 pages soft cover $45
Shipping $7 (U.S.)

Order from Small Farmer's Journal
P.O. Box 1627, Sisters, Oregon 97759

800-876-2893
agrarian@smallfarmersjournal.com
www.smallfarmersjournal.com

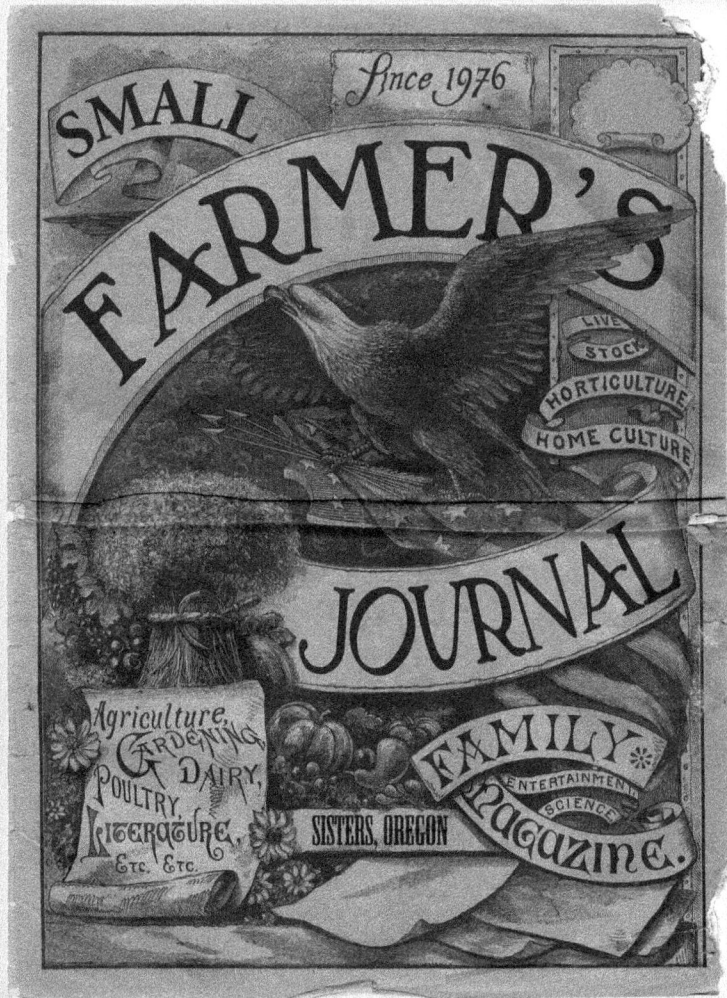

Lynn R. Miller   *founder / editor / publisher*

Since the beginning in 1976 Small Farmer's Journal has
proudly included, in every issue, information on
animal powered lifestyles.

Quarterly $15 per issue $47 year U.S.   otherwise ask

Small Farmer's Journal
PO Box 1627, Sisters, Oregon 97759

agrarian@smallfarmersjournal.com
www.smallfarmersjournal.com

www.ingramcontent.com/pod-product-compliance
Lightning Source LLC
Chambersburg PA
CBHW052349210326
41597CB00038B/6306